T0319660

Global Positioning

Global Positioning

TECHNOLOGIES AND PERFORMANCE

Nel Samama

WILEY-INTERSCIENCE

A JOHN WILEY & SONS, INC., PUBLICATION

For general information on our other products and services or for technical support, please
contact our Customer Care Department within the U.S. at 877-762-2974, outside the United States at
317-572-3993 or fax 317-572-4002.

Wiley also publishes it books in variety of electronic formats. Some content that appears in print,
however, may not be available in electronic format.

Library of Congress Cataloging-in-Publication Data:

Samama, Nel, 1963-
   Global positioning : technologies and performance / Nel Samama
        p. cm.
    Includes bibliographical references.
    ISBN 978-0-471-79376-2 (cloth)
    1. Global Positioning System I. Title.
    G109.5.S26 2008
    623.89′3—dc22   2007029066

CONTENTS

Foreword, xiii
Preface, xv
Acknowledgments, xvii

CHAPTER 1
A Brief History of Navigation and Positioning, 1

1.1 The First Age of Navigation, 1
1.2 The Age of the Great Navigators, 5
1.3 Cartography, Lighthouses and Astronomical Positioning, 11
1.4 The Radio Age, 12
1.5 The First Terrestrial Positioning Systems, 15
1.6 The Era of Artificial Satellites, 19
1.7 Real-Time Satellite Navigation Constellations Today, 23
  • The GPS system, 23
  • The GLONASS system, 24
  • The Galileo system, 25
  • Other systems, 26
1.8 Exercises, 26
Bibliography, 27

CHAPTER 2
A Brief Explanation of the Early Techniques of Positioning, 29

2.1 Discovering the World, 30
2.2 The First Age of Navigation and the Longitude Problem, 30
2.3 The First Optical-Based Calculation Techniques, 33
2.4 The First Terrestrial Radio-Based Systems, 35
2.5 The First Navigation Satellite Systems: TRANSIT and PARUS/TSIKADA, 36
2.6 The Second Generation of Navigation Satellite Systems: GPS, GLONASS, and Galileo, 39
2.7 The Forthcoming Third Generation of Navigation Satellite Systems: QZSS and COMPASS, 40
2.8 Representing the World, 40
  • A brief history of geodesy, 40
  • Basics of reference systems, 42
  • Navigation needs for present and future use, 46

- Modern maps, 53
- Geodesic systems used in modern GNSS, 53

2.9 Exercises, 54

Bibliography, 55

CHAPTER 3

**Development, Deployment, and Current Status of
Satellite-Based Navigation Systems, 57**

3.1 Strategic, Economic, and Political Aspects, 58
- Federal Communication Commission, 58
- European approach, 58
- International spectrum conference, 60
- Strategic, political, and economic issues for Europe, 61

3.2 The Global Positioning Satellite Systems: GPS, GLONASS, and Galileo, 65
- The global positioning system: GPS, 65
- The GLONASS, 72
- Galileo, 76

3.3 The GNSS1: EGNOS, WAAS, and MSAS, 80

3.4 The Other Satellite-Based Systems, 85

3.5 Differential Satellite-Based Commercial Services, 85

3.6 Exercises, 91

Bibliography, 91

CHAPTER 4

Non-GNSS Positioning Systems and Techniques for Outdoors, 95

4.1 Introduction (Large Area Without Contact or Wireless Systems), 96

4.2 The Optical Systems, 97
- The stars, 97
- Lighthouses, 97
- The "ancient" classical triangulation, 98
- Lasers, 99
- Cameras, 100
- Luminosity measurements, 101

4.3 The Terrestrial Radio Systems, 101
- Amateur radio transmissions, 102
- Radar, 102
- The LORAN and Decca systems, 104
- ILS, MLS, VOR, and DME, 107
- Mobile telecommunication networks, 108
- WPAN, WLAN, and WMAN, 114
- Use of radio signals of various sources, 114

4.4 The Satellite Radio Systems, 115
- The Argos system, 115
- The COSPAS-SARSAT system, 117
- DORIS, 119

- The QZSS approach, 119
- GAGAN, 121
- Beidou and COMPASS, 121

4.5 Non-Radio-Based Systems, 123
- Accelerometers, 124
- Gyroscopes, 125
- Odometers, 125
- Magnetometers, 126
- Barometers and altimeters, 126

4.6 Exercises, 127

Bibliography, 128

CHAPTER 5
GNSS System Descriptions, 131

5.1 System Description, 131
- The ground segments, 132
- The space segments, 134
- The user (terminal) segments, 139
- The services offered, 140

5.2 Summary and Comparison of the Three Systems, 142

5.3 Basics of GNSS Positioning Parameters, 142
- Position-related parameters, 143
- Signal-related parameters, 147
- Modernization, 151

5.4 Introduction to Error Sources, 153

5.5 Concepts of Differential Approaches, 153

5.6 SBAS System Description (WAAS and EGNOS), 157

5.7 Exercises, 158

Bibliography, 159

CHAPTER 6
GNSS Navigation Signals: Description and Details, 163

6.1 Navigation Signal Structures and Modulations for
GPS, GLONASS, and Galileo, 163
- Structures and modulations for GPS and GLONASS, 164
- Structure and modulations for Galileo, 168

6.2 Some Explanations of the Concepts and Details of the Codes, 171
- Reasons for different codes, 172
- Reasons for different frequencies, 174
- Reasons for a navigation message, 175
- Possible choices for multiple access and modulations schemes, 178

6.3 Mathematical Formulation of the Signals, 180

6.4 Summary and Comparison of the Three Systems, 182
- Reasons for compatibility of frequencies and receivers, 182
- Recap tables, 183

6.5 Developments, 186
6.6 Error Sources, 187
- Impact of an error in pseudo ranges, 188
- Time synchronization related errors, 189
- Propagation-related errors, 190
- Location-related errors, 193
- Estimation of error budget, 194
- SBAS contribution to error mitigation, 195
6.7 Time Reference Systems, 195
6.8 Exercises, 197
Bibliography, 198

CHAPTER 7
Acquisition and Tracking of GNSS Signals, 201

7.1 Transmission Part, 201
- Introduction, 201
- Structure and generation of the codes, 204
- Structure and generation of the signals, 206
7.2 Receiver Architectures, 208
- The generic problem of signal acquisition, 209
- Possible high level approaches, 212
- Receiver radio architectures, 213
- Channel details, 217
7.3 Measurement Techniques, 223
- Code phase measurements, 223
- Carrier phase measurements, 225
- Relative techniques, 227
- Precise point positioning, 231
7.4 Exercises, 231
Bibliography, 232

CHAPTER 8
Techniques for Calculating Positions, 235

8.1 Calculating the PVT solution, 235
- Basic principles of trilateration, 236
- Coordinate system, 237
- Sphere intersection approach, 239
- Analytical model of hyperboloids, 243
- Angle of arrival-related mathematics, 246
- Least-square method, 248
- Calculation of velocity, 249
- Calculation of time, 251
8.2 Satellite Position Computations, 251
8.3 Quantified Estimation of Errors, 253
8.4 Impact of Pseudo Range Errors on the Computed Positioning, 255

8.5 Impact of Geometrical Distribution of Satellites and Receiver (Notion of DOP), 256

8.6 Benefits of Augmentation Systems, 258

8.7 Discussion on Interoperability and Integrity, 259
- Discussions concerning interoperability, 259
- Discussions concerning integrity, 260

8.8 Effect of Multipath on the Navigation Solution, 262

8.9 Exercises, 269

Bibliography, 270

CHAPTER 9
Indoor Positioning Problem and Main Techniques (Non-GNSS), 273

9.1 General Introduction to Indoor Positioning, 274
- The basic problem: example of the navigation application, 275
- The "perceived" needs, 276
- The wide range of possible techniques, 277
- Comments on the best solution, 279
- The GNSS constellations and the indoor positioning problem, 283

9.2 A Brief Review of Possible Techniques, 284
- Introduction to measurements used, 284
- Comments on the applicability of these techniques to indoor environments, 285

9.3 Network of Sensors, 287
- Ultrasound, 287
- Infrared radiation (IR), 287
- Pressure sensors, 289
- Radio frequency identification (RFID), 290

9.4 Local Area Telecommunication Systems, 291
- Introduction, 291
- Bluetooth (WPAN), 293
- WiFi (WLAN), 294
- Ultra wide band (WPAN), 296
- WiMax (WMAN), 298
- Radio modules, 299
- Comments, 299

9.5 Wide-Area Telecommunication Systems, 299
- GSM, 300
- UMTS, 301
- Hybridization, 302

9.6 Inertial Systems, 304

9.7 Recap Tables and Global Comparisons, 305

9.8 Exercises, 305

Bibliography, 306

CHAPTER10
GNSS-Based Indoor Positioning and a Summary of Indoor Techniques, 309

10.1 HS-GNSS, 310
10.2 A-GNSS, 312
10.3 Hybridization, 314
10.4 Pseudolites, 315
10.5 Repeaters, 319
- The clock bias approach, 320
- The pseudo ranges approach, 323

10.6 Recap Tables and Comparisons, 328
10.7 Possible Evolutions with Availability of the Future Signals, 333
- HS-GNSS and A-GNSS, 333
- Pseudolites, 333
- Repeaters, 334

10.8 Exercises, 341
Bibliography, 342

CHAPTER11
Applications of Modern Geographical Positioning Systems, 345

11.1 Introduction, 345
11.2 A Chronological Review of the Past Evolution of Applications, 346
- TRANSIT and military maritime applications, 346
- The first commercial maritime applications, 347
- Maritime navigation, 347
- Time-related applications, 348
- Geodesy, 350
- Civil engineering, 350
- Other terrestrial applications, 350

11.3 Individual Applications, 353
- Automobile navigation (guidance and services), 354
- Tourist information systems, 357
- Local guidance applications, 357
- Location-based services (LBS), 358
- Emergency calls: E911, E112, 360
- Security, 361
- Games, 362

11.4 Scientific Applications, 362
- Atmospheric sciences, 362
- Tectonics and seismology, 364
- Natural sciences, 364

11.5 Applications for Public Regulatory Forces, 365
- Safety, 365
- Prisoners, 365

11.6 Systems Under Development, 366

11.7 Classifications of Applications, 367

11.8 Privacy Issues, 368

11.9 Current Receivers and Systems, 369

- Mass-market handheld receivers, 369
- Application-specific mass-market receivers, 371
- Professional receivers, 372
- Original equipment manufacturer (OEM) receivers, 374
- Chipsets, 375
- Constellation simulators, 375

11.10 Conclusion and Discussion, 375

- Accuracy needed, 376
- Availability and coverage, 377
- Integrity, 377

11.11 Exercises, 377

Bibliography, 378

CHAPTER12
The Forthcoming Revolution, 381

12.1 Time and Space, 382

- A brief history of the evolution of the perception of time, 382
- Comparison with the possible change in our perception of space, 383
- First synthesis, 385

12.2 Development of Current Applications, 386

- Transportation, 386
- Cartography, 388
- Location-based services, 389

12.3 The Possible Revolution of Everybody's Daily Lives, 389

- A student's day, 390
- A district nurse's day, 392
- Objects, 393
- Ideas in development, 394

12.4 Possible Technical Positioning Approaches and Methods for the Future, 395

12.5 Conclusion, 397

12.6 Exercises, 398

Bibliography, 398

Index, 401

FOREWORD

Positioning is a function that has always been of major importance in sustaining human activities, whether to explore new lands, improve conditions, or to conduct offensive or defensive warfare. In the past six decades, thanks to the arrival of electronic and related technologies, positioning methods have made a quantum leap and can be applied with sufficiently low cost and low power miniature devices to be accessible to individuals. A leap in these technologies occurred with the introduction of the satellite-based Global Positioning System (GPS) in the 1980s. Superior availability, accuracy, and reliability performances outdoors resulted in a rapid and revolutionary adoption of the system worldwide. The development and successful commercialization of GPS methods for indoor applications in the late 1990s is currently resulting in scores of applications and mass-market adaptation. Meanwhile, the introduction of other satellite-based systems has resulted in the use of a more generic label, namely Global Navigation Satellite Systems (GNSS) for a technology that is increasingly considered a public utility.

The author, Nel Samama, has done a wonderful job in compiling this introductory book on *Global Positioning* along the above timeline. Although the focus is on GNSS, as it should be, other earlier and current methods are clearly described in context. A full understanding of GNSS principles can be a frustrating experience for readers that are not familiar with the required fundamentals of celestial mechanics, signal processing, positioning algorithms, geometry of positioning, and estimation. These topics are well treated in the book and are supplemented by an introduction to modern receiver operation, indoor signal reception, and GNSS augmentation. Examples of applications, described in a separate chapter, illustrate well the utilization diversity of GNSS. The book concludes with an entertaining crystal ball gazing into the future...

Global Positioning keeps the mathematical and physical baggage to a minimum in order to maximize accessibility and readability by an increasingly large segment of developers and users who want to acquire a rapid overview of GNSS. The book fits nicely between existing introductory texts for non-technical readers and the more highly technical textbooks for the initiated engineers and will be of value for numerous college courses and industrial use.

PROFESSOR GÉRARD LACHAPELLE

CRC/iCORE Chair in Wireless Location
Department of Geomatics Engineering
University of Calgary

PREFACE

This preface gives some ideas about the way this book has been written: the goals, the philosophy, and how it is organized. Within the "Survival Guide" series, it is intended to provide an overview of geographical positioning techniques and systems, with an emphasis on radio-based approaches.

The idea of this book is to give a summary of the past, the present, and the short-term progress of positioning techniques. In addition, the goal is to make it simple to understand the main trends and reasons for current developments. With the advent of the European constellation of navigation satellites, Galileo, the planned developments of GPS and GLONASS, and the potential arrival of the Chinese constellation, COMPASS, the positioning domain is experiencing a real transformation. It is likely that positioning will enter everyone's lives, transforming then on a wide scale. An understanding of the fundamental principles, realizations, and future improvements will help estimate the real limitations of the systems.

An important part of the book is thus devoted to Global Navigation Satellite Systems (GNSS) — six chapters almost exclusively deal with matters relating to them. The history of navigation, from both the historical and the technical points of view, will assist in assessing the main advantages and disadvantages of the various possible solutions for positioning. Also of prime importance, two chapters are devoted to indoor positioning, which will provide an overview of the approaches currently under development.

The footnotes are designed to provide additional hints or comments to the reader. There is no absolute need to read them at first sight, thus allowing for smoother reading. These notes are often based on personal comments and should be considered accordingly. Although positioning is the main subject, the approaches described are often also applicable to velocity or time determination, which are as important as location in numerous practical cases. Furthermore, the term positioning is used instead of localization because the raw piece of information is in fact positioning, and is the data this book intends to deal with. Localization usually describes the use of positioning for applications: it is a higher level concept.

The book is organized in six parts. The first part includes Chapters 1 and 2 and gives a brief history of navigation from both the historical (Chapter 1) and technical (Chapter 2) points of view.

The second presents an overview of the possible techniques for geographical positioning and includes Chapters 3 and 4. Chapter 3 is dedicated to GNSS, with a comparison of the three main constellations (GPS, GLONASS, and Galileo), their current status, and their short-term modernization. Chapter 4 addresses the other main techniques of positioning, including, for instance, Wireless Local Area Networks or mobile network approaches.

Chapters 5, 6, and 7 form the third part and give details of GPS, GLONASS, and Galileo. In order to allow a direct comparison, each chapter deals with the three constellations. Chapter 5 sets out a description of the above systems, Chapter 6 gives detail of the various satellite signals, and Chapter 7 deals with the acquisition and tracking of these signals.

Chapter 8 constitutes the fourth part and shows how to calculate a position once the measurements are available. The methods of calculation given in this chapter, although related to GNSS, are applicable to any positioning system.

Indoor positioning, as a fundamental current challenge in navigation systems, is dealt with in the fifth part, and includes Chapters 9 and 10. Chapter 9 describes some non-GNSS-based techniques, while Chapter 10 is devoted to those using satellite signals, in one way or another. The first part of Chapter 9 is a global introduction to the "indoor problem." Note also that summary tables are provided at the end of Chapter 10, including all the indoor techniques described.

Applications, either current or future, constitute the last and sixth part of the book. Chapter 11 is a description of the main current applications and devices. Chapter 12, on the other hand, intends to analyze how everyone's lives are likely to be modified in the coming years, with the wide availability of positioning for both people and goods.

Finally, exercises are designed to strengthen the understanding of some specific aspects of the chapters, but also to go one step further. To do this, further reading may sometimes be required. Many exercises are not calculation ones but rather oriented towards the analysis of the chapter contents.

ACKNOWLEDGMENTS

I now understand the reason why all writers thank their families and friends for their patience and abnegation: all these books and documents open on the table, all this time spent in front of the computer, all this energy spent on this piece of paper, these so frequent moments thinking of the best figures that could be used to show a particular aspect, and so on. May all of the people concerned have my real recognition for their continuous efforts.

More specifically, I would like to thank the people who agreed to read the early versions of the book. Their comments and suggestions have really improved the final version in its legibility and pertinence. Thanks to Anca Fluerasu, Muriel Muller, Serge Bourasseau, Nabil Jardak, Marc Jeannot, Jérôme Legenne, Michel Nahon, and Per-Ludwig Normark.

Special thanks and much gratitude to Gerard Lachapelle who, in addition, agreed to write the foreword of this book and to Emmanuel Desurvire who answered my basic questions about writing such a book. Also a great thank you to Günther Abwerzger, Jean-Pierre Barboux, and Alexandre Vervisch-Picois for the very interesting comments they provided.

This book also would not have been possible without the constant understanding of the people of the Navigation Group of the Electronics and Physics Department of the Institut National des Télécommunications (INT), France. They have very often been obliged to carry on their activities alone. I would also like to thank former colleagues, Marc François and Julien Caratori, who helped me, a few years ago, start the positioning-related activities at INT. Also important are all the students who have enriched our reflections with their work and valuable exchanges — thanks also to all of you.

Last but not least, another special thank you to Dick Taylor, who made many corrections to the English of the book — he is certainly the only person who will ever read the book twice!

CHAPTER 1

A Brief History of Navigation and Positioning

In this chapter, we look back at the evolution of geographical positioning, from astronomical navigation of ancient days to today's satellite systems. Major dates are given together with a description of the fundamental techniques. These techniques are essentially still used today. The development phases of modern satellite positioning systems are also provided.

As soon as human beings decided to explore new territories, they needed to be able to locate either themselves or their destination. At first, only terrestrial displacements were of concern, and the issue was to be able to come back home. The "come back" function was achieved by using specific "marks" in the landscape that had to be memorized. Quite quickly, because of the possibility of carrying very large loads by sea, maritime transportation became an interesting way of traveling. New needs arose regarding positioning because of the total absence of marks at sea. Thus, navigators had the choice of following the shore, where terrestrial marks were available, or of finding a technique for positioning with no visibility to the shore. This was the starting point of geographical positioning.

■ 1.1 THE FIRST AGE OF NAVIGATION

The origins of navigation are as old as man himself. The oldest traces have been found in Neolithic deposits and in Sumerian tombs, dating back to around 4000 BC. The story of navigation is strongly related to the history of instruments, although they did not have a rapid development until the invention of the maritime clock, thanks to John and James Harrison, in the eighteenth century. The first reasons pushing people to "take to the sea" were probably related to a quest for discovery and the necessity of developing commercial activities. In the beginning, navigation was carried out

Global Positioning: Technologies and Performance. By Nel Samama
Copyright © 2008 John Wiley & Sons, Inc.

without instruments and was limited to "keeping the coast in view." It is likely that numerous adventurers lost their lives by trying to approach what was "over the horizon."

Hieroglyph inscriptions comprise the most ancient documents concerning ships and the art of navigation. Oars and a square sail mounted on a folding mast were the means of propulsion of the first maritime boats, which followed on from fluvial embarkations (around 2500 BC). The steering was achieved with an oar used as a rudder located at the back of the boat and maintained vertically. This configuration did not allow sailing in all wind conditions. Then came the Athenian *trireme*; about 30 m long and 4 m wide, it was a military vessel with one or more square sails and two or three rows of rowers. The Roman galleys showed no great improvement over it. Compared to these military ships, commercial vessels had more rounded forms and only used sails for their displacement, oars being reserved for harbor maneuvers. Thus, navigation was only possible with restricted wind conditions, leading to the need for a good knowledge of the weather.

Thanks to this knowledge and in spite of the limited size of their boats and the rusticity of their navigation instruments, the Phoenicians, as early as the beginning of the twelfth century BC, had moved all around the Mediterranean Sea. The Carthaginians had even navigated as far as Great Britain and it seems they tried to sail around Africa with no success. Navigation was mainly achieved during the day, and the instruments used were the eyes of the navigators, in order to keep in sight of the coast, and a sounding line. When navigators had to travel by night, they used reference to the movements of the stars as had the Egyptians. Then, the Greek astronomer Hipparchus created the first nautical ephemeris, and built the first known astrolabes (around the second century BC). It has to be noted that the basics of modern navigation principles were already established then and only the technical aspects would be improved. For instance, current global navigation satellite systems use reference stars (the satellites), together with ephemeris (to allow the receiver to calculate the actual location of the satellites), and highly accurate measurements. This latter requirement, together with the fact that current systems carry out distance measurements, are the main differences from ancient navigation.

The astronomical process was quite inaccurate and frequent terrestrial readjustments were required. Localization was even more complex because of the lack of maritime maps. The ancients rapidly drew up documents to describe coasts, landmarks, and moorings — this allowed coastal navigation. For the same purpose, and also for security reasons, lighthouses were built. The most famous is the one in the harbor of Alexandria, built on the island of Pharos during the third century BC. Unfortunately, astronomical positioning was only able to provide the latitude of a point, as can be understood from Fig. 1.1. The longitude problem would remain unresolved for centuries, as described in Section 1.2.

The first known instrument was the Kamal (Fig. 1.2). By making a measurement of the angle between the horizon and a given star, using respectively the bottom and the top of the Kamal, it was possible to obtain directly the latitude of the point of observation. Of course, the accuracy of the Kamal was not sufficient to provide precise navigation, but it was sufficient for coastal navigation. The string seen in Fig. 1.2 had knots tied in it at given positions to note specific locations (i.e., latitudes).

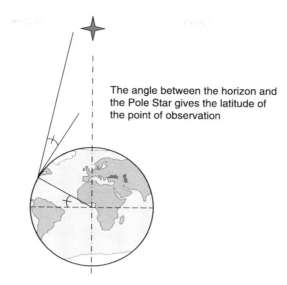

The angle between the horizon and the Pole Star gives the latitude of the point of observation

FIGURE 1.1 Determining latitude with the Pole Star.

The sailor would put a specific knot between his teeth to define the actual "navigation parameters" for the current destination (latitude in this case). This was achieved by varying the distance the Kamal was from the teeth of the sailor, leading to a given angle.

Empirical knowledge of maritime currents has certainly been of great help for the navigators who planned incredible voyages, such as those undertaken by the people of the Polynesian islands. Northern Europeans also participated actively in the field of navigation, the story starting in the third century AD, with the first attempts by the Vikings to explore the northern Atlantic Ocean (the colonization of Iceland and

FIGURE 1.2 The first navigational instrument: the Kamal. (*Source*: Peter Ifland, *Taking the Stars, Celestial Navigation from Argonauts to Astronauts.*)

Greenland happened from the ninth century AD). The first European discovery of northern America, which they called Vinland, also happened in around 1000 AD.

The medieval maritime world was separated into two areas: the *Levant* (the Mediterranean area), where the Byzantines were leaders, and the *Ponant*, from Portugal to the north, where Scandinavian maritime techniques were used. It was only at the end of the fourteenth century that the two types of techniques were joined, in particular by using the stern-mounted rudder (appearing during the thirteenth century).

At around the same time, an instrument indicating absolute orientation appeared in the Mediterranean area, where the first reference to the magnetic needle for navigation purposes was attributed to Alexander Neckam (around 1190). Since the discovery of the properties of a needle in the Earth's magnetic field in the first or second century, it seems probable that the use of magnetism for navigation dates from around the tenth century AD, in China. From then, the development of the compass was continuous, starting with a pin attached to a wisp of straw floating on water, to the mounting of a dial to cancel out the vessel's movements. Other important discoveries include the determination of the difference between the magnetic north and the geographical north (fifteenth century).

Although the compass was a fundamental discovery, it was far from the final answer for ocean navigation. The main empirical characteristics of ancient navigation remain: dead reckoning,[1] which is based on the navigator's expertise, imprecise calculation of latitude by astral observations, and the deduction of the current location using nautical ephemeris established in Spain during the thirteenth century AD (Alphonsine tables).

In any case, it would have been of no interest to establish a precise location without having similarly precise maritime maps, which was assuredly not the case. Following the maps of the world used during the eleventh century, the first maps showing the contours of the coasts associated with compass marks for orientation purposes appeared. These were the first portolan charts (Fig. 1.3), which show a set of crossing lines referenced to a compass rose (a specific representation of the compass). From the ideas of Ptolemy, the Arabians accomplished great cartographical developments. For instance, Idrisi (1099–1165 AD) drew a map that can be considered as the synthesis of the Arabian knowledge of the twelfth century. It included details from Europe to India and China, and from Scandinavia to the Sahara.

Portolans were used to navigate from harbor to harbor. A network of directions referenced to the magnetic north allowed courses to be set. On such a map, information was available concerning the shore, but very little about the hinterland. The portolans are the main medieval contribution to cartography, and the precursors of modern maritime maps. An example of a sixteenth-century portolan of Corsica is given in Fig. 1.3.

One very important geographical area was the Indian Ocean. This zone was a meeting place for sailors from the Mediterranean, Arabia, Africa, India, and the Far East. The ancients well understood the advantage of the monsoon, and the ocean was a great opportunity for commercial exchanges. At the beginning, from the first century AD, the rhythm of the monsoon was well known, and Arabian and Chinese

---

[1]Dead reckoning is the ability to evaluate one's displacement with no absolute positioning instrument.

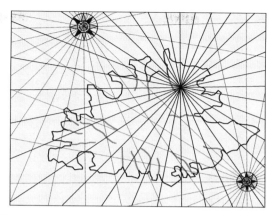

FIGURE 1.3 A sixteenth-century portolan of Corsica.

(ninth century) sailors were already used to navigating in the ocean for commercial purposes. The adventures of Marco Polo (thirteenth century) and Ibn Battuta (fourteenth century) confirm the persistence of such cultural and commercial exchange routes.

The Arabian navigation techniques used in the Indian Ocean were empirical approaches mainly based on the sidereal azimuth rose, which took advantage of the low latitude and navigation in clear skies. Such techniques took about two centuries to reach the Mediterranean region. The principle of such an azimuth rose is to divide up the horizon into 32 sectors using 15 stars scattered through the sky. It seems that Chinese navigators were a little ahead in terms of astronomical navigation, as well as magnetic tools. Their boats were also certainly more advanced concerning their sails, the axial rudder they used, and probably also a stern-mounted rudder. However, at the end of the Middle Ages, the Portuguese techniques of navigation had a definitive advantage.

■ 1.2 THE AGE OF THE GREAT NAVIGATORS

From the middle of the fifteenth century there arose the need to find a route to the East for commercial activities with India, but which avoided the region of Persia. There were two possibilities: the first was to sail around the south coast of Africa and the second was to sail west, under the assumption that the Earth was a sphere. The Portuguese chose the first solution (led by Henry the Navigator) when Bartolomeu Dias reconnoitered the Cape of Good Hope in 1487. Following this first expedition, Vasco da Gama reached India around the aforementioned cape in 1498. The Spanish chose the second route, heading directly to the west across the Atlantic Ocean, and Christopher Columbus finally "discovered" America in 1492. The real west route to India was only discovered later when Magellan found the way through the Magellan strait to the south of South America in 1520. Figure 1.4 gives a global view of the most famous navigation routes. Note that in the Indian Ocean, Zheng He led many expeditions in the fifteenth century.

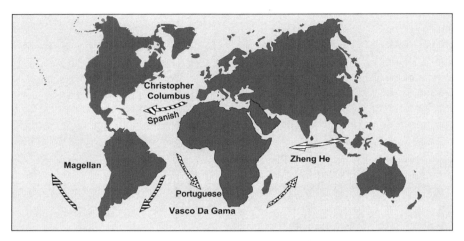

FIGURE 1.4 The great navigators' travels — fifteenth century.

The navigation techniques, however, remained identical to those known and used at the end of the Middle Ages. The progress in navigation was mainly attributable to men's skills and training. The Portuguese opened the Sagres School of navigation and the Spanish the Seville College, where the prestigious Amerigo Vespucci trained many famous sailors. Dead reckoning (with its associated accuracy), evaluation of currents, and the hourglass were the main methods and instruments used. A very important contribution due to Christopher Columbus was the discovery of magnetic declination and its variations.

The Portuguese and Spanish then decided to share the world, thanks to their maritime superiority (Fig. 1.5). Despite the treaties of Tordesillas (1494) and Saragossa (1529), both nations had claims that could not be precisely checked because of the low accuracy of their positioning techniques, even on land. Remember that only the

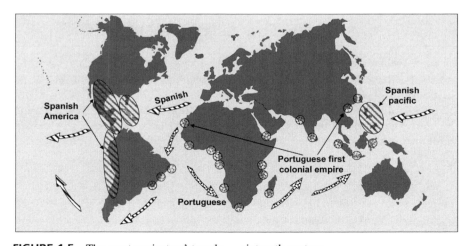

FIGURE 1.5 The great navigators' travels — sixteenth century.

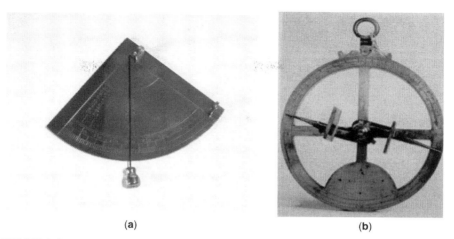

(a)   (b)

FIGURE 1.6 A fifteenth-century quadrant (**a**) and a seventeenth-century astrolabe (**b**). (*Source*: Peter Ifland, *Taking the Stars, Celestial Navigation from Argonauts to Astronauts*.)

latitude of a location could be established, but not the longitude, as no technique was available. So territorial limits were defined through the help of significant land characteristics (river, mountains, and so on), but this was not a simple approach for countries located ten thousand miles away! Determining longitude on land was just about to find a satisfactory answer following Galileo's work on Jupiter's moons, but determining longitude at sea would remain impossible until the late eighteenth century.

The evaluation of latitude was based on the elevation measurement of a reference star over the horizon, which is a simple notion at sea. In the northern hemisphere, the pole star[2] was established as the reference star at the very beginning of navigation, but was not visible once in the southern hemisphere. The Portuguese, who had chosen to investigate the southern route around Africa, faced this problem quite early in the middle of the fifteenth century. A new method was required, and it was found that the course of the Sun over the sky (and more precisely the elevation of the Sun when passing through its apogee), together with astronomical tables (the *regimentos*), allowed the evaluation of latitude worldwide.

To achieve the angle measurements required for both polar or Sun elevation, the instruments used were based simply on technical developments of the Kamal concept. The requirement was to make a double sighting: horizon and star. The first instrument was a quadrant equipped with a sinker, allowing a direct reading of the latitude angle. Figure 1.6a shows the one attributed to Christopher Columbus.

---

[2]The Pole or North Star, Polaris, is at the very end of the Little Bear constellation. It is quite an important star, as it is almost exactly above the North Pole. Thus, its apparent movement is static, unlike the other stars that appear to move during the night due to the Earth's rotation. The Pole Star was found to be a very good "reference light."

The Greeks realized that Polaris was not exactly above the North Pole. We know that the reason for this is the slow movement of the Earth's axis over thousands of years. It appears that 5000 years ago, a star called Thuban was nearest to the polar axis and, in 5000 years, Alderamin will be the one, and back to Polaris in 28,000 years.

FIGURE 1.7 A Davis cross staff — sixteenth century. (*Source*: Peter Ifland, *Taking the Stars, Celestial Navigation from Argonauts to Astronauts.*)

The astrolabe (Fig. 1.6b), whose first use was to define the stars' relative locations in the sky, was also adapted to navigation, allowing angle measurements. The moving part of the astrolabe, the alidade, is much better than the sinker for reading when the sea is rough. The main problem of both the quadrant and the astrolabe is that it is difficult to achieve the pointing of both the horizon and the star simultaneously. In addition, a direct sighting of the Sun is dazzling. Thus, another instrument was designed: the cross staff (Fig. 1.7). Its principle is to carry out a measurement of the length of a shadow cast by the Sun.

The use of reflection mirrors, attributed to Isaac Newton in 1699, allowed simultaneous sighting of both the horizon and the star, using two mirrors (one of which was mobile). The sighting was then achieved in one go. The first such instrument was the octant, using a 45° graduated arc, soon followed by the sextant, using a 60° graduation (Fig. 1.8). The compass was also updated by dividing it into 360°, instead of 32 sectors, and adopting specific mountings in order to ease the plotting of landmarks. Dead reckoning was also improved by the use of a *loch*, a new instrument for measuring speed, simply composed of a hemp line graduated with knots that the navigator let out at sea for a fixed period of time, defined by the hourglass. Thus, he had an evaluation of the distance traveled during this time and consequently the speed of the boat.

Another important point was related to the evolution of maps. In the middle of the sixteenth century, Gerhard Mercator invented the projections that bear his name. This approach consisted of considering a representation in which the distance between two parallels[3] increases with latitude. It was therefore possible to represent the route that follows a constant heading by a straight line because of the conservation of angles. At the same time, the plotting of coasts was becoming better. Nevertheless, the problem of location accuracy remained, as longitude was still not measurable.

---

[3]A parallel is a circle at the Earth's surface being perpendicular to and centered on the North–South axis of the Earth. A parallel defines locations of equal latitude.

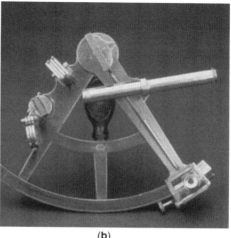

(a)
(b)

FIGURE 1.8 An octant (**a**) and a sextant (**b**). (*Source*: Peter Ifland, Author of *Taking the Stars, Celestial Navigation from Argonauts to Astronauts.*)

The conquest of new territories continued with the British, the French, and the Dutch during the eighteenth century. This was the time of expeditions in North and South America, and also in the Indian Ocean, trying to find the best way to travel towards the East (Fig. 1.9). Cook's famous expeditions to the Pacific Ocean were also great chapters in this era of navigation.

It took almost three centuries for the longitude problem to be solved. During this period, significant progress occurred in the development of instruments and maps, but nothing in determining longitude. As early as 1598, Philipp II of Spain offered a prize to whoever might find the solution. In 1666, in France, Colbert founded the Académie

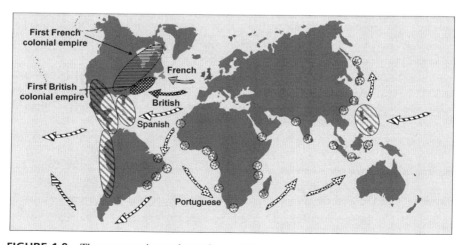

FIGURE 1.9 The great navigators' travels — eighteenth century.

des Sciences and built the Observatory of Paris. One of his first goals was to find a method to determine longitude. King Charles II also founded the British Royal Observatory in Greenwich in 1675 to solve this problem of finding longitude at sea. Giovanni Domenica Cassini, professor of astronomy in Bologna, Italy, was the first director of the French academy, and in 1668 proposed a method of finding longitude based on observations of the moons of Jupiter. This work followed the observations made by Galileo concerning these moons using an astronomical telescope. It had been known from the beginning of the sixteenth century that the time of the observation of a physical phenomenon could be linked to the location of the observation; thus, knowing the local time where the observations were made compared to the time of the original observation (carried out at a reference location) could give the longitude. Cassini established this fact with Jupiter's moons after having calculated very accurate ephemeris. Unfortunately, this approach requites the use of a telescope and is not practically applicable at sea.

In 1707, Admiral Sir Clowdisley Shovell was shipwrecked, with the loss of 2000 men, on the Scilly Islands, because he thought he was east of the islands when in fact he was west of them! This was too much for the British. On June 11, 1714, Sir Isaac Newton confirmed that Cassini's solution was not applicable at sea and that the availability of a transportable time-keeper would be of great interest. It should be noted that Gemma Frisius also mentioned this around 1550, but it was probably too early... On July 8, 1714, Queen Anne offered, by Act of Parliament, a £20,000 prize[4] to whoever could provide longitude to within half a degree. The solution had to be tested in real conditions during a return trip to India (or equivalent), and the accuracy, practicability, and usefulness had to be evaluated. Depending on the success of the corresponding results, a smaller part of the prize would be awarded.

The development of such a maritime time-keeper took decades to be achieved, but finally had an impact on far more than navigation. The history of Harrison's clocks is quite interesting, and time is really the fundamental of modern satellite navigation capabilities. We have seen that Isaac Newton himself confirmed that the availability of a transportable maritime clock would be the solution to the longitude problem. The realization of such a clock, however, was not so easy. The main reason is that the clock industry was fundamentally based on physical principles dependent on gravitation (the pendulum). This was acceptable for terrestrial needs, but of no help in keeping time when sailing. Thus, a new system had to be found.

The reason that time is of such importance is because of the Earth's motion around its axis. As the Earth makes a complete rotation in 24 hours, this means that every hour corresponds to an eastward rotation of 15°. Thus, let us suppose that one knows a reference configuration of stars (or the position of the Sun or Moon) at a given time and for a given well-known location (e.g., Greenwich). If you stay at the same latitude, then you will be able to observe the same configuration but at another time (later if you are eastward and earlier if westward); the difference in times directly gives the longitude, as long as the time of the reference location (Greenwich in the present example) has been kept. The longitude is simply obtained by multiplying this difference by 15° per hour, eastward or westward. The method is

---

[4]This amount is equivalent to more than 15 million dollars today.

very simple and the major difficulty is to "keep" the time of the reference place with a good enough accuracy, that is, with a drift less than a few seconds per day. Pendulums, although of good accuracy on land, were unable to provide this accuracy at sea, mainly because of the motions of the ship and changes in humidity and temperature.

John Harrison built four different clocks, leading to numerous innovative concepts. After almost 50 years of remarkable achievements, in August 1765 a panel of six experts gathered at Harrison's house in London and examined the final "H4" watch. John and William (his son) finally received the first half of the longitude prize. The other half was finally awarded to them by Act of Parliament in June 1773. Certainly more important is the fact that John Harrison was finally recognized as being the man who solved the longitude problem.

One of the most famous demonstrations of Harrison's clocks' efficiency was provided by James Cook during the second of his three famous voyages in the Pacific Ocean. This second trip was dedicated to the exploration of Antarctica. In April 1772, he sailed south with two ships: the *Resolution* and the *Adventure*. He spent 171 days sailing through the ice of the Antarctic and decided to sail back to the Pacific islands. He returned to London harbor in June 1775, after more than 40,000 nautical miles. During this voyage, he was carrying K1, Kendall's copy of Harrison's H4. The daily rate of loss of K1 never exceeded eight seconds (corresponding to a distance of two nautical miles at the equator) during the entire voyage. This was the proof that longitude could be measured using a watch.

■ 1.3 CARTOGRAPHY, LIGHTHOUSES AND ASTRONOMICAL POSITIONING

It is quite obvious that before planning a route, knowing the current location is fundamental. The notion of route also means being able to "draw" the locations on a map. Thanks to Mercator, a graphical representation of the Earth was defined that allowed angles to be preserved whilst projecting the sphere onto a plane (Fig. 1.10). This approach is essential, as it was thus possible to navigate following

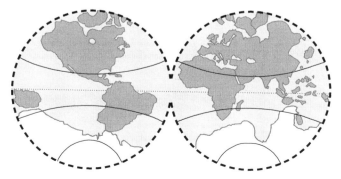

FIGURE 1.10 The great Mercator's projection.

a compass orientation, that is, a constant angle. The projection is carried out on a cylinder tangent to the Earth at the equator, and modifies the projected values in such a way that a route crossing all the meridians with the same relative angle becomes a straight line in the projected plane. The shortest route from one point to another is then represented by a complex concave curve (because the constant-angle route is not the optimal one on a sphere).

Astronomical positioning is not very accurate, so sailors also used landmarks to keep track of their location. These terrestrial landmarks were sighted from the boat, allowing the plotting of a straight line passing through both the landmark's location (which needs to be known, like the satellite locations in today's satellite navigation systems) and the boat. The plotting of three such lines gave the theoretical location of the boat.[5] Of course, because of measurement errors and inaccuracies, this would lead to the creation of a triangle, and the location was usually considered to be the center of the circle that could be drawn inside the triangle. Lighthouses are one of the most noticeable of the landmarks, but others include natural sights, buoys, and so on.

Astronomical positioning consisted of determining the location of the boat by measuring the heights of some given stars above the horizon (by using a sextant). The possible positions on Earth that exhibit the same observed height of a given star are located on a circle that is centered on the vertical line joining the star and the center of the Earth. The radius of this circle is dependent on the heights of the star above the horizon. Of course, the location of the reference star was required — this was obtained through ephemeris, which gave the exact locations of various stars according to the time. Calculating the position (described in Chapter 2) required two measurements to solve the set of two second-degree equations. To be rigorous, the two measurements should be carried out at the same time, otherwise complex recalculations are needed.

Positioning in ancient times was thus typically a discrete event. To be able to follow the location of the boat permanently, or at least to have an idea of its route, required so-called "dead reckoning." The basic principle of this is to measure both the direction and the speed of the boat. In this way, the complete kinematics is defined and the location can be evaluated with accuracy, as long as both the measurements and initial position are accurate. For a long time this was not the case, but the new radio electric techniques would bring about a revolution in positioning, as described in the following.

■ 1.4 THE RADIO AGE

The desire to communicate over long distances was described long before the radio conduction phenomenon was discovered. The first related facts using optical means date from the fourth and fifth centuries BC, when fires on the tops of mountains were used to serve as "communication relays." This approach was still being used by

---

[5]This technique is similar to that still currently used for topography purposes and is called "triangulation."

the first optical telegraphs in the seventeenth century. Of course, the main disadvantage of such a system lies in the fact that transmission is limited to the optical line of sight and requires good "air conditions," that is, no fog. This problem led to the development of the electrical telegraph.

On November 24, 1890, Edouard Branly discovered the phenomenon of "radio conduction," in which an electrical discharge (generated by a Hertz oscillator) had the effect of decreasing the resistance of his "tube." It appeared that electrical propagation was possible without cables. Further work showed that adding a metallic rod to the generator improved the range of the transmission (that is, the detection was also possible further away from the generator). Alexander Popov was in fact just about to invent antennas. The transmission path grew from a few tens of meters to 80 m. In 1896, Popov succeeded in transmitting a message over 250 m (the message being composed of two words, "Heinrich Hertz").[6]

At the same time, Guglielmo Marconi, who was deeply influenced by the publications of Faraday and the life of Benjamin Franklin, felt that it should be possible to establish a transmission over a few kilometers. After a lot of work, he transmitted the letter "S" coded in Morse ("...") over 2400 m at the end of 1895. In September 1896, by using a kite as an antenna, Marconi achieved a 6 km, then 13 km radio path. In May 1897, a transmission of 15 km was demonstrated between two English islands (Steep Holm and Flat Holm), followed by similar performances in Italy in La Spezia harbor. Marconi founded, on July 20, 1897, the Wireless Telegraph and Signal Company. In March 1899, the first trans-Channel message was sent between South Fireland (Great Britain) and Wimereux (France). The addressee was Edouard Branly. With antenna heights of 54 m, this 51 km transmission was achieved with a global performance of 15 words per minute. In July, a 140 km path was achieved between a sea position and the coast. After this new success, Marconi was almost certain that trans-horizon radio paths were possible.

In October 1900, Marconi started drawing up the plans of Poldhu Station (in Cornwall, UK; see Fig. 1.11), which was planned to be the transmission station for the first trans-Atlantic transmission. In April 1901, the construction of the antenna started with the first mast of 65 m in height. In August, 20 masts were aligned in a 56-m diameter circle ... and were destroyed by a storm in September. One week later, a new temporary station was available. Meanwhile, Marconi searched for a reception site in the United States from March 1901. His choice was finally Cape Cod, in Massachusetts. In October, a storm destroyed the antennas, once again. The decision was then made to turn to more simple antenna architectures, such as two masts of 15 m height with wires in between for the transmission site, and a kite-supported wire for the reception.

The new site was Signal Hill (Fig. 1.12) in Newfoundland, still a British colony at this time. This station was ready for experimentation on December 9, 1901. From this date, it was decided that Poldhu would send the letter "S" ("...") each day between 11:30 and 14:30, Signal Hill time (the need for synchronization is definitely a

---

[6]For more details see "Comment BRANLY a découvert la radio," Jean-Claude Boudenot, EDP Sciences (in French!).

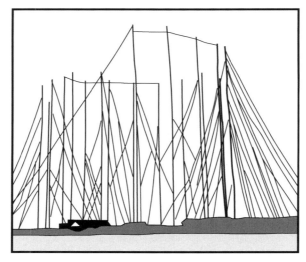

FIGURE 1.11   The Poldhu station.

fundamental aspect). On December 12, the signal was received at 12:30, over a path of 1800 miles (3500 km) including the Earth's curvature!

To return to navigation, it was only about ten years later (1907) that radio electric signals were used, by transmitting time signals. As already described, knowing the time at a specific location is fundamental in calculating the longitude. Until then, this was achieved through the use of Harrison's clocks. Radio transmission was a fantastic improvement, especially in terms of accuracy, because the signal is transmitted at the speed of light, thus greatly increasing the accuracy of the "time transfer."

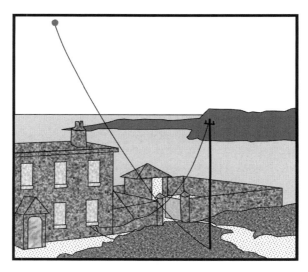

FIGURE 1.12   The Signal Hill station.

The corresponding improvement of positioning is around tenfold. The second application of radio electric waves was to use the signal as a new landmark that no longer needed to be in a visible line of sight. The first such system was implemented on board a ship in 1908, together with a movable antenna that could give an indication of the bearing of the transmitter. This was the first dedicated radio navigation system.

Although a physical understanding of wave propagation was certainly not available at these early stages of development (the wavelength of Marconi's first trial of transatlantic transmission was at around 2000 m, dictated by the wave generator rather than chosen for propagation purposes), new capabilities, and also great similarities, can be highlighted from a comparison of astral measurements and the first navigation approaches with radio electric waves. The first improvement obviously lies in the fact that this new "landmark" can still be used in bad meteorological conditions. This is certainly one of the most important features as it is specifically in these conditions that navigation at sea is at its most dangerous and when positioning is so important. It has often been said that radio electric signals are a modern implementation of a lighthouse, and this is the reason why these beacons are often known as "radio lighthouses." The way the old and new methods were used also showed great similarities. Sailors were used to making angle measurements to obtain distances from a central astral component (the Moon for example) and given stars, or to take sightings from referenced lighthouses to evaluate their location. The new radio beacons allowed the same kind of positioning using measurements based on electrical properties such as the amplitude of currents or voltages. This would simplify the automation of navigation systems as electrical engineering rapidly progressed.

■ 1.5 THE FIRST TERRESTRIAL POSITIONING SYSTEMS

The first systems were based on radio goniometry[7]; by having a rotating antenna and by detecting the maximum power, it was possible to determine the direction of the landmark. The radio compass was one of the most advanced forms of radio goniometrical systems. Another approach was that used for radio lighthouses. Determining both the identification and the orientation of the transmitters had to be easy to obtain, so the technique consisted of having a couple of antennas radiating complementary signals (for instance, the equivalent of A "·–" and N "–·" in Morse). When a receiver is in both main radiating lobes, the signal received is continuous. In 1994, more than 2000 radio lighthouses were available all around the world.

As local time generators (oscillators or atomic clocks) were evolving rapidly, new uses of radio signals were developed. This was the case for hyperbolic systems. The basic principle states that all locations having the same difference of signal travel time to two fixed points, for instance two radio transmitters, lie on a geometrical figure that is a hyperbola. The focal points of this hyperbola are the transmitters. As

---

[7]Goniometry is the way of measuring the angle of rotation of the aerial of a wireless system in order to obtain the direction of arrival of the radio wave.

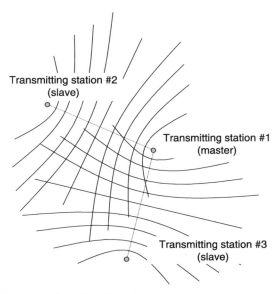

Transmitting station #2
(slave)

Transmitting station #1
(master)

Transmitting station #3
(slave)

FIGURE 1.13 Representation of the hyperbolic approach.

signal-processing capabilities increased, such time difference estimations and measurements became possible. Note that synchronization at the mobile receiver's end is thus avoided as long as time differences are carried out. The basic idea was to obtain two such differences in order to allow the calculation of the intersection point of the resulting two hyperbolae (Fig. 1.13). This approach leads to a theoretical single point in a two-dimensional space.

The first system that used this technique was the Decca,[8] which came into operation at the end of World War II. It worked within a frequency band of 70–128 kHz, allowing for approximately 450 km of operational range. The resulting accuracy was in the range of a few hundred meters, depending on propagation conditions. The new era of radio electric signals allowed for a rigorous evaluation of accuracy — a very important parameter.

The current e-Loran[9] is also a hyperbolic system, but which added new features concerning the modulation scheme, based on pulse trains forwarded by each master and slave station.[10] This approach allowed for a more precise positioning, together with an efficient way of identifying the various stations. The frequency band used was first in the 1750–1950 kHz range for LORAN-A, and is currently in the 90–110 kHz range for eLoran (which is the latest development of LORAN-C). This frequency involves the use of very large antennas (Fig. 1.14) in order to achieve long range and high power transmission.

---

[8]Proposed by the Decca Navigator Company.
[9]Enhanced LOng RAnge Navigation system.
[10]The master station is the one that masters the time. The slave stations have to be synchronized with the master station.

FIGURE 1.14 A typical LORAN antenna. (*Source*: Megapulse.)

These first terrestrial systems provided what is termed "local" area coverage, even though this coverage can be quite large (this is the case for LORAN). However, some people imagined an even more ambitious project that would be the ultimate version of a terrestrial system with a global coverage: the Omega system. It was made up of eight stations using a very low frequency (VLF) band in order to have a complete coverage of the Earth. It was still a hyperbolic approach; each station transmitted sequentially, always in the same order for about 1 s (the duration of emission is specific to each station). The emission consisted of pure continuous waves (no modulation scheme) at respectively 10.2, 11.33, and 13.6 kHz. The sequencing of transmission of these three frequencies was also specific to each station. The total polling sequence lasted 10 s, and the synchronization was required to be better than 1 μs. To calculate an accurate location, it was also necessary to apply propagation corrections. These were based on long-period (typically 15 days) corrections depending on the date, the time of the day, and also on the estimated location. Once more, as we shall see later, the modern principle of satellite constellations was already present, only without the satellite aspects. The global accuracy was generally better than 8 km.

The major reason for the poor accuracy of the abovementioned systems is included in propagation modeling (this point has constantly driven the evolution of modern systems). A new system was designed in order to reduce the propagation errors, and was called the differential Omega. The idea was simply to consider a receiver located in a well-known position, and which monitored the difference between the calculated location and the actual one. This difference was used as an error vector that could be subtracted from the calculated location of any receiver that encountered the same propagation conditions, that is, which was located in the vicinity of the fixed reference receiver. Using this differential approach, the accuracy dropped down to 1.5 km, as long as the receiver remained within about 450 km of the fixed reference; that is, the propagation conditions remained almost identical. The reader should note that this approach was also deployed with the Global Positioning System (GPS) when the U.S. government attached a deliberately generated error, the so-called Selective Availability (which was switched off on May 1, 2000).

Besides these global coverage systems, some specific systems are locally deployed, such as the VOR (VHF Omnidirectional Radio Range) or TACAN (Tactical Air Navigation), as well as others such as ILS (Instrument Landing System) and MLS (Microwave Landing System). All these systems were developed for air navigation purposes, in order to provide the locations of aircraft relative to ground facilities. The VOR is essentially a rotating radio lighthouse and has a medium distance range. The frequency of transmission lies between 108 and 118 MHz, and the signal is modulated in such a way that the transmission is composed of two simultaneous and independent signals at 30 Hz, whose difference of phase characterizes the azimuth of the receiver. In order to achieve this, the VOR radiates at a variable 30 Hz with a symmetric radiating diagram exhibiting a cardioid pattern. Simultaneously, the second signal is a 30 Hz uniform (omnidirectional pattern) signal whose phase is identical in all directions.[11] The onboard receiver calculates the direction of the VOR station, but also selects an azimuth route for a direction chosen by a user. If this direction is the VOR station, then the phase difference signal should be zero. Thus, an indication of the deviation between the real route and the VOR station direction can be provided. Such equipment could be used for positioning, as long as three VOR stations are in radio visibility, by achieving a triangulation from the three directions of arrival measurements. As a matter of fact, this system is mainly used as an alignment device and is usually associated with DME (Distance Measuring Equipment), which gives the distance between the device and a reference ground station. An association between VOR and DME thus gives the plane's location (in polar coordinates) in the ground station referential. The range is typically 200 miles in good conditions and the accuracy lies around 0.2 miles or 0.25% of the measured distance. The military extrapolation of VOR and DME is the TACAN system, which includes both functions on the same carrier frequency.

The accuracy of such a system is nevertheless not enough to provide aviation with an all-weather landing system. Thus, the ILS and MLS were developed. The first is composed of two rotating radio lighthouses that respectively define a direction (the alignment of the runway) and an angle of approach. The first uses frequencies in the range 108–112 kHz and the second 328–335 MHz. Because of the use of these frequencies, multipath effects can occur and trouble the angle measurements. In addition, the system also includes two or three "markers" (radio beacons), which radiate vertically and are distance markers on the approach to the runway. The MLS was seen as the solution to this problem; it used a high-frequency (5 GHz) narrow rotating beam $(1-3°)$ in order to scan space.

---

[11]The reference of all stations is the magnetic north, except in the Polar Regions.

In addition to the abovementioned systems, there are some other local systems that implement either hyperbolic approaches (Hi-Fix, Sea-Fix, Raydist, Lorac, Toran, and so on) or circular approaches (Mini Ranger, Micro-Fix, Trisponder, Tellurometre, Geodimeter, Syladis, Axyle, and so on). The hyperbolic systems almost exclusively use the 1.6–3 MHz band. The corresponding wavelength, when compared to LORAN for instance, allowed much better measurement accuracy, to the order of a few meters. Carrier phase measurements were carried out, and an ambiguity (distance error of half the wavelength) was required to be removed by specific methods (not described here). In the case of circular systems, positioning is obtained using intersecting circles (and no longer hyperbolae) as direct distance measurements are carried out and there is no longer the need for establishing the difference. These systems were called "range–range" and the use of higher frequencies allowed frequency modulation to be developed, notably in relation to code sequences, which permitted the ambiguity problem to be reduced.

■ 1.6 THE ERA OF ARTIFICIAL SATELLITES

In the late 1920s, physicians and mathematicians showed that it was theoretically feasible to imagine artificial satellites launched from the Earth's surface and orbiting the Earth. Of course, a lot of research was still required, but it was thought possible. In 1952, the International Council of Scientific Unions decided that from July 1, 1957, to the end of 1958 would be the "International Geophysical Year (IGY)." The main reason for this choice was that the astrophysical activity of the Sun and a few other stars would be of spectacular importance, thus allowing a large number of valuable research activities. In October 1954, the Council adopted a resolution calling for artificial satellite launches during the abovementioned IGY. One has to remember that this was the time of the Cold War between the United States and the Soviet Union; the proposition was a new area of competition, this time scientific, between the nations. In July 1955, the White House announced its wish to make such a launch and issued a call for projects. In September 1955, the Vanguard project (Fig. 1.15), proposed by the Naval Research Laboratory, was selected among others to represent the United States.

On October 4, 1957, the Soviet Union launched Sputnik-1 (Fig. 1.16), called the "basket ball," weighing 183 lb, on an elliptic orbit with a 98-min revolution period. The Soviet Union also launched Sputnik-2 on November 3, 1957, with the dog Laika onboard. On October 23, 1957, tests were carried out on the Vanguard launcher, which broke down on 6 December at the time of launching.

On 31 January, 1958, the United States launched Explorer-1 (Fig. 1.17), the new project, using a Jupiter C launcher, developed by a U.S. Army team led by Wernher von Braun. Explorer-1 had a mass of 13.9 kg, and would discover the Van Allen radiating belts.[12]

---

[12]The Van Allen belts are charged particle belts linked to the presence of the Earth's magnetic field and are located around the Earth.

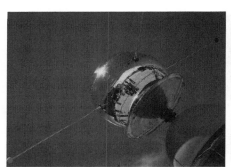

FIGURE 1.15 The Vanguard project (the satellite was called the "grapefruit"). (*Source*: NASA.)

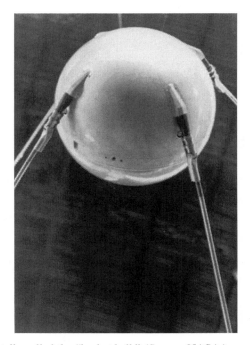

FIGURE 1.16 Sputnik, called the "basket ball." (*Source*: NASA.)

FIGURE 1.17 The Explorer 1 project. (*Source*: NASA.)

Let us come back to the first Sputnik launch. To prove that a satellite was actually orbiting the Earth, it was planned that it should transmit a signal. Sputnik used a 400 MHz carrier frequency with sound modulation data. In such a way, once demodulated, it was possible to "hear" Sputnik.[13] Nothing was really known about this flight — the orbit, the speed of the satellite, the duration of the transmission, and so on. So, it was a fantastic opportunity to carry out some tests. Among others, George C. Weiffenbach and William H. Guier, members of the Applied Physics Laboratory of the Johns Hopkins University, carried out such investigations. They succeeded in determining Sputnik's orbit by analyzing the Doppler shift[14] of the signal while the satellite was in radio visibility, that is, for about 40 min of the 108 min of a complete revolution of Sputnik.

The method they used to achieve such a goal was of fundamental importance as it is the starting point of all modern satellite navigation systems. The measurement was of the Doppler shift, the unknown variable was the orbit of the satellite, and another piece of data was the actual location of the place of observation (that is, the laboratory). After about three weeks of observation and a few calculations, they finally showed that it was possible to calculate the orbit, knowing both the Doppler shift

---

[13]What was then "hearable" can be listened to at www.amsat.org/amsat/features/sounds/firstsat.html.

[14]The Doppler shift is the physical phenomenon that shifts the frequency of any wave transmitted, depending on the relative speed between the transmitter and the receiver. Let us define D as the distance between the transmitter and the receiver: the frequency received is increased when D decreases and decreased when D increases. Note that this phenomenon is a physical time compression of the signal and applies to all waves (sound, radio, light, and so on).

and the exact location where the measurements were carried out. It has to be remembered that, in 1957, this was at the height of the Cold War between the Soviet Union and the United States. The U.S. Army, and more specifically the U.S. Navy, had a problem concerning the positioning of its fleets cruising in northern oceans. These ships were equipped with missiles to which precise missions could be allocated. The problem was that, although the guidance of such a missile was controlled by an inertial system of high quality, the starting location of the flight was still obtained through the use of terrestrial systems, and was therefore not very accurate. A more accurate system would be of great help for this specific purpose.

Frank McClure, of the Applied Physics Laboratory, made a suggestion: would it be possible to invert this problem? That is, would it be possible to be able to calculate the location of the observation point, knowing the orbit of the satellite, by carrying out the same measurements as those achieved to define the Sputnik orbit, that is, the Doppler shift of the received signal. Thus, the problem of satellite positioning was solved, thanks to Sputnik, and led to the Navy Navigation Satellite System (NNSS), or "TRANSIT" program, which was launched in 1958, directed by Richard Kershner.

The first satellite was launched in September 1959 and, before the end of 1964 (an amazingly short time for anyone working on modern projects), 15 launches had been carried out, with 8 more for research purposes. These eight were related to the program and concerned

- The establishment of a network of terrestrial surveillance stations;
- The determination of terrestrial gravity, which is of primary importance in order to predict the orbits of a satellite over a long period (12 h in the case of TRANSIT); and
- The definition of terrestrial and maritime receivers.

The TRANSIT system became operational for the Navy in 1964. The mean accuracy obtained was typically in the range of 200–500 m.

In the meantime, with some delay compared to the United States, the Soviet Union was preparing a program that was comparable to TRANSIT: TSYKLON. Their data were obtained from the measurement of the Doppler shift of the signals transmitted in the VHF band, typically at 150 and 400 MHz, together with orbital data. The acquisition of a few signals allowed the calculation of a location with a typical accuracy similar to that of TRANSIT. The time to fix depended on both the location and the constellation. TSYKLON was made up of a total of ten satellites divided into two different constellations: Parus (first launch in 1974), sometimes referred to as Tsikada–Military and Tsikada (first launch in 1977). A location was typically obtained in one or two hours.

The Soviet Union then decided to deploy the GLONASS system, but with the revolution of 1989 and the collapse of the Berlin Wall, they experienced financial difficulties in maintaining the GLONASS program in accordance with the initial plans.[15]

---

[15]Some launches in 2000 and 2001 seem to have been related to low orbital satellites, comparable to that of the Tsikada system.

The improvement of positioning due to satellite systems was real in terms of availability, coverage, and accuracy when compared to terrestrial systems, and more specifically to LORAN. However, there were still some limitations that prohibited the spread of TRANSIT to aviation or terrestrial forces, for example. The first was limited coverage due to the limited constellation, followed by the fact that the Doppler shift of the receiver was not taken into account, making it necessary to remain static while measuring (which is quite difficult to achieve when traveling by plane), and also the fact that only two-dimensional positioning was possible. This last point is of no real importance when at sea, but is absolutely crucial with a plane. New specifications were therefore required to provide improvements.

The main specifications of TRANSIT can be summarized as follows:

- *Availability.* A fix can be obtained every 40 min for a static user.

- *Accuracy.* Two-dimensional positioning, speed and altitude are not available.

- *Coverage.* A user on the equator must be more patient (35–100 min depending on latitude).

Of course, these limitations were the starting point for the specifications for the second generation of United States satellite-based positioning system, which were as follows:

- *Availability.* Twenty-four hours a day, 365 days a year, for all the covered locations and whatever the meteorological conditions (we mentioned this point concerning the terrestrial systems, for which the propagation conditions are of great concern). This last point has fundamental implications and modern systems still require a great deal of effort on improvements to propagation-related matters.

- *Accuracy.* Three-dimensional positioning with delivery of speed (real speed vector in three dimensions too) and precise time (one has to remember that time delivery was the first application of radio signals to maritime and navigation domains in the early years of the twentieth century).

- *Coverage.* The whole planet should be covered, with an extension to space (low and medium Earth orbit satellites usually position themselves by using GPS signals).

Section 1.7 deals with the new infrastructure deployed in order to meet these requirements.

◼ 1.7 REAL-TIME SATELLITE NAVIGATION CONSTELLATIONS TODAY

The GPS System

As the TRANSIT system was made operational in 1964 for the U.S. Navy, the early works on what would become, in 1973, the Global Positioning System (GPS) program started with tests on both the CDMA (code division multiple access) scheme and the PRN (pseudo random noise) code approach. These two techniques, widespread nowadays in radio systems, and more specifically in wireless telecommunications, were then quite innovative concepts. In 1967, the U.S. Navy started the TIMATION program to assess the effect of relativity, both special and generalized, on a satellite-based atomic clock.[16] In 1973, the programs related to satellite navigation from both the U.S. Navy and the U.S. Air Force merged into an official "Navigation Technology Program" called "NAVSTAR GPS" sometimes referred to as "Navigation Satellite with Time and Ranging Global Positioning System".

After the first stage of research programs, phase II started, in 1978, with the launching of the first four NAVSTAR satellites. From 1978 to 1985, 11 satellites were launched (called block I), and from 1989 to 1997, the 28 block II/IIR operational satellites followed. In 1985, seven satellites were available allowing about 5 h a day of positioning. The 24 nominal satellites were in orbit in 1994 and the GPS system was declared operational in 1995.

The major difference of this system over the TRANSIT system is that it is now based on a trilateration technique; that is, multiple distance measurements are carried out in order to allow the receiver to calculate a fix (TRANSIT was based on Doppler shift measurements).

The GLONASS System

In December 1979, the Council of Ministers of the Soviet Union took the decision to deploy its own global navigation satellite system. The scheduling, presented in August 1979, was accepted. The technical aspects of the system were consolidated in November 1978 by the Minister of Defense of the Soviet Union. GLONASS appeared to be the equivalent of GPS: a military-driven system. The technical aspects were the responsibility of both the armed forces and the civil aviation minister.

The various phases of the development of GLONASS were as follows:

- Experimental tests, system improvements and the first orbital constellation of a few satellites, from 1983 to 1985;
- Twelve satellites in orbit, flight tests, and initialization of the system, from 1986 to 1993;
- Deployment of the nominal 24-satellite constellation from 1993 to 1995.

It can be noted that, although the green light was given to the GLONASS program a few years later than GPS, the final availability of the system was roughly the same. In fact, the technology used for GLONASS did not involve modern multiple access schemes and so allowed a quicker development.

In March 1995, following a decision of the ex-Soviet Union countries,[17] GLONASS was declared available for civilian users "for the long term."

---

[16]GPS is the first widespread system that must implement both theories of relativity in order to obtain accurate positioning. Neglecting these effects would lead to a 10 km error per day!

[17]The Soviet Union was dismantled in 1991, leading to the transfer of this program to the Russian Federation.

Unfortunately, the life-time of the satellites was very short, of the order of 2.5 years, compared to the seven years forecast. In comparison, the GPS satellites exhibited a life-time of about ten years in comparison to the seven years forecast. Of course, this meant launching satellites at short intervals and required a significant financial budget. Thus the constellation declined. Some improvements in the life-time of the satellites, from 2.5 to 4.5 years, were not enough to solve the problem. In 2002, the constellation was composed of only seven operational satellites. We are currently experiencing a renewal of GLONASS as there are now about 19 satellites in orbit,[18] and the full constellation is planned to be ready in 2012/2013.

The Galileo System

As navigation-based applications attracted great interest in a large range of domains, from scientific to highly commercial, through military and telecommunications fields, they became a strategic issue for all the great nations. Indeed, if one can imagine that the use of such a system could be almost everywhere in a few years, let us imagine the impact of a system failure on economic activities. Boats could no longer travel the world because of the lack of satellite navigation capabilities, and this would lead to no navigation at all. The synchronization of the Internet network would no longer be under control and it would collapse within a few hours. Terrestrial displacements, mainly based on automatic systems using GPS, are limited to a few specific categories, and so on. The economies of large communities would really be affected. Even if the real situation is far from being the one described above, this was a risk. The European Community, as one of the major political entities, decided to be a real partner in satellite navigation.

The program was divided into two parts, GNSS 1 and GNSS 2.[19] The first is the European Geostationary Overlay Service (EGNOS), consisting of an "augmentation" to GPS designed to allow a (relative) European independence.[20] The second, called Galileo, is a completely autonomous satellite constellation able to provide positioning and time services.

As the cost of GNSS 2 would be huge, the European authorities in charge of this program decided to go ahead with a complete review of the possibilities. At that time,[21] the only "available" constellations were GPS and GLONASS. Discussions started first with the Americans in order to find a way to cooperate. The idea of the European Union was to share the monitoring of the constellation and its developments. The European part could have provided both funding and technical contributions. The real problem with this kind of approach was that GPS is really a military controlled system. Thus, it is not possible to share the control of such

---

[18]On March 25, 2007, there were eight operational satellites, three satellites in the commissioning phase and eight satellites temporarily switched off (http://www.GLONASS-ianc.rsa.ru).

[19]GNSS stands for Global Navigation Satellite Systems and is intended to include all the systems aimed at navigation purposes and based on satellites.

[20]The idea is to be able to warn users if the GPS becomes unusable, whatever the reason. Furthermore, its capabilities are far more useful than only monitoring GPS, as described in the following chapters.

[21]The official document relating to these aspects appeared on February 10, 1999, "GALILEO–Involving Europe in a new generation of Satellite Navigation Service" (http://europa.eu.int).

equipment with any other country, as it would clearly lead to sharing some of the strategic plans of the United States. It appeared quite quickly that, even if a financial contribution was possible (leading to European action on the GPS program), it was not possible to consider any share of the system's control. This was not acceptable for the European Union.

Some discussions then took place between the European Union and Russia concerning the GLONASS constellation. The limited life-time of the GLONASS satellites was identified and the idea for Europe was to provide both financial and technical aspects. The answer was once again negative, officially due to differences related to financing. The fact that the Russian GLONASS was also a military program and that the Americans declined to cooperate probably also played a role. Thus, Europe decided to build its own constellation, currently planned to be operational in space in 2012.

Other Systems

There are now many programs all around the world that intend to provide either full navigation capabilities or local or regional augmentations to the global positioning systems. This is the case for COMPASS (the Chinese satellite-based worldwide positioning system), but also for GAGAN (Indian augmentations) or QZSS (Japanese regional augmentation). Many other systems should be available in the coming years and are described in this book (see Chapters 3 and 4).

✔ 1.8 EXERCISES

Exercise 1.1 Assuming an uncertainty of ε_a degrees in the angle measurement obtained using the Pole Star elevation measurement technique, determine the accuracy of the latitude ε_{lat} in meters. Let us imagine that longitude was also available, estimate the accuracy of the longitude ε_{long}. Note that this latter accuracy depends on the latitude.

- Give the literal expression first.
- Then carry out calculations for $\varepsilon_a = 1, 2, 5$, and 10 degrees for various latitudes.
- Finally, draw the curve $\varepsilon_{long} = f(\text{latitude})$ considering ε_a as a parameter.

Exercise 1.2 For a trip from London (UK) to Dakar (Cameroon) passing through Lisbon (Portugal), Agadir (Morocco), and Cap Verde, define the location of the knots on the Kamal rope that is being used. Let us assume that the Kamal is 10 cm high.

Exercise 1.3 What is the real distance Christopher Columbus would have had to sail to reach the east coast of India (approximate value) from the Canary Islands? Compare this to his approximation of around 2400 km. How long would it have taken assuming the same speed as his first voyage to America? Note that you need to do some research to estimate the duration of his first trip.

Exercise 1.4 What were the main advantages of the Davis cross staff over the Kamal or the quadrant measurements? Same question for the sextant over both Kamal and quadrant?

Exercise 1.5 The antennas, both at the transmitting and the receiving sites, of the first transatlantic radio link were huge. What were the reasons for such dimensions?

Exercise 1.6 Explain in detail the reason why the first terrestrial radio navigation systems were based on the intersection of hyperbolae instead of the intersection of circles. Answering this question is fundamental for a good understanding of positioning systems.

Exercise 1.7 What was the main investigation carried out by the members of the Applied Physics Laboratory of the Johns Hopkins University while Sputnik was transmitting? How did the measurements then carried out lead to the concept of a satellite-based positioning system?

Exercise 1.8 Write a concise chronological description of the evolution of satellite-based positioning systems, together with their advantages and limitations.

■ BIBLIOGRAPHY

Boorstin DJ. The discoverers. New York: Random House; 1983.

Gardner AC. Navigation. UK: Hodder and Stoughton Ltd.; 1958.

Guier WH, Weiffenbach GC. Genesis of satellite navigation. Johns Hopkins APL Technical Digest, January–March 1998;19(1).

Ifland P. Taking the stars, celestial navigation from Argonauts to astronauts. The Mariners' Museum Newport News, Virginia, and The Krieger Publishing Company, Florida; 1998.

Kaplan ED, Hegarty C. Understanding GPS: principles and applications. 2nd ed. Artech House; 2006.

Kennedy GC, Crawford MJ. Innovations derived from the transit program. Johns Hopkins APL Technical Digest, January–March 1998;19(1).

Parkinson B. A history of satellite navigation. Navigation: Journal of The Institute of Navigation 1995;42(1).

Parkinson BW, Spilker JJ Jr. Global positioning system: theory and applications. American Institute of Aeronautics and Astronautics; 1996.

Pisacane VL. The legacy of transit: guest editors introduction. Johns Hopkins APL Technical Digest, January–March 1998;19(1).

Sobel D. Longitude. London: Fourth Estate Limited; 1996.

Sobel D. A brief history of early navigation. Johns Hopkins APL Technical Digest, January–March 1998;19(1).

Web Links

http://grin.hq.nasa.gov. Nov 7, 2007.

http://history.nasa.gov/hhrhist.pdf. Nov 7, 2007.

http://history.nasa.gov/nara/nara1.html. Nov 7, 2007.

http://pwifland.tripod.com/historysextant/. Nov 7, 2007.

http://techdigest.jhuapl.edu/td1901/index.htm. Nov 7, 2007.

http://www.mat.uc.pt/~helios/Mestre/Novemb00/H61iflan.html. Nov 7, 2007.

CHAPTER 2

A Brief Explanation of the Early Techniques of Positioning

In this chapter, we describe the evolution of the techniques used for positioning and the major historical events relating to them. The approach presented here is more technical than in Chapter 1. We will compare ancient and modern calculations to show that it is mainly the management of radio electric waveforms that has enabled improved accuracy, the main principles having been established centuries ago. In addition, an introduction to cartography and two-dimensional representations of the Earth are also given.

When dealing with techniques used for positioning, one can consider a few fundamental approaches based on physical measurements. Measuring angles was the first natural technique, using either specific stars or a terrestrial point of interest. Then two successive improvements became possible by introducing time into the problem. The first improvement involved an increase in complexity, and hence of accuracy, of positioning using angle measurements and well-established ephemeris (this requires knowing a reference location time). The second was related to the new possibility of carrying out distance measurements. This latter approach required synchronization between a few transmitters. The satellite era opened the way to even more accurate measurements, thanks to electronics. The first satellite measurements were based on the Doppler shift effect of a moving transmitter–receiver system (see Chapter 1 for explanation). A significant improvement was made possible with the availability of time measurement, and hence distance estimation (this feature was obtained with satellites thanks to the GPS program).

Global Positioning: Technologies and Performance. By Nel Samama
Copyright © 2008 John Wiley & Sons, Inc.

■ 2.1 DISCOVERING THE WORLD

Coastal navigation was well established at the beginning of the fifteenth century, and commercial exchanges were economically important for European countries. The same coastal navigation had also been developed in the Indian Ocean, and even though boat construction was very different, navigation was roughly at the same level. Positioning and displacement measurement techniques allowed quite a good estimation of routes, but it was increasingly apparent that coastal navigation was no longer adequate. In addition, the spice and silk routes passed through Persia and attacks on traders were frequent. The need to find other routes was thus of prime importance for those countries that wanted to develop commercial activities with distant countries.

Maritime routes also presented the advantage of allowing the transportation of huge loads at reasonable cost, for instance, the movement of military squadrons and equipment. Unfortunately, positioning was only possible when in view of the coast. Traveling out of sight of the coast was dangerous, as no visual references were available and only latitude could be evaluated (from stars), not longitude.

■ 2.2 THE FIRST AGE OF NAVIGATION AND THE LONGITUDE PROBLEM

The historical side of the longitude problem has been described in Chapter 1. Let us come back to the technical aspects. Many theories were proposed, among which three directions of investigation were significant (the modern techniques used for positioning purposes, whatever the technology deployed, have their mathematical and physical origins here):

1. Variation of the terrestrial magnetic field;
2. Measurements of distances from specific stars to the Moon;
3. Maritime clocks.

The first area is based on the observation that the magnetic terrestrial field varies from one point to another. The idea was then to imagine that it would be possible, by drawing a "map" of the magnetic field all over the globe, to have a direct correspondence between the local field measured and the location where this measurement was carried out. We now know that this is not possible and that, for instance, the magnetic "north" is subject to variation over time. Nevertheless, although John De Castro demonstrated that it was untrue as early as at the beginning of the sixteenth century, experimental works were carried out until the middle of the seventeenth century before this avenue of research was abandoned. In 1638, the mathematician Henry Gellibrand showed that there were extensive local variations in the magnetic field over time; that is, the variation at a given fixed location is huge compared with that needed for positioning. Furthermore, Edmund Halley undertook a measurement campaign in 1698 that finally came to the conclusion that longitude determination was not possible with this approach. It has to be noted that this activity was in fact before the Queen Ann challenge, and probably has a link to it, as a solution was neither available nor foreseeable at this time.

The second method was part of the logical evolution of the nautical navigation art at the end of the seventeenth century. Measurement instruments were good enough to allow acceptable navigation, and angle resolution improved somewhat. Galileo and Cassini had shown that angle measurements could lead to finding a location by observing Jupiter's moons. Although not applicable at sea, research was carried out to show that transposition to the Earth's Moon was possible (seventeenth and eighteenth centuries), requiring the use of both measurements and tables. Unfortunately, the accuracy of angle measurement (two to three degrees) was not good enough. Thus, new techniques appeared, together with improved accuracy. Tables referred now to the angular distances from 15 well-known stars to the Moon, and the advent of the sextant made location-finding possible. Accuracy nevertheless remained inadequate, and the calculations were complex. The first use of this approach is quoted from around 1767, concomitant with important improvements in sextant technology (from 1770 onwards).

The third approach, the maritime clock, took decades to achieve, but finally had an impact on much more than just navigation. The history of the Harrison clocks is quite interesting and was the starting point of modern satellite navigation. We know about the story (see Chapter 1), but let us discuss briefly the technical aspects.

John Harrison was born in Foulby in 1693. His first clock, built in 1713, had a mechanism that was made entirely from wood. He also worked for a while with his younger brother James, and their first major project was a turret clock that required no lubrication. In 1726, John and James designed two precision clocks, to see how far they could push the capabilities of their design. By inventing a pendulum rod made of alternating wires of brass and steel, Harrison solved the problem of the pendulum's length varying with temperature, slowing or accelerating the clock in the process. Thus, Harrison's clocks achieved an accuracy of one second in a month. A maritime time-keeper had to exhibit the same kind of accuracy in a very different environment . . . During this period, being unable to meet the Board of Longitude, he contacted Edmund Halley, who facilitated a meeting with Georges Graham, a member of the Royal Society. The reason Graham agreed to meet with Harrison was probably linked to the fact that Graham's earlier works were carried out in the watch-making domain.

Four trials were required for John Harrison to achieve the final so-called H4 (Fig. 2.1), which was quite different from the previous time-keepers (H1, presented in Fig. 2.2, and H2 and H3), because it was of the size of a large pocket watch. William, John's son, with H4, sailed for the West Indies aboard the ship *Deptford* on November 18, 1761. The watch appeared to be only 5.1 s late on its arrival in Jamaica. A second trial was carried out on a journey to Barbados aboard the *Tartar* on March 28, 1764, and demonstrated an error of less than 40 s (the journey lasted 47 days). These two results were judged as excellent, close to being enough to win the prize, yet not quite enough to allow the Board of Longitude to award the prize. Harrison was asked to disclose his entire design to the Astronomer Royal in order to allow him to make and test such a time-keeper. This was the condition

FIGURE 2.1   Harrison's H4 clock.

to allow half the prize to be released, the second half being potentially awarded when the other time-keepers had exhibited similar performances at sea.

Harrison finally accepted to disclose the inner mechanism of H4 and received the first half of the longitude prize. A lot of stories concerning innovation are very similar

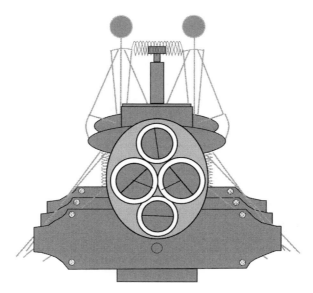

FIGURE 2.2   Harrison's H1 clock.

to this, requiring huge efforts from the inventor to prove his truth against the well-established authorities — but that is another story.

H1 (1730–1735) was a portable version of the previous wooden clocks. It was based on the use of springs in order to allow the effects of gravity to be dealt with, unlike in a pendulum clock. (Note that overcoming the effects of gravity is the second difficulty at sea, the accuracy problem being the other.)

H2 and H3 (1737–1759) were not successful in resolving the longitude problem. In fact, the accuracy required to win the prize was not reached, despite the numerous innovations. Harrison requested additional funding from the Board to continue his efforts. At the end of this period, Harrison was convinced that the design should be entirely new; this led to the H4.

■ 2.3 THE FIRST OPTICAL-BASED CALCULATION TECHNIQUES

Angle measurements, taken from observations of the Pole Star, were the starting point. These provide a direction, when thinking in terms of positioning, but were first used to evaluate latitude, as the angle from the horizon to the Pole Star gives the latitude of the point of observation. This is not yet positioning. When considering terrestrial landmarks such as lighthouses, it really is positioning that is being sought as long as two or more such landmarks are available, as described in Fig. 2.3. Angle measurements are still used, but in order to define the directions of the arrival of light signals. If two such directions are measured, then it is possible to draw the intersection of these two directions in order to plot a location on a two-dimensional map. For maritime positioning, this is quite acceptable. Of course, one can easily understand that the locations of the landmarks are essential and that positioning is thus obtained in a reference frame that is relative to the landmarks.

Once the angles from a boat to any reference points are defined, by any means, calculations need to be carried out to obtain the location of the boat. When dealing

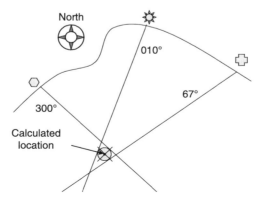

FIGURE 2.3 Landmark-based location evaluation.

with angle measurements from landmarks, the method is quite simple to understand as it only involves drawing straight lines on a map (as long as the angles are conserved by the projection), knowing the location of the landmarks. Owing to measurement errors, the intersection of three such lines leads to a triangle. The center of the greater circle included in this triangle is considered as the location (see Fig. 2.3).

When taking angle measurements of stars, the sextant is the primary instrument. The calculations to find the location are then much more complex. As explained in Chapter 1, the possible positions on Earth that have the same observed height of a given star are located on a circle that is centered on the vertical line joining the star and the center of the Earth. The radius of this circle is dependent on the heights of the star above the horizon. Two measurements are required to be able to mathematically calculate the geometrical intersection of two circles on the Earth (this in fact leads to two points, which are far apart, thus allowing the one relevant point to be determined if the location is roughly known). The techniques used were either to make two simultaneous measurements from two different stars, or to make two measurements of the same star at different times. The latter required additional mathematics to correct the time shift. Thus, tables were used, and navigators had to be quite good at mathematics to obtain the correct location.[1]

Dead reckoning was also of great importance because it was not possible to carry out the measurements and the calculations on a continuous basis. The navigator therefore needed to estimate the boat's speed and direction, using *lochs* or a compass. Many factors are bound to degrade these estimations. For compass measurements, thanks to Christopher Columbus, one knew that there was a difference between the magnetic north and the geographical north (the magnetic declination). Furthermore, the metallic parts of the boat, even if partially compensated, remained of concern. Although technological improvements occurred, the achievable accuracy remained in the 1° range. Speed determination was obtained through the use of *lochs*, which allowed a measure of the relative velocity of the boat over the sea. It uses a rope graduated with knots, where the number of knots unwound during a fixed amount of time (measured by an hourglass) gives the required value. This measure gave a typical accuracy of half a knot,[2] depending on the technology.

■ 2.4 THE FIRST TERRESTRIAL RADIO-BASED SYSTEMS

When using the stars, and especially the Pole Star or the Sun, to achieve latitude determination, we have already seen that the time of the reference location where ephemeris were carried out is fundamental. The first use of radio electric waves

---

[1] The resolution is based on the fundamental formula of the spherical triangle formed by the pole, the zenith, and the star observed (called the position triangle): $\sinh = \sin (lat) \sin d + \cos (lat) \cos d \cos (ta - long)$ where d and ta are respectively the declination and the time angle of the star, obtained from ephemeris. *Lat* and *long* are the latitude and longitude of the boat.

[2] A knot is equivalent to one nautical mile covered in one hour. One nautical mile, 1852 m, is the equivalent in distance to one minute of arc of a meridian (or the distance equivalent to one minute of a latitude arc).

was achieved with the transmission of a time signal in order to synchronize a boat's clocks. In such a way, there was no longer the need for maritime clocks and the error of positioning could be reduced. In 1910, for example, the Eiffel Tower in Paris was equipped with a large wire (the antenna) and transmitted time signals.[3]

With the advent of radio signals, new technical possibilities arose such as direct time measurements. In fact, it was then possible to use time propagation as the primary measurement, leading to the distance from a transmitter to a receiver being obtained by simply multiplying the time by the speed of the wave, that is, the speed of light. Note that, although based on time availability, such as for longitude determination, this is clearly a new approach as distances are now considered. However, not everything can be solved by using radio waves. In order to carry out time measurements, there is the need for synchronization between the transmitter and the receiver. The first use of radio signals was for time synchronization purposes, but when dealing with radio wave propagation, the need for time measurements was even more stringent, because just 1 ns of error in synchronization means 30 cm of corresponding length error.[4]

Because of this difficult synchronization problem, two successive approaches were deployed for global systems:

- Hyperbolic systems, where time synchronization is only required at the ground facilities and can be achieved through wired architectures. The idea is to carry out differences of time of arrival that eliminate the time synchronization requirement at the receiver's end.

- Spherical systems, where time synchronization must be carried out from ground facilities to receivers. No such systems were developed before the GPS because of the complexity of this synchronization. Indeed, for GPS, as described in the following chapters, synchronization between satellites and receivers is achieved in a clever way.

Note that some local systems implemented a spherical approach — the main idea was either to use an atomic clock at the receiver, or to measure propagation time from a transmitter to a receiver and back. In this latter case, the time is measured in a unique reference frame, and synchronization is no longer required.

The first multiple transmitter hyperbolic system deployed was the Decca. It comprised a master station associated with three slave stations. Each couple of master and slave stations (there are three such couples) gave a network of hyperbolae (as shown in Fig. 1.13). The frequencies transmitted were different for each station, and are a multiple of a given reference frequency (we shall see that this principle, in a modified form, is still used in modern satellite constellations). As the signals of the slave stations are under the control of the master station signal, the time difference was obtained by establishing the phase difference between the master station signal and the three slave station signals.

---

[3]The street in front of the Eiffel Tower was named Quai Edouard Branly in 1943 in memory of these first experiments, which were possible thanks to the receiver technology of Edouard Branly.
[4]Considering 3×10^8 m/s as the speed of light, 1 ns (10^{-9} s) is equivalent to 0.3 m.

First, the receiver was set up with the reference numbers of the calculated hyperbolae, for each master–slave pair. Then, the navigator had to use a specific Decca map to locate the boat at the intersection of the various hyperbolae. With advances in electronics, receivers were then immediately able to give the location of the receiver.

The LORAN[5] (LOng RAnge Navigation) system, still in use today, is also a hyperbolic system, which adds a pulse-based modulation to the synchronization of the master and slave signals. The number of slave stations varies from two to four. Each station transmits pulses in sequence. Once a slave station receives the master station's transmission, it transmits its own sequence with a known delay (which characterizes the station). Some pulses are in opposite phase compared to the carrier, so that identification of the station is simplified. The calculation of the location of the receiver is achieved in two steps: (1) A rough acquisition is obtained by measuring the time difference between the reception of the master and slave signals; (2) A more precise time measurement is made by analyzing the phase differences. As with modern satellite-based systems, the resulting accuracy greatly depends on parameters such as the quality of time synchronization between the stations and the quality of the propagation correction models. The accuracy is then in the range of 800 m for a distance up to 1300 km (where the propagation is mainly achieved through ground waves), and less than 8 km for distances up to 3000 km (where it is the sky waves that have to be considered).

As we shall discuss in more detail in Chapter 4, eLoran is currently widely deployed in the United States, but also in the north of the Pacific Ocean shores, in the Persian Gulf, and in Europe.

■ 2.5 THE FIRST NAVIGATION SATELLITE SYSTEMS: TRANSIT AND PARUS/TSIKADA

As has already been described in Chapter 1, the advent of Sputnik was the real starting point for satellite navigation systems. By following Sputnik's Doppler shift, assuming a fixed location of observation, it was possible to deduce the orbit of the satellite. Let us explain how this is possible. When considering a fixed place of observation, that is, the laboratory, the Doppler shift of Sputnik had the typical form given in Fig. 2.4. When crossing the x-axis (time), the satellite was at its nearest point to the laboratory. In fact, it was moving towards the laboratory before this point and would be moving away afterwards.

If the transmission frequency is not known, there is the possibility of defining this closest point by finding where the slope of the curve above is the steepest. So far, as no distances are available, knowing the shortest distance does not allow the location of the satellite to be determined. It was therefore necessary to wait for many successive records of Doppler curves to obtain enough data. Then, by orbital modeling, calculations showed it was possible to define the orbit of Sputnik.

---

[5]The current version is eLoran. The first version was LORAN-A, which used very low frequencies. LORAN-C is the latest version, but is currently being replaced by eLoran. The other versions, from LORAN-B to LORAN-E, were designed either for specific purposes or were not developed to an operational level.

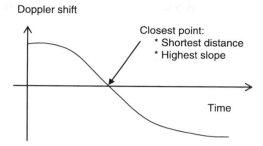

FIGURE 2.4   A typical satellite Doppler shift curve obtained at a fixed location.

It is interesting to invert the problem in order to obtain the location of the observation point, given the orbit of a satellite and assuming a receiver is able to measure the Doppler shift. The idea was that satellites could be launched and used for positioning purposes. The observations would be identical, that is, measures of Doppler shifts. If, in addition, one knows the location of the satellite at the time it is at its closest point, then it is possible to draw the curve of all possible corresponding locations of observation. Figure 2.5 (left) gives such a typical curve. It is clear that it is not enough to define a single location. One has to wait for another measurement (Fig. 2.5 right, second track) corresponding, for example, to another satellite.

The TRANSIT system was based on this approach and was able to provide users with a two-dimensional location. This was quite enough for the purposes of this system, as long as the goal was mainly to give an acceptable (accurate enough) starting location for missiles launched from sea.

The TRANSIT system was made up of six satellites (see Fig. 2.6), orbiting at an altitude of 1100 km, with a revolution period of 108 min. Note that these values are quite close to the corresponding values for Sputnik (98 min of revolution at an altitude of 985 km). The system worked in three steps:

1. Transmission from a satellite enabled a tracking station to digitize and record Doppler signals of the satellite in order to transmit some data to a computing center.

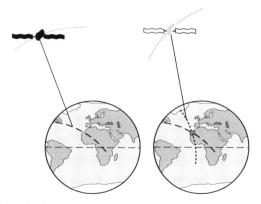

FIGURE 2.5   Location finding using Doppler shift measurements.

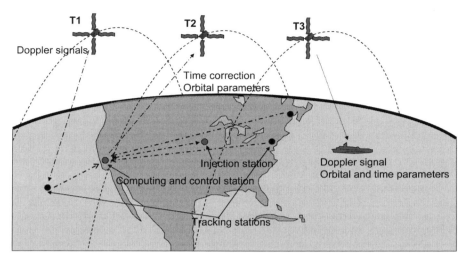

FIGURE 2.6 The TRANSIT system architecture.

2. The computing center prepared orbital modeling data for the satellite (this introduces the notion of ephemeris, already discussed in the light of ancient navigation techniques concerning astral movements), together with time corrections.

3. The receiver made the Doppler shift measurements in order to solve the set of equations to output the location of the receiver. It also received ephemeris data, which is essential for the calculation of the satellite location and time synchronization.

Note that the notion of ephemeris was required in order to improve the accuracy of the measurement. If the receiver can accurately evaluate the actual location of the transmitter, that is, the satellite, then it can more accurately use the Doppler measurements to find its location. This point, although involving slightly different parameters today with modern satellite systems, is still of major significance when improving accuracy. A typical diagrammatic representation of the TRANSIT system is given in Fig. 2.6, and images of satellites in Fig. 2.7.

As it is based on Doppler shifts, TRANSIT required at least two measurements from different satellites (or more precisely from two different ground tracks). Thus, the user would wait for some time either to make measurements from two satellites or to use two ground tracks from the same satellite at different times. In this case, the two resulting ground tracks should show good geometrical characteristics in order to provide an acceptable positioning accuracy. Six satellites were available for an acceptable coverage of the northern part of the world (remember that this system was designed for military purposes during the Cold War between the United States and the Soviet Union).

Each transmission was achieved through two frequencies, 150 and 400 MHz. The ground segment was composed of one control center and three ground

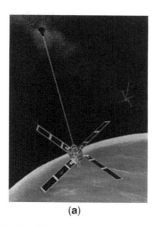
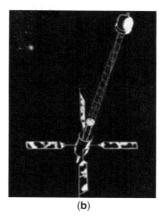

(a)　　　　　　　　　　　　(b)

FIGURE 2.7 Satellites used in the TRANSIT System: Oscar (**a**) and Nova (**b**). (*Source*: Applied Physics Laboratory, Johns Hopkins University.)

stations, all located in the United States. The TRANSIT system was stopped on December 31, 1996.

In the meantime, the TSYKLON system was deployed by the Soviet Union. In many ways it is quite similar to TRANSIT. There are two constellations: PARUS is the military one, composed of six satellites positioned on two orbital planes separated by 30°, at about 1000 km of altitude. The second one, TSIKADA (which means "dragonfly") is dedicated to commercial and civilian use and is composed of four satellites positioned on two orbital planes separated by 45°, complementary to those of PARUS.

■ 2.6 THE SECOND GENERATION OF NAVIGATION SATELLITE SYSTEMS: GPS, GLONASS, AND GALILEO

The first generation of systems was based on Doppler shift measurements. In order for these systems to provide users with 24-hour, 365-day permanent coverage of the Earth with three-dimensional (3D) positioning, the number of satellites would have been too high, mainly because one needs to wait for the nearest distance point in order to obtain one measurement. Thus, the satellites needed to be either at a very low Earth orbit or the user had to wait for a long time. Thus, real-time 3D 365-day positioning requires another approach.

Time-based measurements were one solution, but required a great deal of research in order to learn about the clock running modes in space. This was achieved through extensive research studies concerning atomic clocks, for example, for GPS. GLONASS and Galileo implement the same global technical approaches. The basic idea is to carry out propagation time measurements from satellites at the receiver end. The major difficulties are to cope with the many error sources in the traveling path of the waves and the problem of time synchronization between the transmitters (satellites) and the receiver.

Various models have been studied and proposed, together with specific measurement techniques (see Chapters 5 to 7 for details) in order to cope with error sources. For synchronization, the approach has been to consider time de-synchronization as a variable in the calculations, like the spatial positioning variables. Thus, there is the need for an additional measurement over what is needed from a geometrical point of view. This clever method allows an important reduction in the cost of receivers, as there is no need for very stable oscillators (the important point is only to have a stable oscillator over a very short period of time).

■ 2.7 THE FORTHCOMING THIRD GENERATION OF NAVIGATION SATELLITE SYSTEMS: QZSS AND COMPASS

The success of the second-generation systems has been greater than could have been imagined. It has also led to new requirements in a wide variety of applications. One of the most important needs is based on communication. Positioning data are one thing, but for many applications, the most important aspect is the transmission, by any possible means, of these positioning data to a server or whatever, in order to allow processing and related services to be provided. Thus, the next generation, here called the third generation, has to take these aspects into account at the system level, that is, with embedded telecommunication capabilities in the satellites and ground facilities.

The Japanese QZSS is planned to provide such a system, although the first satellite to be launched will probably not exhibit the telecommunication features (because of economic and industrial reasons). The Chinese COMPASS (also named Beidou[6]) global navigation satellite system intends also to incorporate such features.

■ 2.8 REPRESENTING THE WORLD

Positioning is interesting, but has no sense without a common representation of the world. In other words, there is a need for a global coordinate referential in order to allow a single representation. The study of the shape of the world is called geodesy.

A Brief History of Geodesy

The first attempt at defining the perimeter of the Earth, while considering it to be spherical, was carried out by Eratosthene.[7] He made a direct calculation by considering the size of the shadow of the Alexandria lighthouse when the bottom of

---

[6]Beidou is the Chinese name of the Great Bear constellation.
[7]Greek (276 BC/194 BC).

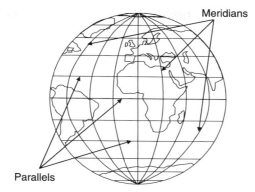

Meridians

Parallels

FIGURE 2.8 *Source*: Meridians and parallels.

the Aswan well was totally lit by the Sun. In order to evaluate the real distance separating the sites, he used the cadenced step of a military infantry division; the final result of his evaluation gave around 40,000 km (39,690 km). This incredibly accurate value was not believed by Christopher Columbus when evaluating the distance separating Europe and India: he estimated that about 2400 km was the correct value, corresponding to more or less three weeks in favorable navigation conditions. No significant progress was achieved, despite many attempts, until the end of the seventeenth century, with the works of Father Picard and Sir Isaac Newton.

During the eighteenth century, two opposing theories existed. The first, defended by Newton, stated that the Earth was flattened at the poles and the second, defended by Cassini, stated that it was flattened at the equator. Two expeditions were carried out in Lapland (Maupertuis and Clairaut), and on the equator (Peru; Condamine and Bouguer). Newton's theory happened to be the better one and, in 1799, the French Science Academy established a definition of the meter as being a fourth of one ten-millionth of a terrestrial meridian. This was the official birth of the metric system.[8] The definitions of equator, meridian, and parallel are given as follows (see Fig. 2.8 for details):

- The equatorial plane is the plane containing the greatest circle of the Earth perpendicular to the polar axis.
- The great circles surrounding the Earth and including both poles are the meridians.
- The small circles perpendicular to the pole axis are the parallels.

The development of a number of geodesic networks occurred throughout the nineteenth century. This was also the time of the birth of the ellipsoid representation of the Earth. In order to allow exchanges between different countries, the International

---

[8]Note that with the current definition of one meter, given by the distance traveled by light during 1/299792458 of a second, a terrestrial meridian is 40,074 kilometers.

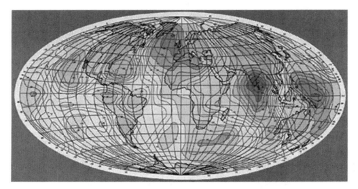

FIGURE 2.9　A graphical representation of the equigravity field surface (Courtesy of Applied Physics Laboratory, Johns Hopkins University).

Geodesic Council was created in 1886. Meanwhile, it appeared that the Earth was not exactly an ellipsoid and a new concept was proposed, the geoid, corresponding to the equipotential surface to the field of gravity (the one of equal gravity value, perpendicular to the gravity vector at each point). Unfortunately, it also appeared to be not exactly a "simple" surface, as shown in Fig. 2.9.

Basics of Reference Systems

The orbit of the Earth is an ellipse of low eccentricity ($e = 0.01673$), the Sun being at one focal point. The terrestrial orbit plane is called the ecliptic plane. Furthermore, the Earth rotates from west to east in 24 h, and the rotation axis deviates from the perpendicular of the ecliptic plane by an angle equal to $23°27'$. The Earth is also rotating around the Sun in the same direction. The North Pole lies in the northern hemisphere, defined as being the hemisphere containing Paris, France. Figure 2.10 is a graphical representation of this.

The vernal point is defined as being the point where the Sun crosses the equator when going from the southern hemisphere to the northern hemisphere. The intersection line of the equatorial plane and the ecliptic plane indicates the vernal point (Fig. 2.11).

Considering the vernal point, a reference in the astronomical domain, a terrestrial location can be defined using both right ascension and declination, defined as follows (see Fig. 2.12a for graphical details):

- The right ascension α is the angle between the meridian that crosses the location and the vernal point. It is calculated in the equatorial plane counterclockwise from 0 to $360°$.

- The declination δ is the angle of the direction of the location with the equatorial plane, evaluated on the local meridian, from 0 to $90°$ north or south.

The referential associated with this representation of a location is called the Celestial Equatorial referential.

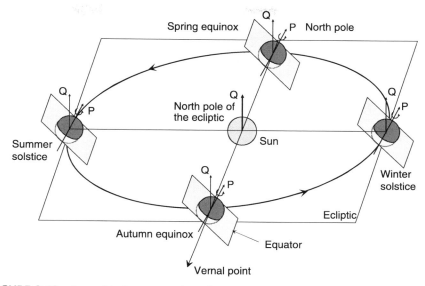

FIGURE 2.10 A graphical representation of astronomical basic definitions.

From this referential, it has been possible to extract a Cartesian referential. It is called the Earth-Centered Inertial (ECI) referential, commonly used in astronomical calculations. It is defined as follows (see Fig. 2.12b):

- The origin O is the center of the Earth.
- The z-axis is the polar axis.
- The x-axis lies in the equatorial plane and crosses at the vernal point.
- The y-axis is such that Oxyz is a direct trihedron.

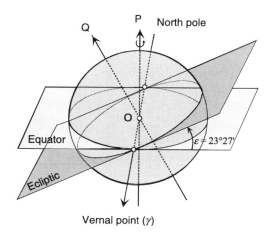

FIGURE 2.11 Definition of the vernal point.

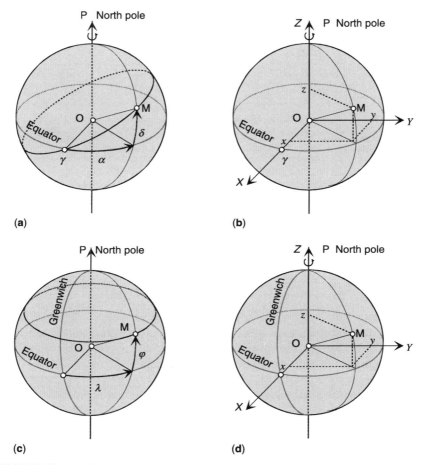

FIGURE 2.12 (**a**) The Celestial Equatorial Referential; (**b**) The ECI (Earth-Centered Inertial) Referential; (**c**) The Geographical Referential; and (**d**) The ECEF (Earth-Centered Earth-Fixed) Referential.

This referential is not really practical for positioning, because the vernal point is a fixed point. Thus, a given location on Earth does not keep the same coordinate values, as the Earth is rotating on itself. This referential is mainly used to define star locations.

For positioning purposes, it is better to choose a referential that rotates with the Earth, allowing a single coordinate value for a given location. The geographical coordinates (longitude and latitude) were designed for this purpose (see Fig. 2.12c).

- The longitude λ is the angle between a reference meridian (Greenwich for instance) and the local meridian, calculated counterclockwise from 0 to 360°.
- The latitude φ is the angle between the direction of the location with the equatorial plane, evaluated from 0 to 90° north or south.

The associated system of reference is the Geographical Referential (with reference to a meridian of origin).

From this referential, it is also possible to define a Cartesian referential, called the Earth-Centered Earth-Fixed (ECEF). The main parameters are as follows (see Fig. 2.12d):

- The origin O is the center of the Earth.
- The z-axis is the pole axis, oriented north.
- The x-axis is in the equatorial plane and passes through the Greenwich meridian.
- The y-axis is such that Oxyz is a direct trihedron.

This ECEF referential is very practical for positioning and is the one commonly used for satellite navigation systems.

So, many different references are in use and one needs to know which one is currently being used by the navigation system concerned. Of course, this is also of uppermost importance when comparisons between different positioning systems are required. This is particularly the case when a positioning system is an integration of various techniques that are referenced to different referential systems.

As well as the definition of geographical referential systems, it is necessary to define a model of the Earth. As satellite-based positioning systems use spatial signals, there is theoretically no need for such a representation. Nevertheless, it is interesting to propose a visual display to the user, on a map for example. The need for an accurate correspondence between the representations of the Earth and reality is thus an important topic.

The shape considered as the model for the Earth is an ellipsoid (Fig. 2.13). It is flattened at the poles and exhibits symmetry of revolution. This mathematical model has been chosen in order to be as close as possible to the geoid and is defined by its longer axis a and another parameter such as the small axis b, its flatness f, or the first eccentricity e.

Note that an interesting feature is the fact that the perpendicular to the local tangent does not cross the center of the ellipsoid, unless on the equatorial plane

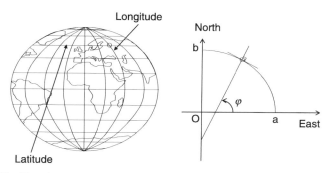

FIGURE 2.13   The ellipsoid representation.

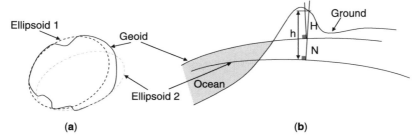

FIGURE 2.14 (a) Ellipsoid: a local approximation; (b) The problem of altitude.

and at the poles (Fig. 2.13). This shows clearly that it is only a mathematical representation of the geoid.

Any given location on the Earth surface can be defined by one of the following:

- Its Cartesian geocentric coordinates (x, y, z) relative to the three axes of a reference frame whose origin is located at the center of the Earth.
- Its geographical coordinates longitude λ, latitude φ, and ellipsoidal height h.
- Its plane coordinates obtained when making a plane projection onto a local Earth surface area.

For historical reasons, many geodesic reference systems exist, each of which is a local geoid approximation. Both an ellipsoid and a meridian of reference are associated to each geodesic reference frame. A user must take care in the way these geodesic systems are used, as they can be quite different from one place to another, as described in Fig. 2.14. Furthermore, the geoid is a complex surface, and noticeable differences can be observed between ellipsoids and the geoid (as much as 200 m, or even more). As the geoid is also defined as being the mean sea level, it is considered as the origin of altitudes. Unfortunately, the geoid surface is very difficult to actually measure because of both the variation of ocean heights (leading to averaging) and the presence of continents.

Thus, the altitude H of a given location M is the difference between the ellipsoidal height h and the so-called undulation N ($H = h - N$). The undulation is usually available through modeling (or a conversion grid).

Navigation Needs for Present and Future Use

In practice, a plane representation of the Earth is used, in order to:

- Provide us with a plane surface representing a part of the ellipsoid;
- Transform angular metrics into a more practical metric system; and
- Allow metric measurements of distances.

Unfortunately, such a projection cannot be achieved without deformations and distortions. Usually, the points are projected on a plane using the center of the sphere as the origin. This leads to incorrect values for distances (Fig. 2.15).

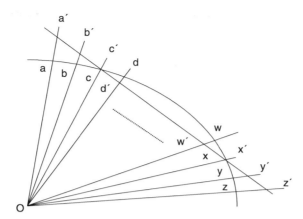

FIGURE 2.15 A plane projection form [a ... z] to [a′ ... z′].

Nevertheless, it is possible to define the parameters required in order to minimize deformations and distortions for a given projection. This is essentially achieved through calculation. Usually, one can choose to preserve either

- The surfaces;
- The angles;
- The distances from a specific point; or
- None of the metric characteristics.

Note that the basic projection shown in Fig. 2.15 is a typical projection that preserves the angles. Usually, the projection surface is either

- A cylinder tangent or secant to the Earth (see Fig. 2.16a, b);
- A cone tangent or secant to the Earth (see Fig. 2.16c, d); or
- A plane tangent to the Earth (see Fig. 2.16e).

The Universal Transverse Mercator (UTM) Projection

This famous and most commonly used projection is a transverse tangent cylindrical projection that covers the whole Earth in 60 zones, each of 6° of longitude, from 80° south to 84° north. Each zone is divided into 20 bands of 8° of latitude amplitude. Figure 2.17 shows the global UTM reference grid of the world.

The UTM reference grid for Europe is presented in Fig. 2.18. As an example of its use, let us consider the UTM representation of France. It is divided into three zones:

- UTM North: zone 30, from −6° west to 0° (Greenwich), zones T and U;
- UTM North: zone 31, from 0° west to +6°, zones T and U;
- UTM North: zone 32, from +6° west to +12°, zones T and U.

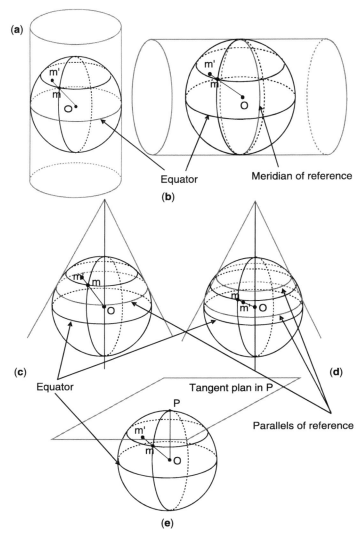

FIGURE 2.16   Various types of projections: (**a**) tangent cylinder; (**b**) secant cylinder; (**c**) tangent cone; (**d**) secant cone; (**e**) tangent plane (bottom).

Within a given zone, the UTM coordinates are given in meters relative to

- The horizontal origin X0 (called false easting), which equals 500,000 m and corresponds to the central meridian;
- The vertical origin Y0 (called false northing), which equals 0 m and corresponds to the equator.

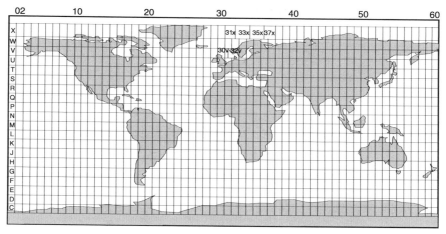

FIGURE 2.17 UTM reference grid of the world.

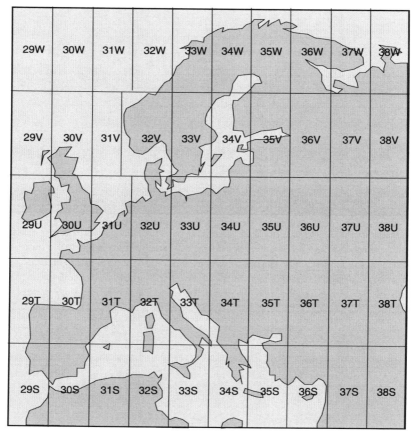

FIGURE 2.18 UTM reference grid of Europe.

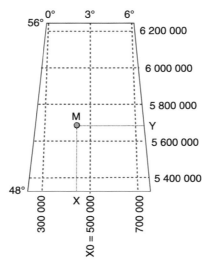

FIGURE 2.19 Example of UTM coordinate.

An example is given as an illustration in Fig. 2.19, for UTM zone 31U, which exhibits a scale factor of 0.9996. Let us consider the point of the following coordinates:

- 31U 0,445,600 m with reference to "zero,"
- 31U 5,691,400 m with reference to "equator."

Note that it is common to use the UTM projection with other geodetic reference frames, such as WGS84 (GPS). In such a case, the projection can be associated with any ellipsoid (IAG GRS 1980 for WGS84).

The Lambert Projection

The Lambert projection (Fig. 2.20) is a conical secant projection associated, in the case of France, with the NTF[9] geodetic system (which uses the Clarke 1880 IGN ellipsoid). In order to minimize distortions, the country is divided into four zones. An additional projection is defined in order to provide a single map for the whole country (France in the present case). The latter is characterized by larger distortions than the four others, but allows a single reference frame for the whole country. The user should know about this and decide which projection to use.

The longitude of origin is the Paris meridian and corresponds to $\lambda = 0$ grade. The longitude used for Paris[10] is $2°20'14.025''$ east of Greenwich. Table 2.1 gives the details of the three main zones, together with characteristics of another projection used for the whole country. Note that standard[11] parallels are also given.

---

[9]NTF: Nouvelle Triangulation de la France.
[10]At a place called La croix du Panthéon.
[11]The standard parallels are those where the cone is secant to the Earth's ellipsoid.

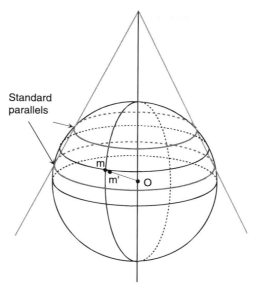

FIGURE 2.20 The Lambert projection.

TABLE 2.1 Parameters used for the Lambert projection in France.

	Lambert Zone			
	I	II	III	II extended
Range (gr)	53.5–57	50.5–53.5	47–50.5	France
Latitude of origin (gr)	55	52	49	52
Standard parallels	48°35′54.682″ 50°23′45.282″	45°53′56.108″ 47°41′45.652″	43°11′57.449″ 44°59′45.938″	45°53′56.108″ 47°41′45.652″
X0 (m)	600,000	600,000	600,000	600,000
Y0 (m)	200,000	200,000	200,000	2,200,000

Source: IGN, Institut Géographique National, the French National Institute of Geography.

In comparison with the UTM projection, it gives horizontal bands rather than vertical ones. France is divided into the three abovementioned zone, as shown in Fig. 2.21.

The coordinates of point M described in the UTM section are now (zone II) 2N 187,200 m with respect to the origin and 2E 545,600 m, also with respect to the origin.

Change of Coordinate Reference Frame

It is of prime importance to be able to translate from one reference frame to another, especially for global navigation systems that are likely to provide a very

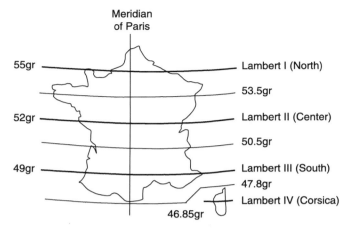

FIGURE 2.21 France divided into three Lambert zones (plus one for Corsica).

large coverage of the Earth. Furthermore, it is necessary to have the ability to achieve any projection required by the user. Such a change often also requires a change in the geodetic referential. Such a change is usually achieved in Cartesian coordinates and uses a seven-parameter modeling system, as described in Fig. 2.22.

The standard transformation between Cartesian coordinate reference frames is a three-dimensional similitude (also known as the "Helmert transformation") with seven parameters, as follows:

- Three translations Tx, Ty, and Tz;
- Three rotations Rx, Ry, and Rz;
- A scale factor D.

The conversion approach is fully described in Fig. 2.23.

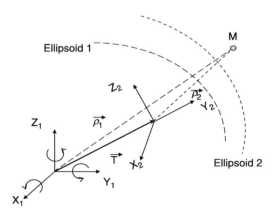

FIGURE 2.22 The "standard" transformation.

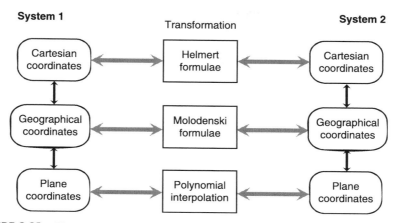

FIGURE 2.23 The general conversion approach.

Modern Maps

For some modern maps, it is possible to have access to different scales in different reference frames. For example, the TOP25 series of IGN exhibits a GPS compatibility and gives two scales in degrees or grades and two kilometric scales, as follows:

- Inside, the NTF system: longitude and latitudes (grades from the Paris meridian), in black, and Lambert kilometric indications (zone I, II, III, or IV depending on the location) also in black, and extended Lambert II in blue.

- Outside, the WGS84: longitude and latitudes (degrees from the Greenwich meridian), in black, and UTM kilometric indications in blue with a 1 km step.

It is now obvious that a user should be aware of the real coordinate system used in order to be able to properly use a map. This explains the discrepancies that sometimes occur between a GPS receiver's output and the feeling of a user who knows perfectly well that the positioning is not good. The receiver does not give the right location. Very often, it actually does, but the tuning of the receiver is not compatible with the map and comparisons are not carried out on the right basis.

Geodesic Systems Used in Modern GNSS

Usually, one distinguishes between the referential system and its implementation, which is the referential frame. The geodesic coordinate frames used by GPS, GLONASS, and Galileo are respectively WGS84 (World Geodetic System 1984), PZ-90, and GTRF (Galileo Terrestrial Reference Frame).

In the WGS84 model, the plane sections parallel to the equatorial plane are circular. The equatorial radius equals 6378.137 km. The plane sections perpendicular to the equatorial plane are ellipsoidal. The section containing the Z-axis has its larger semi-axis equal to 6378.137 km (the equatorial radius) and its smaller semi-axis is equal to 6356.752 km (flattening[12] equal to 1/298.257223563).

The PZ-90 model is similar to the WGS84, with a larger semi-axis of the plane section of the equatorial plane being 6378.136 km and a smaller semi axis of 6356.751 km (flattening equal to 1/298.257839303).

The GTRF model is also similar (based on GRS80) with the larger semi axis of 6378.137 km and the smaller semi axis of 6356.752 km (flattening equal to 1/298.257222101).

✅ 2.9 EXERCISES

Exercise 2.1 Describe the importance of the maritime clock in order to solve the longitude problem at sea. For instance, why were astronomical observations, such as those carried out by Galileo, not applicable? What time does one have to keep once at sea and why? How is this time information to be used?

Exercise 2.2 Considering a typical landmark-based positioning system, such as the one described in Fig. 2.3, define

- Qualitatively, the error of location induced by a measurement error of 2° and 5°.
- The resulting location obtained with only two landmarks (assuming respectively 2° and 5° of measurement error).
- The resulting location obtained with three landmarks and the approach of Fig. 2.3 (also assuming 2° and 5° of measurement uncertainty).

Exercise 2.3 Explain the typical form of the Doppler curve given in Fig. 2.4. Describe the different possible forms (qualitative comments on the amplitudes) depending on the actual satellite trajectory. Conclude concerning this Doppler curve for users at various latitudes (for the same satellite ground track on Earth).

Exercise 2.4 Do the same as exercise 2.3 in the specific case of Sputnik for the observations carried out at the Johns Hopkins University (approximate latitude of 36°25' North).

Exercise 2.5 Describe the major difference between the three generations of satellite-based navigation systems.
Explain the basic principles of the TRANSIT and GPS systems. Highlight the technical major differences in the physical measurements and the location calculations.

Exercise 2.6 Considering Fig. 2.15, describe diagrammatically the places where the deformation induced by the projection is the most important. Empirically, find the best tuning of the projection plane (orientation and position) in order to minimize the deformation for points "a" to "d". What is the effect on the projection of points "x" to "z"? What conclusion do you come to?.

Exercise 2.7 Explain the reasons for the need for a high number of UTM zones in order to cover the world. How are the junctions dealt with between two successive zones?

---

[12]The flattening is the ratio $a/(a - b)$, a being the larger semi-axis and b the smaller.

Exercise 2.8   Why consider different types of projections: tangent cylinder, secant cylinder, tangent cone, secant cone, tangent plane, and so on?

Exercise 2.9   Why are reference frame conversion systems of such importance in positioning systems?

■ BIBLIOGRAPHY

Adams WS, Rider L. Circular Polar Constellation providing continuous single or multiple coverage above a specified latitude. The Journal of Astronautical Sciences 1987;35(2): 155–192.

Boorstin DJ. The discoverers. New York: Random House; 1983.

Bostrom CO, Williams DJ. The space environment. Johns Hopkins APL Technical Digest, January–March 1998;19(1).

Danchik RJ. An overview of transit development. Johns Hopkins APL Technical Digest, January–March 1998;19(1).

Gardner AC. Navigation. UK: Hodder and Stoughton Ltd.; 1958.

Ifland P. Taking the stars, celestial navigation from argonauts to astronauts. Virginia: The Mariners' Museum Newport News; Florida: The Krieger Publishing Company; 1998.

Kaplan ED, Hegarty C. Understanding GPS: principles and applications. 2nd ed. Artech House; 2006. Norwood, MA, USA.

Kennedy GC, Crawford MJ. Innovations derived from the transit program. Johns Hopkins APL Technical Digest, January–March 1998;19(1).

Merrignan MJ et al. A refinement of the World Geodetic System 1984 reference frame. In: ION GPS 2002: Proceedings; Portland (OR); 2002. p 1519–1529.

Parkinson BW, Spilker Jr. JJ. Global positioning system: theory and applications. American Institute of Aeronautics and Astronautics; 1996.

Sobel D. Longitude. London: Fourth Estate Limited; 1996.

Yionoulis SM. The transit satellite geodesy program. Johns Hopkins APL Technical Digest, January–March 1998;19(1).

Web Links

http://grin.hq.nasa.gov. Nov 7, 2007.

http://history.nasa.gov/hhrhist.pdf. Nov 7, 2007.

http://history.nasa.gov/nara/nara1.html. Nov 7, 2007.

http://pwifland.tripod.com/historysextant/. Nov 7, 2007.

http://techdigest.jhuapl.edu/td1901/index.htm. Nov 7, 2007.

http://www.mat.uc.pt/~helios/Mestre/Novemb00/H61iflan.html. Nov 7, 2007.

Development, Deployment, and Current Status of Satellite-Based Navigation Systems

Among the numerous technical solutions to positioning, Global Navigation Satellite Systems (GNSS) have a special place. This is mainly due to the fact that they have brought such simplicity of use and low cost that many applications and domains have taken advantage of positioning. Examples include civil engineering or the observation of wild animals. This chapter describes the current status of the three main constellations, either existing ones such as GPS and GLONASS, or those soon to be available (Galileo). In addition, the other satellite-based facilities are also described, both public and commercial. At the end of the chapter, a table summarizes all the current satellite systems available for positioning.

Initially designed for military purposes, the American Global Positioning System (GPS) has found a large public, both in professional and mass market sectors. This success has largely exceeded the best hopes of its founders and has mainly been due to the incredible performances provided to users. Positioning rapidly became a new way to carry out many tasks that had previously demanded considerably more effort. This success, especially concerning GPS, has led to a strategic problem — if one imagines the deployment of positioning in so many domains such as transport, telecommunications, or safety, then it is of vital importance to share the management of the global positioning system. This was indeed the problem in the European Union, and the decision to launch Galileo was closely related to strategic options.

Global Positioning: Technologies and Performance. By Nel Samama
Copyright © 2008 John Wiley & Sons, Inc.

■ 3.1 STRATEGIC, ECONOMIC, AND POLITICAL ASPECTS

The spectrum allocated to GNSS, which needs allocating for the whole planet, is decided by international bodies. The International Telecommunication Union (ITU) is in charge of organizing such debates, and frequency bands are then allocated to services. The 1559–1610 MHz band was thus reserved for satellite-based navigation systems. The first "user" can make use of this band within the delimited constraints of power levels and the spectrum shape defined by the ITU. GPS, within this band, was the only user for many years. When Galileo appeared in the scope of GNSS, new discussions took place in order to define both the way this band could be used by both GPS and Galileo, and the possibility of assigning new bands to GNSS. These discussions are always complicated, as there is a need for international consensus and the satisfaction of specific national demands. This process is usually very long and demanding in order to meet all the divergent interests.[1] ITU conferences are held every three years or so.

National regulation bodies also exist to cope with local affairs. The United States Federal Communication Commission (FCC), for instance, is the organization that decided to make it compulsory to give the terminal location for any E911 emergency call. Let us give a quick description of these bodies for the United States and the European Union.

Federal Communication Commission

To understand the complexity of spectrum allocation, a quick look at the United States frequency allocation table, limited to the bands relating to GNSS, is instructive (Fig. 3.1). Services in capital letters are primary services allocated to the corresponding bands. The GPS bands are clearly identified in the 1215–1240 MHz and 1559–1610 MHz bands. The third GPS frequency, L5, is intended to lie in the 1164–1188 MHz band and is shared with the Aeronautical Radio Navigation Service (ARNS). This chart, specific to the United States, is a good illustration of the potential difficulty of finding "room" for new services or systems.

When considering worldwide systems, one can easily imagine that it is even more complicated, especially when a single band is being searched for. The International Telecommunication Union (ITU) is in charge of this allocation at the World Radio Conferences (WRC), which take place every three years or so, and where states try to find agreements for the allocation of bands and orbits.

European Approach

The European authorities in charge of the frequency allocations devoted to Galileo have worked within the ITU frame. The allocations of the WRC held in 1997 in the L2 and L1 bands are respectively shown in Figs 3.2 and 3.3 (note that the future L5 was already planned). The Galileo program had to find a way to be

---

[1]The typical time this process takes is about ten years. For Galileo, spectrum allocation was finally settled in 2004, with a final agreement between the United States and the European Union.

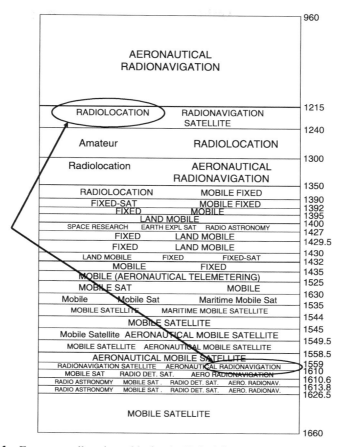

FIGURE 3.1 Frequency allocation table for the United States.

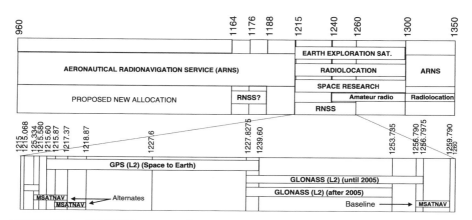

FIGURE 3.2 WRC'97 frequency allocation in the L2 band.

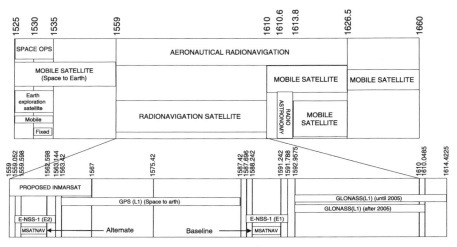

FIGURE 3.3 WRC'97 frequency allocation in the L1 band.

included within this global frame. It is interesting to see that the ARNS occupied quite a lot of bandwidth in the vicinity of the Radio Navigation Satellite Services (RNSS), partly for air control systems. Demonstrating the interest of new constellations and frequencies for aeronautical services was probably of great help in the negotiations to obtain part of the ARNS spectrum.

Tough discussions took place at the WRC conference in 2000 and led to further fruitful bilateral discussions between the United States and Europe in order to find an acceptable compromise for the 2003 WRC. At this conference, a final agreement was found; Fig. 3.4 summarizes the frequency and bandwidth allocations for GPS, GLONASS, and Galileo. Note that only two bands, L1 and L5/E5a, are common to both GPS and Galileo.

International Spectrum Conference

When a government signs an agreement with the ITU, it actually agrees to only a limited commitment. It agrees with the radio regulations of the ITU, which have been agreed by successive WRCs. This includes that Member States accept that spectrum allocations do not cause interference with other Member States' allocations. However,

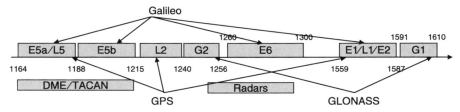

FIGURE 3.4 WRC'03 frequency allocation for GNSS (2003).

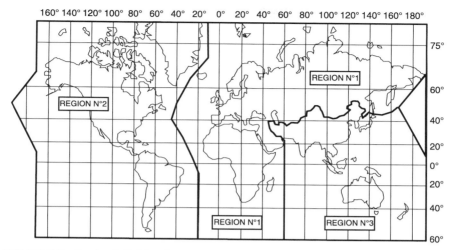

FIGURE 3.5 Regional division for spectrum allocation.

Member States are free to draw their own national frequency tables, but which have to follow the WRC's tables. Nevertheless, they have a certain flexibility in allocations, taking into account some regional or local constraints. Thus, national allocations can differ greatly from one to another.

The world of frequency allocation is divided into three regions, as shown in Fig. 3.5. This historical division means that the WRC and agreements for all the worldwide systems, such as the Universal Mobile Telecommunication System (UMTS) or GNSS are crucial. Another example of worldwide systems is WiFi, where national regulations are not uniform; the industry took this difficulty into account by implementing a dynamic channel allocation protocol in order to be able to develop one single radio architecture. The final multiband approach of the Ultra Wide Band (UWB) radio for Wireless Personal Area Networks (WPAN) is also a demonstration of the same aspect — in order to enable deployment in almost all countries, the designers have chosen a "versatile" approach, that is, one that easily adapts to local frequency allocations.

Strategic, Political, and Economic Issues for Europe

The decision to build a new global satellite navigation system in Europe was taken after careful analysis of both economic and strategic aspects. As already stated in Chapter 1, after the economic analysis of the domains related to satellite navigation, either for positioning or for time-based applications, the European Union had to make a choice from the following strategies (1998–1999):

- Formation of a global joint system with all major players;[2]

---

[2]These are the exact terms used in the official publication of the European Union in the document "GALILEO — Involving Europe in a new generation of Satellite Navigation Services" dated February 10, 1999.

- The European Union developing a GNSS with one or more international partners, namely the United States or Russia;
- Independent development by the European Union of its own system.

Although it was obvious that a joint development would be the most efficient and the least expensive, some issues had to be considered; for instance, a guarantee against disruption of the signals, a participation for future development, operation, and control of the constellation, or an opportunity for European industry to participate in the market segments. Thus, contacts were made, specifically with the United States and the Russian Federation.

At that time, the United States was not in favor of sharing any aspects of the control of either the current GPS constellation or its future modernization, mainly for military-related reasons. On the other hand, the United States had a positive attitude to cooperation in certain technical domains. As a matter of fact, once it was understood that Europe was willing to (and actually would) enter the GNSS field, there was a real interest in remaining in the discussions. Furthermore, as GPS was the first system in operation within projected shared bands, a careful analysis of European propositions was compulsory. The 2000 WRC discussions on this matter were really tough and led to the planning of many meetings carried out over the following years in order to reach an agreement that would be fully acceptable to both parties. Thus, the 2003 WRC, which settled the main principles of Galileo, was much easier.[3]

In 1999, discussions with the Russian Federation seemed likely to lead to a real full partnership in developing a new international system based on GLONASS. The main idea was to design a new system, interoperable with GPS and also using the GLONASS frequency spectrum allocated. In 2001, cooperation was no longer so obvious and slowed down, due mainly to differences related to financing.[4] Meanwhile, the Russian Federation decided to implement a renewal program for GLONASS, and the participation of Europe was then envisaged only as a financial help in order to facilitate this renewal. Note that the current GLONASS constellation is composed of 15 or so satellites that already allow partial positioning.

Once these observations had been carried out, a choice had to be made, and Europe decided to design, build, and launch a new constellation with some stringent initial constraints. Among others, the technical one was that it had to be interoperable with GPS. At first, the meaning of this term was not really clear: would it mean that the two systems could run independently without interference (which is required in any case to comply with the ITU standards), or that technical specifications such as codes, frequencies, and so on should be shared. The real interoperability concept, as described in more detail for example in Chapter 8, consists in allowing a receiver to use satellites from different constellations in order to compute a position. This fact means a real technical link between both systems, and is the final approach followed.

---

[3]From the point of view of an outside party, it looked a little bit like the Americans, at first sight, were willing to rule Galileo out. Once they understood that Galileo would be a reality, cooperation started on a good basis . . .
[4]Reported by the "Progress report on the Galileo program" official document of the European Commission dated December 5, 2001.

Besides these strategic and political aspects, the economic one had also to be carefully considered. Galileo should be a civilian-controlled system, unlike GPS and GLONASS, based on the definition of primary services to be offered to the user communities. However, as long as GPS (and in a less crucial way GLONASS) was free of charge for the users in its Standard Positioning Service (SPS), the Galileo program had to invent a new financing approach to show that the system would be able to generate revenues and that industry would be interested and involved in the program (also for the financial aspects). As the program was presented as a civil one, it was necessary to tell the public why such a huge amount of money would be spent building a system while there was already another system (GPS) in place. It seems that this is, among others, one of the reasons the United States decided on May 1, 2000, to withdraw the so-called Selective Availability (SA).[5] The argument of the European Union that GPS signals were intentionally corrupted was no longer a reality, and it was then more difficult to convince European public opinion to go ahead with Galileo.

The financing strategy of Europe was to establish a Private Public Partnership (PPP) where both public funding (notably during the development phase) and private investments could be joined. This PPP could, in addition to providing complementary finance, ensure overall value for money. Such a private sector commitment would certainly encourage the different services to generate income and to ensure that users' needs were placed at a central position during the development. Many different public structures were successively engaged in the process: the Galileo Joint Undertaking (GJU) until the end of 2006, and then the Galileo Supervisory Authority (GSA) from January 1, 2007. This latter organization is in charge of public interests for the GNSS programs and also the regulatory authority concerning this field. The private counterpart is the so-called Concessionaire, which is in charge of deployment and exploitation of the program. It should have taken over from the public structures after the development phases. The final consortium is composed of eight partners: AENA, Alcatel, EADS, Finmeccanica, Hispasat, Inmarsat, Thales, and TeleOp. Note that the final offer from private partners is the result of the merging of the two remaining consortia of the first round of the bidding, following the European demand. Negotiations took longer than originally planned and difficulties arose concerning the risks, both on technical and economic aspects. At the time of writing, a communication[6] from the commission "invites the Council and the European Parliament to take note of the failure of the current concession negotiation and to conclude that, on this basis, the current PPP negotiations should be ended."

In order to facilitate the revenue part of the program, and hence its viability, some European directives have already introduced the use of GNSS in various sectors such as maritime traffic, waterways transport, and the interoperability of road tolling systems. A recommendation has also been issued regarding the emergency call number E112. Note that Europe has only issued a recommendation for E112 where

---

[5]The SA was set up in order to provide civil users with degraded performance. Noise on both the orbital parameters and the satellite's clock synchronization was applied for this purpose.
[6]"Galileo at a cross-road: the implementation of the European GNSS programmes" on May 6, 2007.

the United States and Japan have mandated the operator to provide the user's location when using the E911. This probably makes a difference in terms of general investments.

The initial (2000) planning of Galileo included the following stages:

- Development and validation phase from 2001 to 2005;
- A deployment phase in 2006 and 2007;
- An operating phase from 2008 onwards.

The schedule is now a little delayed due to discussions by Member States concerning the first round of financing (because of European Union rules concerning the return on investment of European projects), and it took more than a year to settle this problem. In addition, the process of selecting the Concessionaire did not go ahead as quickly as previously planned. Thus, the new global planning was officially as follows[7]:

- The development and validation phase will continue until the beginning of 2009 (from 2007 until the end of this phase, the GSA is in charge of pursuing the tasks and activities of the former GJU);
- The deployment phase in 2009 and 2011;
- The operating phase in 2012.

Following the fact that the PPP negotiations have been stopped, alternative scenarios have been proposed by the European Commission. The various dates for the operating phase are 2012, 2013, or 2014, depending on the scenario.[8]

Another essential aspect of Galileo is international cooperation. At first, this cooperation was seen as a means to increase European know-how and to limit political risks, but it was also the affirmation that the program was really based on a service-oriented approach to offer users the best possible system. As the program was of world importance, relationships with third-world countries were encouraged. A number of such countries expressed an interest in Galileo. In practice, the problems to be solved when thinking of cooperation are related to security, technology transfer, intellectual property, or control of exports. The main countries at first were, as already described, the United States and the Russian Federation, but many others were concerned. With the United States, the main difficulties arose when the Public Regulated Service (PRS) of Galileo was revealed, because it was no longer possible for Americans, with their new military M-Code, to jam the PRS signals without impacting their M-Code, thus jamming themselves. With the Russian Federation, although early discussions on a potential cooperation took place before the definition of Galileo, they slowed down a little afterwards, despite the mutual interest in expanding cooperation. Other countries that have signed cooperation agreements are the People's Republic of China, Israel, Ukraine, India, Morocco, and South Korea. Other cooperation agreements are in discussion with Norway and Argentina. Finally, initial discussions are under way with Switzerland, Canada, Australia, Saudi Arabia, and Brazil.

---

[7]Official document of the European Commission "Taking stock of the GALILEO program" dated June 7, 2006.
[8]See the document of the Services of the Commission SEC(2007)624 of May 16, 2007.

It is important to note that technical discussions with the United States led to the signature of a final agreement defining the structure of the codes, modulations, and bandwidths that ensure the interoperability of GPS and Galileo in June 26, 2004.[9]

■ 3.2 THE GLOBAL POSITIONING SATELLITE SYSTEMS: GPS, GLONASS, AND GALILEO

The first system developed and deployed was the Global Positioning System (GPS), but GLONASS, the Soviet Union's equivalent constellation, followed quickly after. Nowadays (2007), with the so-called "modernization" of both systems, the future European Galileo constellation is foreseen to become available in 2012/2013. Other systems are also either already available or under development in various parts of the world, such as Japan, India, and China. The Galileo program is different in that it tried to bring together and to integrate a large community of countries; agreements exist, in addition to the European Union, with Africa, South America, China, Israel, and so on. In this chapter, we shall briefly describe the current status and the foreseen improvements of the major Global Navigation Satellite Systems (GNSS).

The Global Positioning System: GPS

Current Status

The GPS is based on a three-segment architecture (Fig. 3.6): the ground segment, the space segment, and the user segment. As their names suggest, the ground segment is the part of the system that allows the monitoring, upload, and/or update of the system, and is currently completely based on Earth stations (in the GPS III evolution, scheduled for 2013 onwards, there are plans to implement a spatial version of the ground segment too). The space segment is made up of the satellites. The user segment is composed of all the possible receivers. The fundamental feature of the system, as was also the case with TRANSIT, PARUS, and TSIKADA, is the unidirectional link between the space segment and the user segment, making it impossible to saturate the system; that is, there can be as many receivers as one wants. In addition, this feature also allows receivers to be undetectable, which is very important for military applications. Of course, a link exists between the ground segment and the satellites. This is shown graphically in Fig. 3.7.

One has to note that the two other constellations, GLONASS and Galileo, also use this principle of operation.

The nominal constellation (Fig. 3.8) is made up of 24 satellites, orbiting in quasi-circular orbits, tilted at 55° from the equatorial plane. Six such orbital planes are used, each having four satellites (non-uniformly distributed, because the coverage had to be better for northern areas for military purposes). The altitude of the satellites is around 20,200 km (with an orbital radius of approximately 26,600 km), and the revolution

---

[9]The official document is available at http://ec.europa.eu.

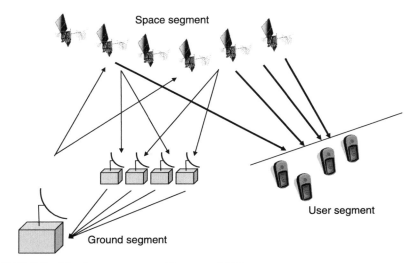

FIGURE 3.6   The three-segment architecture of GPS.

period is 12 h. The United States' policy of replenishment is to wait for the highly probable failure of a satellite, and to keep a "reasonable" margin. Thus, the real number of satellites in orbit, and really used for positioning, is generally between 28 and 30.

One of the most important aspects of a navigation constellation relies on signals. From their form, their number, and their main characteristics, the performances of positioning can be assessed. The current GPS signal structure available on the latest satellites is diagrammatically represented in Fig. 3.9 (note that there is also a "data message" that is not represented).

From a very high quality fundamental frequency of 10.23 MHz, all the required signals are generated. In fact, as will be seen in Chapter 7, this frequency is slightly different from 10.23 MHz because of relativistic effects. The high stability is achieved

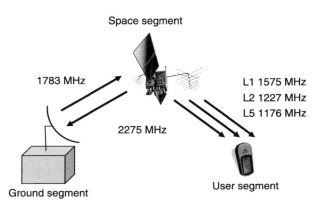

FIGURE 3.7   Bi- and uni-directional links in the GPS system.

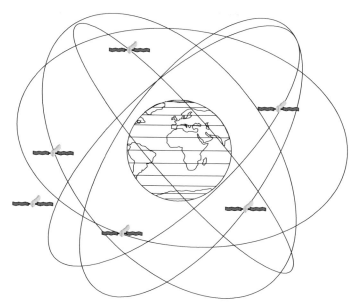

FIGURE 3.8 GPS constellation.

through the use of four atomic clocks in each satellite. By taking respectively the 120th and 154th harmonics of this fundamental frequency, both transmitted central frequencies of 1227.60 MHz (called L2) and 1575.42 MHz (called L1) are reached. For measuring as well as identifying purposes, the GPS signals are in fact codes that modulate these carrier frequencies. As codes are generated at rates of 1.023 MHz for the "civil" signal (called C/A for Coarse/Acquisition) and 10.23 MHz for the "restricted access" signal (called P for Precise) using a binary phase shift keying (BPSK) scheme, the corresponding spectral occupation is twice the code rate for the main lobe, leading to a figure of 3.10.[10] Although more details are given about this figure in Chapter 7, one can observe that the C/A code is delivered on L1, in phase, and code P is delivered on both L1 and L2, in phase quadrature. This feature is of prime importance when dealing with performances of the various signals, as is discussed in depth in Chapter 6.

Another challenging aspect of GPS signals is the very low value of the power levels of C/A and P codes. Received by a 3 dBi[11] antenna, these levels can be as low as -128.5 dBm[12] for C/A and -131.5 dBm for P. Physically considering the white noise within the main lobe for both codes, it appears that a receiver that deals with 2.046 MHz of bandwidth (C/A) is subject to a noise power of

---

[10]Note that for all the spectrum representations, in-phase refers to sine waves, whereas in-quadrature refers to cosine waves.

[11]The dBi is a measure of antenna gain with respect to an isotropic antenna that exhibits a unity gain in all directions.

[12]The dBm is a power value expressed in decibels ($10 \times \log(\text{power}/\text{reference})$), relative to a reference power of 1 mW. Thus, 0 dBm is a power of 1 mW and -128.5 dBm is a power of $10^{-12.85}$ mW (as -131.5 dBm is a power of $10^{-13.15}$ mW).

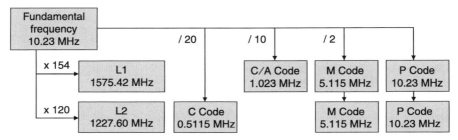

FIGURE 3.9 GPS signal architecture.

−111 dBm, whereas one dealing with 20.46 MHz of bandwidth (P) is subject to −101 dBm of noise power. In such cases, it appears clearly that the GPS C/A signal is −17.5 dB (a factor of more than 50) under the noise level and the GPS P signal is −30.5 dB (a factor of 1000) under the noise level. From these rapid calculations it is easy to understand that special care must be taken in order to detect the signals; this case was devoted to both the CDMA (code division multiple access) scheme and PRN (pseudo random noise) codes.

The current positioning performances depend greatly on the positioning techniques used. There are essentially two techniques: stand-alone and differential (there are numerous differential approaches, see Chapter 5 for details). The first technique relates to stand-alone receivers, which benefit from no outside help for location finding. Depending on the nature of the algorithms and the electronics used (and more specifically the fact that one or both frequencies are tracked[13]), stand-alone techniques provide accuracy in the range of a few meters. If one wants to achieve better accuracy, ranging from 1 to 2 m down to the centimeter, differential schemes are required. Differential stands for "making a difference," whatever it might be (details on differential methods are given in Section 5.5).

From these signals, two services are provided: the Standard Positioning Service (SPS), currently using the C/A code, and the Precise Positioning Service (PPS), which uses P codes and the two available frequencies.

Modernization

Although GPS has outstanding global performances, in comparison with former terrestrial systems or the first satellite-based ones, some interesting features that are not yet available with GPS will be provided by the Galileo constellation. This is the case, for example, for the civilian availability of two frequencies or the ability for monitoring the integrity of the signals. As stated before, GPS provides only one civil frequency and could not compete with Galileo in some aspects ("time to alarm" for integrity[14] is an example). However, the United States system has entered a modernization[15] that started with its launch on September 26, 2005. All

---

[13]Note that current civilian applications usually do not make use of stand-alone dual frequency receivers.
[14]See Section 3.3 for details.
[15]Moreover, this might be due to the Galileo program.

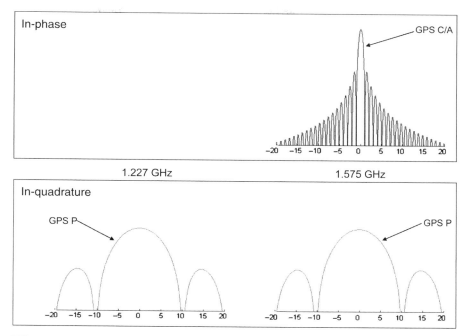

FIGURE 3.10 Typical spectral representation of current GPS signals in L1 and L2.

satellites launched before 2005 have a civil signal L1 C/A and two restricted access signals P on L1 and L2 (see Fig. 3.10).

The satellite launched on September 26, 2005, was the first of a series of 12 so-called Block IIR-M (R for Replenishment and M for Modernized) satellites. They are characterized by a new civil C/A code on L2 (see Fig. 3.11) and also an M-code on both frequencies, featuring the new military capabilities of GPS. The advantage of a second frequency for civilian use appears obvious when considering propagation effects,[16] and the M-code is almost compulsory for the United States to be able to keep the possibility of running GPS in potentially hostile environments (in terms of radio signals). The M-Code also allows the United States to implement a new approach: selective denial (replacing selective availability). As long as GPS is no longer alone, there is the need to be able to deny access to navigation signals (by jamming for instance) other than the M-Code.

An additional L1C signal is planned (2012), using a better modulation scheme and power level than the old C/A one. Typically, the level of L1C could be -155.5 dBW[17] (as compared to -158.5 dBW for the current C/A signal), and the foreseen modulation is

---

[16]From the satellite to the receiver, the signal has to travel through the ionosphere (at least when the receiver is on Earth), which is a non-linear medium. The effect on signal propagation is a speed reduction of the wave depending on many factors and thus affecting the measurements. The fact of having two frequencies allows the real-time measurement of the induced delay, and then its suppression.

[17]The dBW is identical to the dBm with a reference power of 1 W instead of 1 mW.

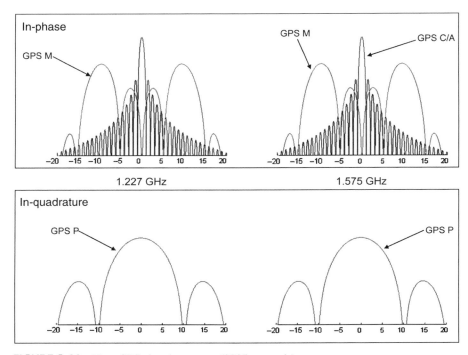

FIGURE 3.11 New GPS signal structure (2005 onwards).

a binary offset carrier (BOC), similar to Galileo signals. This L1C is intended to provide the best interoperability features with Galileo in the L1 band.

In addition to these new signals, it is planned to deploy a third civil signal in the 1176 MHz ARNS (Aeronautical Radio Navigation Service) band (Fig. 3.12) to enhance the system for aviation applications (in order to provide a dual-frequency system for civil aviation because L2C is not in an ARNS band). The power level is configured to be about $-157.9\,dBW$, a dB above L1 C/A. This signal will be available with the Block IIF satellite generation, expected from 2008 onwards. Note that Galileo is a three-frequency system and should be available in 2012/2013.

The official accuracies imagined are summarized in Table 3.1.

The GLONASS

Current Status

As for GPS, the system is composed of three segments: spatial, ground, and user. In GLONASS, a rocket–space complex is added as a fourth component. The 24 nominal satellites are distributed in three orbital planes (as compared to six with GPS), at an altitude of 19,100 km and an inclination of 64.8° in order to allow a good coverage of the northern regions of Russia.[18] The revolution is equal to 11 h

---

[18] In fact, the current GLONASS constellation, even though not complete, is much better for positioning in northern regions than for lower latitudes.

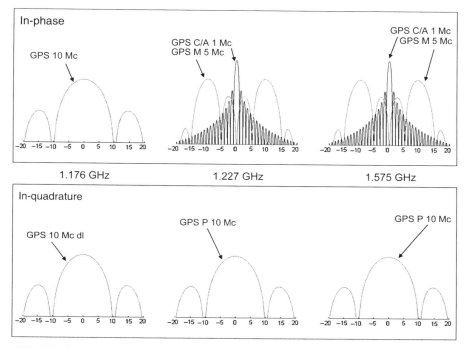

FIGURE 3.12 New GPS signal structure (2008/2009 onwards). Note that L1C is not represented in the figure (2012 onwards).

15 min 44 s. The orbital planes are spaced 120° apart from each other, and the satellites are regularly spaced every 45° in each plane. The planes are switched 15° apart from each other in latitude. The coverage is then of good quality all over the world and in space.

The radio interface for GLONASS is slightly different from GPS, as it uses a frequency division multiple access (FDMA) scheme, and not a CDMA one. Nevertheless, the radio transmission principles remain with two frequencies, L1 and L2, and two types of signals, civil and military. The corresponding "services," as they are called in GPS and GLONASS, are the Standard Precision Service (SPS) and the High Precision Service (HPS). GPS equivalents are respectively the SPS and the PPS. Thus, each satellite is characterized by a couple of frequencies,

TABLE 3.1 Foreseen accuracy of modernized GPS.

	Mass Market Receivers	Professional Receivers
Horizontal (m)	2.5	0.5
Vertical (m)	4.5	1.1
Time (ns)	5.7	1.3

one in L1 and the other in L2, which allows both transmission and identification. Noise considerations are of reduced concern with GLONASS, compared to GPS, due to the multiple access scheme used. To reduce the frequency spectrum used, the possibility of using the same frequency on opposite sides of the Earth was decided and was implemented from 2005 onwards.

A factor of ten is applied between SPS and HPS rates of modulated signal, respectively, at 0.511 MHz and 5.110 MHz. The signals are quite different, a standard code for SPS and a military code for HPS, and are coded through C/A and P codes, as with GPS. The modulation is a bipolar phase-shift keying (BPSK) one, as with GPS. L1 is composed of the modulo-2 addition of C/A code, navigation message data, and an auxiliary meander sequence. L2 is composed of the modulo-2 addition of the P code and auxiliary meander sequence. All the above components are generated using a single onboard time/frequency oscillator, as with GPS.

The main difference with GPS signals is that C/A and P codes are identical for all the satellites. The achievable accuracy of a complete GLONASS constellation is of a few tens of meters in the horizontal and vertical planes (slightly degraded performances vertically because of the geometrical distribution of the satellites, discussed in Chapter 8 concerning the dilution of precision concept). The speed accuracy is about a few centimeters per second. The new GLONASS-M satellites have also transmitted a standard code on L2 since 2003. This makes GLONASS the first dual-frequency system to be available for civilians.

The frequency values for L1 and L2 are defined by the following formulae

$$f_{K1} = f_{01} + K\Delta f_1$$

$$f_{K2} = f_{02} + K\Delta f_2$$

where K characterizes a given satellite. Furthermore, the fundamental frequency values are given by

$$f_{01} = 1602 \, \text{MHz} \qquad \Delta f_1 = 562.5 \, \text{kHz}$$
$$f_{02} = 1246 \, \text{MHz} \qquad \Delta f_2 = 437.5 \, \text{kHz}$$

Before 1998, values of K between 1 and 12 were used without any limitation (channels 0 and 13 being reserved for technical purposes). From 1998 until 2005, the satellites used filters in order to limit side-band transmission, particularly in the upper part of the L1 band, shared with radio astronomy[19]). Values of K were between -7 and 13. From 2005, values of K have shifted from -7 to $+6$ (channels 5 and 6 being reserved by the Russian Federation for technical purposes).

Modernization

In 2001, the Russian government adopted some directives in order to undertake alone the maintenance and the development of the system. The Federal Dedicated Program "Global Navigation System" was approved by the government in August 2001 for the period from 2001 to 2011. The main goal of this program was to preserve

---

[19]CCIR recommendation no. 796 for the 1610.6–1613.8 MHz band.

the Russian role in the satellite navigation community. That meant providing sufficient funds for the development of GLONASS.

To achieve the goal of having GLONASS play a role in the international community, modern satellites are needed. Eighteen GLONASS-M (for "modernized") satellites are planned, as is the development of the following generation of satellites, the GLONASS-K.

The GLONASS-M satellites, whose first launch occurred in December 2003, exhibit a shift in frequency (as described above), but also a power level doubled on L2, and extended codes. In addition, complementary bits of data are added in order to give access to information relative to the system time discrepancy between GPS and GLONASS. Further improvements make it possible to obtain satellites with a working life of about seven years, which is of tremendous importance. As time stability is also improved, achievable accuracy comes down to a level about one-and-a-half that of GPS. Note that power levels are in the range of $-161\,dBW$ for L1 and $-167\,dBW$ for L2.

The evolution of the signals is graphically described in Fig. 3.13 for signals transmitted before 2003, in Fig. 3.14 for signals transmitted after 2003, and Fig. 3.15 for the addition of a third civil signal, L3, in the 1205 MHz band (2008 onwards).

The frequency value for L3 is defined following the formula $f_{K3} = f_{03} + K\Delta f_3$, where K characterizes a given satellite. Furthermore, the fundamental frequency values are $f_{03} = 1201.5\,MHz$ and $\Delta f_3 = 421.875\,kHz$. Note that these values still need to be validated.

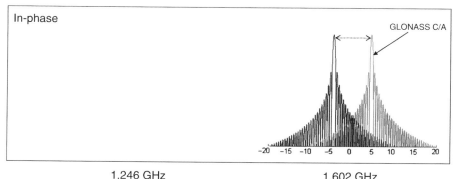

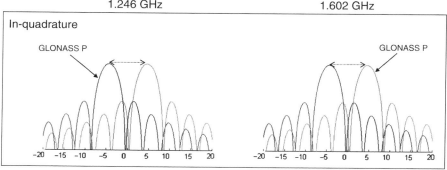

FIGURE 3.13 Diagram of signals before 2003.

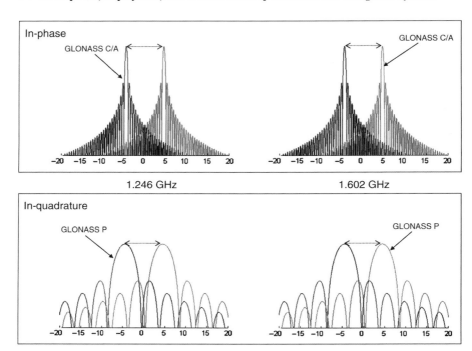

FIGURE 3.14   Diagram of signals after 2003.

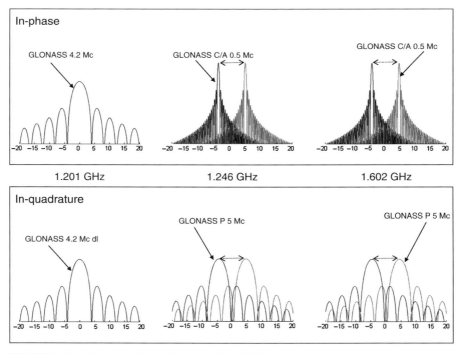

FIGURE 3.15   Diagram of signals planned for 2008.

TABLE 3.2 **Comparison of main parameters between GPS and GLONASS.**

Parameters	GLONASS	GPS
Satellites		
Number of satellites	19 (March 2007)	30 (March 2007)
Number of orbital planes	3	6
Inclination	64.8°	55°
Altitude	19,100 km	20,180 km
Ground track repetition	8 days	1 day
Revolution period	11 h 15 min	12 h
Ephemeris	9 parameters (position, velocity, acceleration)	Keplerian elements of the orbit and their first derivatives
Update rate of ephemeris	30 min	2 h
Geodesic reference frame	PZ 90	WGS 84
Time reference frame	UTC (SU)	UTC (USNO)
Transmission time of almanacs	2.5 min (7500 bits)	12.5 min (37,500 bits)
Signal		
Signal identification	Own frequency	Own code
Multiple access scheme	FDMA	CDMA
Carrier frequencies		
L1	1598–1606 MHz	1575.42 MHz
L2	1243–1249 MHz	1227.60 MHz
Codes	Same for all satellites	Different for each satellite
Code frequency	C/A: 0.511 MHz	C/A: 1.023 MHz
	P: 5.11 MHz	P: 10.23 MHz
Selective availability	No	No
Anti-spoofing	No	Yes

The latest development of GLONASS will be achieved (2008) through the use of GLONASS-K and in 2015 with GLONASS-KM satellites. The GLONASS-K implements a third civil frequency, as with Galileo and the modernized GPS, in order to improve the accuracy but mainly the reliability of the system for "life-saving" applications (which are the most demanding ones in this domain). The life-time is also planned to be increased to up to ten years. Further improvements in synchronization are bound to provide even better accuracy.

A first comparison of global parameters with GPS can be carried out. Table 3.2 provides some elements (note this table is based on L1 and L2 only).

Galileo

The only possibility for Europe was to develop a third global positioning satellite constellation. Once again, this was not the original goal of European countries, but it is difficult to have imagined that a compromise with real military systems could have been found. Of course, the new problem of the program was to convince European public opinion of the need for such expenditure, estimated at some 3.2 billion Euros.[20] So it was decided to make this program the first big achievement of European industries and research activities. Thus, Galileo is the first joint program between the European Union and the European Space Agency (ESA), financed by both parties for the research and development phase. The program is civilian-driven and controlled, has to be implemented by an industrial consortium, and is expected to generate revenues. Thus, the notion of services was at the forefront of the thinking. The signals required are then only derivatives of these services.[21] A lot of work has been carried out over many years, involving industry, government, research centers, and universities with a view to develop the services. As the system should be profitable, "killer" applications have been sought, but have not yet been found. Furthermore, knowing about some obvious limitations of the GPS was of great help in designing a "better" system, at least at the time of the specifications. For example, the major drawback of satellite-based navigation systems is certainly poor urban coverage, where modern applications are bound to be mainly deployed and where most of the population lives. Thus, the "local elements" concept, initially designed with this goal, is interesting; unfortunately, it is easier to state a wish than to succeed in finding real solutions. Although some concepts have arisen from this "local elements" part of the program, definitive answers have not been provided for this difficult problem (further discussions on indoor positioning are given in Chapters 9 and 10).

A total of four navigation services and one to help the search and rescue services have been defined in order to answer the users' needs. The communities targeted are mass market, scientific, commercial, and governmental ones. The Galileo constellation can be used autonomously, even if the obvious goal is to provide interoperable constellations. The above mentioned services are the following:

- The Open Service (OS), which is composed of corresponding open signals and is free of charge, like the corresponding GPS and GLONASS signals. It is mainly designed for mass market applications.

- The Safety of Life Service (SoL), which provides integrity data in addition to the OS. The main domain of applications is transportation where lives could be in danger.

- The Commercial Service (CS), characterized by two additional signals, allows faster data rates and potentially improved accuracy. It will also provide added value services on payment.

- The Public Regulated Service (PRS) provides positioning and timing information to government authorized users, such as police or customs.

- The Search and Rescue (SAR) Service is a support of Galileo to the international COSPAS-SARSAT[22] system (humanitarian search and rescue).

---

[20]About four billion US dollars.

[21]The reader will certainly be able to see that although this approach is quite interesting in itself, it has not led to fundamentally new concepts: the finally adopted structure of the signals remains quite "classical." The constraint to be interoperable with GPS was certainly a limitation.

[22]COSPAS-SARSAT stands for Cosmicheskaya Systyema Poiska Avariynich Sudov — Search and Rescue Satellite Aided Tracking.

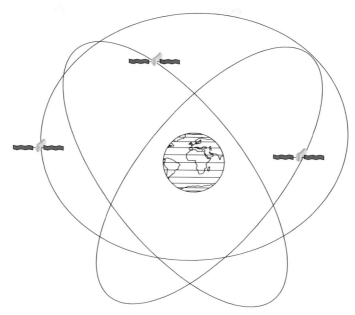

FIGURE 3.16 The Galileo constellation.

Current Status of the Program

The constellation is made up of 30 satellites on a Medium Earth Orbit (MEO) at an altitude of 23,222 km and an orbital period of 14 h. There are three orbital planes, as with GLONASS, each composed of ten satellites. The nominal constellation is 27 satellites, the last three being spare. Thus, each orbital plane is composed of nine "in use" satellites plus one that is in reserve in case of a failure. The inclination of each orbital plane is 56°, comparable to GPS but at a higher altitude, which leads to a better coverage than GPS for polar regions, where European countries still have population and territories. See Fig. 3.16 for a graphical representation of the constellation.

A first experimental satellite was successfully launched on December 28, 2005: GIOVE-A.[23] The main goals of GIOVE-A were to transmit on the frequencies allocated to Galileo during the international radio communication conference held in 2003. A first representative transmission should have been demonstrated before June 2006 to "secure" the frequencies for Galileo: this was definitely achieved in March 2006. The second goal, as important as the first, was to carry out many experimental tests and validations on signals, modulation schemes, in order to prepare the real Galileo constellation intended to be launched from 2009 until the end of 2012.

---

[23]GIOVE-A: Galileo In-Orbit Validation Element-A. A stands for the first satellite. Two were planned in case of the failure of the first launch of GIOVE-A. As it was successful, the launch of GIOVE-B was delayed in order to re-define its space mission.

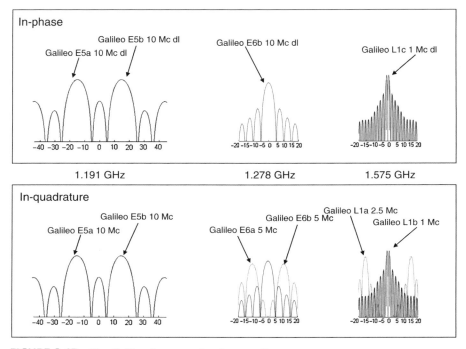

FIGURE 3.17   The Galileo frequency band occupation.

The Galileo signals essentially share three frequency bands, two of which are in common with GPS for interoperability purposes.[24] Figure 3.17 gives a diagrammatic view of the 11 signals distributed throughout the spectrum, from 1.164 GHz to 1.591 GHz.

As already described, the original principle of Galileo was based on services and not on signals. Thus, for each service (OS, SoL, CS, PRS, and SAR), a subtle combination of bands and signal arrangement has been developed. In fact, the signals described in Fig. 3.17 result from the analysis of the needs for the various services in terms of performances, and their translation into physical signal requirements.

In addition to Table 3.3, which gives details of the bands, it can be noted that E5 has a bandwidth of 50 MHz, E6 40 MHz, and E2-L1-E1 (also called L1) 8 MHz for both E2 and E1, in addition to the previous 24 MHz allocated to L1. This gives a total of 122 MHz for satellite navigation purposes, just for the Galileo system. It is obvious that tremendous possibilities are bound to appear, in comparison with existing capabilities.

Of the 11 signals, the distribution to services is defined as follows (see Fig. 3.18): 6 signals are allocated to OS and SoL, 2 signals are solely allocated to CS, and 2 signals are solely allocated to PRS. The 11th signal is dedicated to the SAR service (and is not represented in Fig. 3.18).

---

[24] At least with GPS III including L5.

TABLE 3.3 Frequencies allocated to Galileo.

Carrier	Central Frequency (MHz)
E5a (L5)	1176.45
E5b	1207.14
E6	1278.75
E2-L1-E1	1575.42

Modulation schemes and more technical parameters are discussed in Chapter 6, but an interesting feature of the Galileo signals lies in the presence of so-called "pilot tones." These signals, present in all the bands, do not include navigation data. We shall see that navigation data are a limitation toward long-time coherent integration that could greatly help in increasing the sensitivity of receivers (see Chapter 7 for details). Pilot signals are in quadrature phase with signals incorporating navigation data. Such pilot tones are also planned as a possibility on L2 for GPS satellites from Block-IIR-M onwards. Figure 3.17 gives the representation of these signals in the quadrature phase plane.

A comparison can be drawn with GPS and GLONASS. First, Figs 3.19 and 3.20 show the complete spectrum of navigation signals that will appear in a few years with the advent and modernization of the three constellations. Note that the current situation (2007) is completely different: a few GPS and GLONASS satellites transmit L2C, but the majority is restricted to C/A on L1 and P on L1 and L2. Further discussions are carried out on this point in Chapter 7.

Table 3.4 gives a summary comparing the three constellations' main parameters.

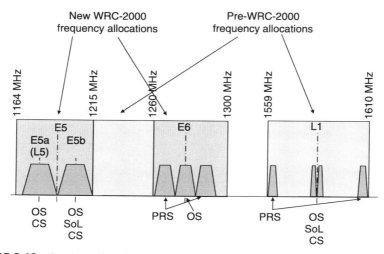

FIGURE 3.18 Service allocation versus frequencies in Galileo. (*Source*: Galileo Joint Undertaking.)

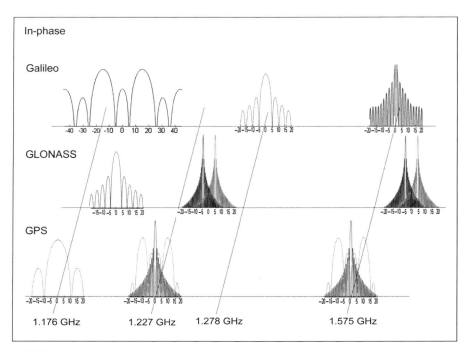

FIGURE 3.19 In-phase GPS, GLONASS, and Galileo signal spectrum as it will appear in a few years (note that L1C GPS is not included).

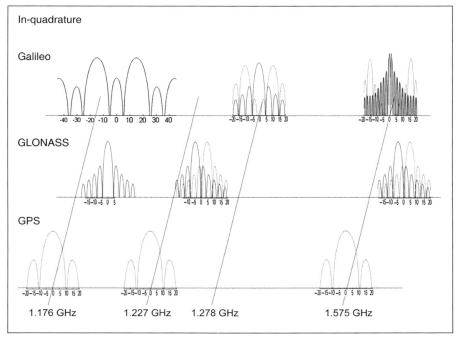

FIGURE 3.20 In-quadrature GPS, GLONASS, and Galileo signal spectrum as it will appear in a few years (note that L1C GPS is not included).

TABLE 3.4 Comparison of main parameters for GPS, GLONASS, and Galileo.

Parameters	Galileo	GLONASS	GPS
Satellites			
Number of satellites	30 (2011)	19 (03/2007)	30 (03/2007)
Number of orbital planes	3	3	6
Inclination	56°	64.8°	55°
Altitude	23,616 km	19,100 km	20,180 km
Ground track repetition		8 sun days	1 sun day
Revolution period	14 h 4 min	11 h 15 min	11 h 58 min
Ephemeris	Keplerian elements of the orbit and first derivative	9 parameters (position, velocity, acceleration)	Keplerian elements of the orbit and first derivative
Update rate of ephemeris	3 h	30 min	2 h
Geodesic reference frame	GTRF	PZ 90	WGS 84
Time reference frame	UTC (BIPM)	UTC (SU)	UTC (USNO)
Transmission time of almanacs	10 min	2.5 min	12.5 min
Signal			
Signal identification	Own code	Own frequency	Own code
Multiple access scheme	CDMA	FDMA	CDMA
Carrier frequencies			
E1/L1	1559–1591 MHz	1598–1606 MHz	1575.42 MHz
L2		1243–1249 MHz	1227.60 MHz
E5	1164–1214 MHz		
E6	1260–1300 MHz		
Code	Different for each satellite	Same for all satellites	Different for each satellite
Code frequency	E1: 1.023 MHz E5: 10.23 MHz E6: 5.115 MHz	C/A: 0.511 MHz P: 5.11 MHz	C/A: 1.023 MHz P: 10.23 MHz
Selective availability	No	No	No
Anti-spoofing	No	No	Yes

■ 3.3 THE GNSS1: EGNOS, WAAS, AND MSAS

Many converging factors encouraged the European Union to enter into the satellite navigation community, most importantly, because of its skillfulness in the satellite and launcher technologies and because over the few years after the GPS constellation had been declared operational, the civil signal was intentionally disturbed[25] by the U.S. government. In such a way, the civil receiver performances were degraded compared to the real capabilities of the original signal. Some user communities therefore developed new techniques to overcome these perturbations: the so-called differential techniques (see Chapter 5 for details). The basic idea was that these errors are not random and so are identical for two receivers close to each other. Thus, it is certainly possible to remove the effects of these errors by positioning a receiver at a perfectly known fixed location and observing the difference between the computed location and the actual one. Then, removing the positioning error vector thus obtained from the positioning computed by another receiver, located at an unknown location, could probably help in increasing the resulting positioning accuracy. This is exactly what happened, and differential approaches gained in popularity as they allowed an accuracy of 10–20 m instead of the 100 m allowed at that time, thanks to selective availability. Approaches that deal with range measurements sent to the mobile receiver, rather than positions, are also available.

Unfortunately, this technique requires a data transmission link, either in postprocessing (early systems) or in real-time (current systems). The fear of the community was that the "links" could be developed in a noncoordinated manner, leading to multiple technical and technological solutions. The normalized RTCM[26] American format existed for such differential transmissions, but it dealt only with the format and not with the radio link. As the European Union was not yet involved in satellite navigation, this was the moment to join.

The GPS system is totally driven by the U.S. government and is thus under no "impartial" control. This means that if one satellite happens to show some difficulties, in any way, there is no possibility of knowing about it, unless the U.S. authorities agree. This has been the case on some occasions, leading to positioning errors that sometimes lasted quite a long time (a few hours) without any information from anyone.

All these reasons led the European Union to decide to participate in satellite-based navigation, under the form of a "satellite-based differential station" and "integrity signals." The corresponding program was called GNSS 1, standing for Global Navigation Satellite System, phase 1. This first step is clearly identified as being the premise for GNSS 2, namely Galileo.

In order to achieve both integrity and differential data transmission, the main requirement is for a large ground infrastructure to gather all the necessary information.

---

[25]This perturbation was called "selective availability" and consisted of some errors on both the ephemeris data (specifically those data associated with orbital parameters) and satellite clock synchronization parameters.
[26]RTCM stands for Radio Technical Commission for Maritime Services.

The program name is EGNOS (European Geostationary Navigation Overlay Service). This is effectively an overlay system that heads the GPS and GLONASS systems. The approach consists simply in gathering information, just like a standard local differential station, but over a large area (typically the European area[27]). Once this has been achieved (see Chapter 5 for details), the differential and integrity data are sent to a geostationary satellite for transmission over a large area. These kinds of complementary systems to satellite navigation systems are called SBAS (Satellite-Based Augmentation Systems), as opposed to Local Area Augmentation Systems (LAAS). The transmission is also another problem. Indeed, a transmission radio data link requires the use of a frequency band that will certainly increase the cost of a receiver. The shrewdness of EGNOS is to use the same band as GPS, and furthermore a similar code to the GPS satellites' C/A codes. This means that no hardware changes are required in order for every GPS receiver to access the differential mode.[28] Of course, as the best C/A codes are reserved for the GPS satellites (36 C/A codes with the best correlation characteristics are thus reserved for the constellation), EGNOS uses other Gold codes, exhibiting slightly degraded correlation performances. This quite clever approach has led to the fact that "EGNOS enabled" receivers were available as soon as the program was technically defined; only a software modification was required and all manufacturers carried out this modification early on.

Together with EGNOS, two other SBAS programs were developed: the Wide Area Augmentation System (WAAS) in the United States and the Multi-transport Satellite-Based Augmentation System (MSAS) in the Japanese area. A fourth is also under development in India, called GAGAN. The coverage areas of these programs (see Fig. 3.21) are dictated by the ground infrastructure, and not by the coverage of the transmitting geostationary satellites. There is a need for ground stations to collect the data required in order to prepare the differential and integrity messages. Discussions are in progress with some countries for the development of the EGNOS coverage, in particular in the Mediterranean regions. Note that WAAS is already operational and EGNOS should be in early 2008.

The EGNOS integrity concept relies on the transmission of various corrections. Although the concept of integrity is very complex, there are essentially two levels of integrity: usable and nonusable, for all visible satellites. In order to achieve this goal, EGNOS is capable of making corrections to orbital and clock errors of each satellite (through the so-called "slow corrections") and to ionosphere time delays through the use of a separated set of corrections. At the time of activation of selective availability, "fast corrections" were also allocated to rapidly varying clock errors; these corrections are still transmitted.

---

[27]In fact, the area covered depends mainly on the ground deployment. EGNOS correction data are available in Europe and North Africa.
[28]The only difference lies in the fact that the navigation message of EGNOS is in fact specific to the differential mode and has to be decoded in accordance with EGNOS message definition rather than GPS navigation message definition.

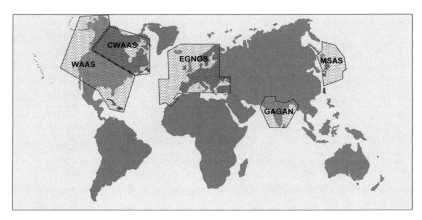

FIGURE 3.21 SBAS coverage areas.

EGNOS typically offers three services intended to be provided to different user communities:

1. *The Open Service (OS)*: Freely available, it allows accuracy with GPS that can reach 1–3 m horizontally and 2–4 m vertically.
2. *The Commercial Data Distribution Service (CDDS)*: Offered through the Internet or through cellular phones on a controlled access basis for professional users needing enhanced performances. It provides in particular

   a. Correction messages (integrity, clocks, orbit, and ionosphere data);
   b. Raw data gathered at ranging and integrity monitoring stations;
   c. Other "products," such as ionosphere computations and troposphere maps.

3. *The Safety of Life (SoL)*: Available in 2007, it offers enhanced and guaranteed performances to the transportation communities. The integrity performance of EGNOS is characterized by

   a. A time to alarm of 6 s;
   b. A horizontal alert limit of 40 m;
   c. A vertical alert limit of 20 m;
   d. An integrity risk of 2×10^{-7} per 150 s.

The way the integrity concept is implemented within EGNOS is described in Chapter 8. Note that this integrity concept is clearly applicable to (and driven by) the aviation domain, although probably not so easily transposable to other environments such as city centers or indoors. The concepts of WAAS, MSAS, and GAGAN are comparable, although not identical.

EGNOS was developed under a tripartite agreement whose members are the European Space Agency (ESA), the European Commission (EC), and Eurocontrol (the European Organization for the Safety of Air Navigation).

■ 3.4 THE OTHER SATELLITE-BASED SYSTEMS

With regard to the many advantages of a satellite-based navigation approach, and relative to the overall cost of a global system, some "local" developments have been carried out throughout the world in order to implement partial systems. This is the case for constellations such as GAGAN in India or QZSS in Japan, but also for some commercial services that intend to provide a better accuracy by providing differential corrections. Let us start with the review of navigation satellite programs. Table 3.5 summarizes the three main constellations around the world. Note that it has been mainly in Asia that developments have been carried out, as other parts of the world have their own constellation, for example Russia, the Unites States, and Europe. African and South American continents have no known projects, but are associated with Galileo.[29]

Some more detail about these systems is given in Chapter 4 in the paragraph about satellite-based positioning radio systems.

■ 3.5 DIFFERENTIAL SATELLITE-BASED COMMERCIAL SERVICES

There are also commercial services based on transmitting differential corrections to end users, to improve availability, integrity, and accuracy of the navigation satellite systems. Table 3.6 summarizes the main characteristics of OmniSTAR™ and StarFire™.

The OmniSTAR system and the StarFire system are comparable concepts; differential corrections are provided through a geostationary satellite transmission link to the users. They both send correction data to the user's receiver, which computes the positioning. Various techniques allow accuracy in the range from 10 cm to around 1 m.

There are also some commercial tracking services based on the use of geostationary telecommunication satellites for fleet management. Table 3.7 summarizes the main characteristics of OmniTRACS® and EutelTRACS™.

The EutelTRACS system is a positioning and communication satellite system for mobiles. It provides message and report features in real time between fleet and master control centers. The various messages use satellites to link the central station of Eutelsat to distribution centers all over Europe. The EutelTRACS system operates

---

[29]More precisely, Brazil, Argentina, and Morocco.

TABLE 3.5 Other satellite-based navigation capabilities.

<table>
<tr><td colspan="2" align="center">BNS/China</td></tr>
<tr><td>Main
  characteristics</td><td>Three satellites on geostationary orbits (140°E, 80°E, 110.5°E)
Interactive navigation experimental services
Typical accuracy of 30–50 m</td></tr>
<tr><td>Development</td><td>Three satellites launched in October and December 2000, and in May 2003</td></tr>
<tr><td>Current status</td><td>Agreement to contribute to the Galileo program for the in-orbit validation phase by the development of services and products related to synchronization and positioning</td></tr>
<tr><td>Perspectives</td><td>Essential use for governmental and military applications
Use of Galileo for civilian applications
The evolution of Beidou towards a fourth worldwide constellation (COMPASS) is planned</td></tr>
<tr><td colspan="2" align="center">GAGAN/India</td></tr>
<tr><td>Main
  characteristics</td><td>A geostationary relay (82°E) of GPS and GLONASS signals</td></tr>
<tr><td>Development</td><td>Increased accuracy for GPS and GLONASS constellations (1.5–2.5 m)</td></tr>
<tr><td>Current status</td><td>Use of GPS and GLONASS signals. Agreement with the Galileo program to participate on a financial basis in the European program</td></tr>
<tr><td>Perspectives</td><td>Payload in the L-band onboard an experimental satellite Gsat-4.
Fifteen reference stations
Under study: a regional constellation (IRNSS) of eight satellites</td></tr>
<tr><td colspan="2" align="center">QZSS/Japan</td></tr>
<tr><td>Main
  characteristics</td><td>Three satellites in an inclined orbit (between 35,000 and 42,000 km), combining navigation (accuracy down to 1 m), communication, and broadcasting capabilities</td></tr>
<tr><td>Development</td><td>Use of the GPS constellation and satellite relays for data transmission</td></tr>
<tr><td>Current status</td><td>Ongoing negotiations between the Japanese government and spatial industry in order to establish a public–private partnership</td></tr>
<tr><td>Perspectives</td><td>Three satellites to be launched in 2008–2009 and a fourth one in reserve (on ground)
The first satellite will not have communication capabilities</td></tr>
</table>

BNS = Beidou Navigation and Positioning System (CAST/Chinese Academy of Space Technology); *GAGAN = Geo-Augmented Navigation (ISRO/Indian Space Research Organization, AAI/Airports Authority of India); *IRNSS = Indian Regional Navigation Satellite System (ISRO/Indian Space Research Organization); *QZSS = Quazi Zenith Satellite System (with ABSC/Advanced Space Business Corporation and NICT/National Institute of Information & Communications Technology, in a private/public partnership).

on the frequency bands 10.7–11.7 GHz and 12.5–12.75 GHz for the space-to-Earth link and 14–14.25 GHz for the Earth-to-space link.

The OmniTRACS system is also a wireless communication and satellite positioning system. It uses QUALCOMM's U.S. two-way satellite wireless link in

TABLE 3.6 **Differential correction commercial satellite services.**

	OmniSTAR/USA (global coverage)
Main characteristics	Differential GPS services on board geostationary satellites, offering various possible accuracies: sub-meter (VBS), 20 cm (XP) and 10 cm (HP)
Development	Use of the Virtual Base Station Solution (VBS) to optimize the corrections for each user's position
Current status	Deployment of approximately 100 ground reference stations that track the GPS satellites and compute corrections every second
Perspectives	Integration of the future GNSS signals

	StarFire/USA (global coverage)
Main characteristics	Differential GPS services on board geostationary satellites, offering real-time decimeter accuracy
Development	Use of the StarFire Network to consider each of the GPS satellite signal error sources independently
Current status	Deployment of more than 60 ground reference stations that compute GPS satellite orbit and clock corrections
Perspectives	Integration of the future GNSS signals

order to allow fleet management. The positioning technique implemented is based on GPS.

As a conclusion to this chapter, Table 3.8 summarizes the various satellite systems used for navigation.

TABLE 3.7 **Commercial tracking satellite services.**

	OmniTRACS/USA (global coverage)
Main characteristics	C-band and Ku-band relays on board geostationary satellites, developed by Qualcomm (∼100 m)
Development	Since 1988, first message and localization of mobile system. Mainly used for real-time monitoring of truck, ship, container fleets
Current status	Deployment of more than 540,000 terminals all over the world
Perspectives	Integration of the future GPS signals as well as the Galileo ones

	EutelTRACS/Eutelsat (Europe/Mediterranean/Middle-East)
Main characteristics	Ku-band relay on board two geostationary satellites (W1 and Sesat-1)
Development	Closed network of secured message and localization, providing for fleet management, since 1990
Current status	Diversification of services for the 25,000 terminals already in use (developed by Qualcomm and Alcatel)
Perspectives	Combined use of modernized GPS and Galileo signals for the next generation

TABLE 3.8 Summary of satellite-based navigation systems.

GPS/United States

Main characteristics	24 satellites/six orbital planes/55°/20,200 km
	Typical accuracy of 3–5 m (based on real observations; official documents give 13 m horizontally and 22 m vertically, 95% of the time)
Development	11 Block I (launched from 1978) & IIA, all non-functional now, and 28 Block II/IIA
Current status	30 operational satellites (15 Block IIA, 12 Block IIR, and 3 Block IIR-M)
	Improvements of the ground segment with a new control center
Perspectives	Conversion of 8 Block II-R into II-RM
	Building of 12 Block IIF (first launch planned in 2008); Block II-RM and Block II-F to be deployed in 2005–2012; GPS III (first launch in 2013): 18 satellites to be deployed from 2013 to 2018

GLONASS/Russia

Main characteristics	24 satellites/three orbital planes/64°/19,100 km
	Typical accuracy of 20 m (current)
Development	87 satellites launched
	2 GLONASS-M satellites launched in 2003
Current status	15 operational satellites
	9 GLONASS-M to be launched from 2005 to 2009
	Development of GLONASS-K satellites
Perspectives	18 satellites in 2008, 24 satellites in 2010–2011
	Launching of 27 GLONASS-K from 2008 onwards
	GLONASS-KM planned for 2015

Galileo/European Union

Main characteristics	30 satellites/three orbital planes/56°/23,600 km
	Typical accuracy of 1–2 m
Development	Two experimental satellites GIOVE-A and GIOVE-B to secure the frequencies and orbits
Current status	First four satellites planned for 2008 in order to validate the complete concept
Perspectives	Full constellation deployed in 2009–2010

WAAS/United States

Main characteristics	L-band geostationary relay for GPS and GLONASS differential corrections
	Typical accuracy of 1–2 m
Development	Two geostationary satellites
Current status	25 ground stations in the United States
Perspectives	Two additional satellites (Anik and PanAm)
	EGNOS/European Union
Main characteristics	L-band geostationary relay for GPS and GLONASS differential corrections
	Typical accuracy of 1–2 m

(Continued)

TABLE 3.8 *Continued*

Development	Three geostationary satellites (Inmarsat 3 + Artemis)
Current status	34 ground stations in Europe and North Africa
	Open Service and Commercial Service available
	Safety of Life Service under certification (available in 2007)
Perspectives	Extension of the services to other areas
	Integration of EGNOS into the Galileo program

MSAS/Japan

Main characteristics	L-band geostationary (140°E) relay for GPS and GLONASS differential corrections
	Additional features such as meteorology and mobile communications
Development	Two geostationary satellites
	Improvements of GPS positioning
Current status	MTSAT-1R launched in 2005 for the Ministry of Transport and the Japan Meteorological Agency
Perspectives	MTSAT-2 under construction (launching planned in 2007)
	Synergies with QZSS

BNS/China

Main characteristics	Three satellites in geostationary orbits (140°E, 80°E, 110.5°E)
	Experimental interactive navigation services
	Typical accuracy of 30–50 m
Development	Three satellites launched in October and December 2000, and in May 2003
Current status	Agreement to contribute to the Galileo program for the in-orbit validation phase by the development of services and products related to synchronization and positioning
Perspectives	Essential use such as governmental and military applications
	Use of Galileo for civilian applications
	The evolution of Beidou towards a fourth worldwide constellation (COMPASS) is planned

GAGAN/India

Main characteristics	A geostationary relay (82°E) of GPS and GLONASS signals
Development	Increased accuracy for GPS and GLONASS constellations (1.5–2.5 m)
Current status	Use of GPS and GLONASS signals. Agreement with the Galileo program to participate on a financial basis in the European program
Perspectives	Payload in the L-band on board an experimental satellite Gsat-4
	Fifteen reference stations
	Under study: a regional constellation (IRNSS) of eight satellites

QZSS/Japan

Main characteristics	Three satellites in an inclined orbit (between 35,000 and 42,000 km), combining navigation (accuracy down to 1 m), communication, and broadcasting capabilities

(Continued)

TABLE 3.8 *Continued*

Development	Use of the GPS constellation and satellite relays for data transmission
Current status	Ongoing negotiations between Japanese government and space industry in order to establish a public–private partnership
Perspectives	Three satellites to be launched in 2008–2009 and a fourth one in reserve (on ground) The first satellite will not have communication capabilities

OmniSTAR/United States (global coverage)

Main characteristics	Differential GPS services on board geostationary satellites, offering various possible accuracies: submeter (VBS), 20 cm (XP) and 10 cm (HP)
Development	Use of the Virtual Base Station Solution (VBS) to optimize corrections for each user's position
Current status	Deployment of approximately 100 ground reference stations that track the GPS satellites and compute corrections every second
Perspectives	Integration of the future GNSS signals.

StarFire/United States (global coverage)

Main characteristics	Differential GPS services on board geostationary satellites, offering real-time decimeter accuracy
Development	Use of the StarFire Network to consider each of the GPS satellite signal error sources independently
Current status	Deployment of more than 60 ground reference stations that compute GPS satellite orbit and clock corrections
Perspectives	Integration of the future GNSS signals

OmniTRACS/United States (global coverage)

Main characteristics	C-Band and Ku-Band relays on board geostationary satellites, developed by Qualcomm ($\sim$100 m)
Development	Since 1988, first message and localization of mobile system. Mainly used for real-time monitoring of truck, ship, container fleets
Current status	Deployment of more than 540,000 terminals all over the world
Perspectives	Integration of the future GPS signals as well as the Galileo ones

EutelTRACS/Eutelsat (Europe/Mediterranean/Middle-East)

Main characteristics	Ku-Band relay on board two geostationary satellites (W1 and Sesat-1)
Development	Closed network of secured message and localization, provided for fleet management, since 1990
Current status	Diversification of services for the 25,000 terminals already in use (developed by Qualcomm and Alcatel)
Perspectives	Combined use of modernized GPS and Galileo signals for the next generation

✔ 3.6 EXERCISES

Exercise 3.1 What was the original goal of GPS? Describe the link between the major technical aspects and the purpose for which the system was set up. Among others, discuss the following specifications:

- Frequency range;
- Number of frequencies;
- Civil code.

Exercise 3.2 Analyze the reasons and the impact of a regulation such as the E911 for the development of GPS and related industries. Comment on the European E112 approach.

Exercise 3.3 Do personal research in order to determine the main reasons only the Galileo program has slipped so far behind schedule (initially planned for 2008, it is now delayed by at least 3 or 4 years).

Exercise 3.4 Make a comparison of Galileo and GPS in terms of technical, political, and economic aspects. In this comparison, try to include the industrial aspects and outline the future of GNSS (keep to a strategic point of view — the technical aspects are dealt with in subsequent chapters).

Exercise 3.5 Why is the allocation of the frequency spectrum such an issue for GNSS?

Exercise 3.6 Describe the philosophy of the satellite-based augmentation systems and the way they have been implemented by EGNOS. In particular, provide the main issues that allowed standard receivers to cope with EGNOS signals long before it had been declared operational.

Exercise 3.7 SBAS allow for integrity for GPS. Find further documents and readings and describe the real definition of integrity, the context of acceptable use, and hence the real limitations of this integrity concept. In addition, give details on the way it is implemented.

Exercise 3.8 Some commercial SBAS services are available. Can you explain the interest these services provide in comparison to standard free performance. Can you find domains where these services are highly valuable?

■ BIBLIOGRAPHY

Commission communications to the European Parliament and the Council on Galileo, COM(2000) 750 Final, Brussels; 22 November 2000.

Commission staff working paper — progress report on the GALILEO programme. SEC(2001) 1960; 5 December 2001.

Communication from the Commission to the European Parliament and the Council — taking stock of the GALILEO programme. COM(2006) 272 Final, Brussels; 2004.

Communication from the Commission to the European Parliament and the Council — GALILEO at a cross-road: the implementation of the European GNSS programmes. COM(2007) 261 Final, Brussels; 16 May 2007.

Council regulation on the establishment of structures for the management of the European satellite radio navigation programme. COM(2003) 471 Final, Brussels; 31 July 2003.

Council Regulation (EC) No 876/2002 of 21 May 2002 setting up the Galileo Joint Undertaking. Official Journal of the European Parliament 2002.

Council resolution of 5 April 2001 on Galileo. Official Journal of the European Parliament 2001;C157(01).

Erhard P, Armengou-Miret E. Status and description of Galileo signals structure and frequency plan. European Space Agency Technical Note, April 2004.

Galileo — involving Europe in a new generation of satellite navigation services. COM(1999) 54 Final, Brussels; 10 February 1999.

Galileo study phase II. Executive Summary, PriceWaterHouseCoopers; 17 January 2003.

Gibbons G. GPS, GLONASS and Galileo — our story thus far. InsideGNSS 2006;1(1): 25–31, 67.

Hatch RT, Sharpe T, Galyean P. StarFire: a global, high-accuracy, differential GPS system. In: ION NTM 2003: Proceedings; Anaheim (CA).

Heinrichs G, et al. To locate a phone or PDA-GNSS/UMTS prototype for mass-market applications. GPS World 2006;17(1):20–27.

Hofmann-Wellenhof B, Lichtenegger H, Collins J. GPS theory and practice. Springer Verlag; 2001. Springer Wien, New York.

Inception study to support the development of a business plan for the GALILEO programme. Executive Summary, PriceWaterHouseCoopers; TREN/B5/23–2001; 20 November 2001.

Kaplan ED, Hegarty C. Understanding GPS: Principles and applications. 2nd ed. Artech House; 2006. Norwood, MA USA.

Küpper A. Location based services — fundamentals and operation. England: John Wiley and Sons; 2005.

Ladetto Q, Merminod B. In step with INS: navigation for the blind tracking emergency crews. GPS World 2002;13(10):30–38.

Mezentsev O, et al. Pedestrian dead reckoning: a solution to navigation in GPS signal degraded areas? Geomatica 2005;59(2):175–182.

Mission high level definition. European Commission, 23 September 2002.

Oehler V, et al. The Galileo integrity concept. In: ION GNSS 2004: Proceedings; Long Beach (CA).

Parkinson BW, Spilker Jr. JJ. Global positioning system: theory and applications. American Institute of Aeronautics and Astronautics, 1996.

Regulation of the European Parliament and the Council on the implementation and the deployment and commercial operating phases of the European programme of satellite radio navigation. COM(2004) 477 Final, Brussels; 14 July 2004.

Yi Y, et al. GPS + INS + Pseudolites: an integrated positioning system. GPS World 2003;14(7):42–49.

Web Links

http://www.fcc.gov/. Nov 14, 2007.

http://igscb.jpl.nasa.gov. Nov 14, 2007.

http://gps.faa.gov/. Nov 14, 2007.

http://tycho.usno.navy.mil/gps.html. Nov 14, 2007.

http://www.navcen.uscg.gov/gps. Nov 14, 2007.

http://www.ngs.noaa.gov/. Nov 14, 2007.

CHAPTER 4

Non-GNSS Positioning Systems and Techniques for Outdoors

Geographical positioning has been enhanced during recent years with the advent of GPS. The original goal of this system was to allow positioning in environments with no infrastructure and where the deployment of an infrastructure would be difficult, for example, at sea or in the desert. Of course, the problem of positioning has not arisen with satellites and, as described in Chapter 1, many earlier techniques were developed. The success of GPS in many applications, mainly transportation, means that satellite-based navigation has entered the public's consciousness, although it still has many limitations. Some limitations have already been mentioned, such as coverage, availability, or even integrity. This chapter is intended to review the geographical techniques, both ancient and modern, whose purpose was to provide positioning in a way other than GPS, GLONASS, or Galileo.

This chapter is devoted to positioning techniques. For example, the triangulation method used in geodesy is briefly described, although it means knowing the base line and is not intended to provide the same kind of positioning as does GNSS. In other words, not just autonomous techniques are described. Large-area positioning techniques are also highlighted. This means that sensor networks are not dealt with in this chapter, but are (very briefly) in Chapter 9, in relation to indoor techniques. Of course, the description is not complete and the environments considered are not exclusive of others, but are presented in order to counterbalance the omnipresence of GNSS. For example, acoustic systems are not included.

Global Positioning: Technologies and Performance. By Nel Samama
Copyright © 2008 John Wiley & Sons, Inc.

◼ 4.1 INTRODUCTION (LARGE AREA WITHOUT CONTACT OR WIRELESS SYSTEMS)

The limitation of the systems presented here arises from the fact that positioning is achieved through contactless techniques, either radio or optical. The only exceptions are the inertial systems based on physical measurements (which could also be seen as contactless). The systems we are going to briefly describe can be classified as in the following.

1. Optical measurement systems:

    a. Stars,

    b. Lighthouses,

    c. Classical triangulation,

    d. Lasers,

    e. Cameras,

    f. Luminosity, and so on.

2. Terrestrial radio measurement systems:

    a. Radar,

    b. Amateur radio,

    c. LORAN,

    d. ILS, MLS, etc.,

    e. Mobile telecommunication networks,

    f. WPAN, WLAN, WMAN,

    g. Radio signals of various sources like TV, and so on.

3. Satellite radio measurement systems:

    a. Argos,

    b. Cospas-Sarsat,

    c. Doris,

    d. QZSS,

    e. GAGAN,

    f. COMPASS, and so on.

4. Physical measurement systems:

    a. Accelerometers,

    b. Gyroscopes,

    c. Odometer,

    d. Magnetometers,

    e. Barometers,

    f. Thermographs, and so on.

■ 4.2 THE OPTICAL SYSTEMS

The first navigation systems used the stars, followed by the Sun when the Portuguese realized that the Pole Star was not available when south of the equator. After this came artificial stars — the lighthouses — the most famous certainly being the one located in the harbor of Alexandria and which is one of the Seven Wonders of the World. The familiar drawback of optical systems concerns the propagation characteristics of this very high frequency electromagnetic wave; it can be quite directional but is easily stopped by obstacles such as clouds (for the stars and the Sun), fog or rain (for lighthouses), or hills or mountains when considering terrestrial systems. So radio systems are now widely spread, but optical systems still present some interesting problems.

The Stars

The story of the Pole Star and the Sun in navigation is briefly described in Chapter 1. Just note that the Pole Star is really an easy way to approximate the latitude of a location. Recall also that following the observations of Galileo concerning the moons of Jupiter,[1] complete[2] positioning has been possible. This is certainly the reason the European Union has given the name of the famous Italian scientist to its future satellite navigation system.

Lighthouses

In early navigation, lighthouses had the role of warning sailors of the presence of a piece of ground. This was somehow the characterization of the interface between the sea and the land. Some lighthouses were identifiable and hence were able to give additional information. Furthermore, these seamarks were used for years, and are still used, for positioning purposes, using the intersection of base lines. The idea is simply to measure the compass bearing of two lighthouses and to use these angles in order to draw the corresponding lines on a map to evaluate the location of the observation point. Figure 4.1 gives an illustration of this technique. Of course, the accuracy of the positioning depends greatly on the accuracy of the compass bearing (and of the map).

Nowadays, lighthouses still have the main function of warning sailors when approaching land, but the navigation aspects, little by little, have been abandoned. Many lighthouses, which are really beautiful pieces of architecture, have now been converted into museums, and their primary purpose of warning of a transition from sea to land is somehow inverted: it is now a warning for people on land that the sea is close.

Compared to these constructions, GNSS presents a real advantage. As lighthouses are significant constructions, they require financial capacities that not all countries are able to provide. This is the same for airport infrastructures for instance. The modern GNSS systems, with their very good accuracy combined with the low cost of receivers, are a real answer to safety in countries under development . . . but this is another story.

---

[1]Although not a star but a planet.
[2]"Complete" means latitude and longitude.

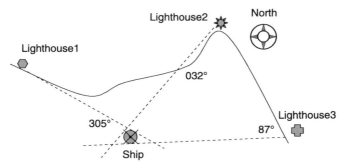

FIGURE 4.1   The compass bearing technique.

The "Ancient" Classical Triangulation

When the need arose to measure the Earth's shape, the so-called triangulation method was developed. The name is closely derived from the properties of triangles. There are many such properties that have led to the use of this geometrical structure in many domains. To measure the Earth, the property that is used is the following:

A triangle is fully defined by one side and the two adjacent angles.[3]

The idea of triangulation was first to define a base line,[4] thus defining two points of the triangle and a segment length (precisely the one corresponding to the base line, of course). The next step was to carry out two angle measurements (see Fig. 4.2 for illustration):

- Angle α from the first point A of the base line to the measured point M;
- Angle β from the second point B of the base line to the measured point M.

Applying the rule described above, knowing the base line and the two angles, allows the complete determination of the triangle AMB, thus leading to the location of M. This method can be applied further considering the MA segment as the new base line, for example. Gradually, it is then possible to triangulate a whole country.

Note that this technique is quite different from the compass bearing technique, because it consists of sighting a targeted point (the measured one, M in our example) from two known points, whereas the compass bearing is almost the opposite approach, consisting of receiving the signal at the measuring location from two lighthouses.

Lasers

One of the most impressive optical components is certainly the laser. Its very clear transmission can be used for many different applications, ranging from

---

[3]There are many other similar properties, hence many other ways to achieve "triangulation."

[4]The approach often implemented is to use an invar cable of 17 m as first base line. The material used, invar, has the property of being very stable, and it presents an extremely low expansion coefficient, thus leading to an accurate base line length value in all weather, pressure, or temperature conditions.

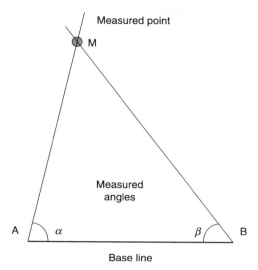

FIGURE 4.2 The triangulation technique.

telecommunications to medicine. One positioning-related application of the laser is telemetry. The basic principle of telemetry is to achieve distance measurement by measuring the flight time of a pulse. In the case of laser telemetry, this pulse is an optical one, generated by a laser and is thus at a very precise frequency. The principle is identical to that of radar, with an accuracy that lies easily in the range of a few millimeters. Modern laser systems also apply "carrier phase" measurements, often using different carrier frequencies in order to achieve approximate and fine measurements. The two difficulties associated with laser telemetry for positioning are pointing and sensitivity to the environment (physical occultation), as for all optical systems.

In some specific environments, knowing the location of obstacles can enable complex laser-based systems to carry out self-positioning. Let us imagine a system composed of three laser beams, in a closed environment indicated by the polygon in Fig. 4.3. The laser telemetry system allows d_1, d_2, and d_3 to be obtained with an accuracy of up to a few millimeters, even when the rays are not perpendicular to the reflected surfaces.[5] Knowing the shape of the confined zone, it is possible to carry out computations in order to define the location of the laser system and its orientation. Of course, this example is only in two dimensions, but a similar approach can be taken with additional measurements in order to achieve three-dimensional positioning and orientation.

The main difficulty is of course that potential obstacles can certainly lead to incorrect distance measurements. This is also the case in indoor environments where there are open doors or windows (although not always the case in this latter instance). A solution could be to have the system pointing at the "sky," that is, up to the ceiling for example, like radio satellite-based systems. In such a case, more

---

[5]Note that this is a major difference from ultrasound telemetry systems.

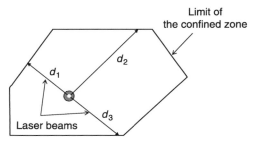

FIGURE 4.3 Possible laser positioning system and environment.

measurements are required as the "sky" is a perfect plane. This approach could be of some interest in static environments where the positions of objects and structures are well defined.

Cameras

The camera is a passive optical system in the sense that it transmits no signal. The scene is captured and forms an image that can then be processed. Therefore, such a system cannot be saturated and is also silent, unlike telecommunication systems.

For positioning purposes, the idea is to carry out specific form recognition within the image. This can be either as markers or "natural" forms like buildings, roads, doors, or characteristic monuments (churches, stadia, transportation systems, and so on). With a certain level of knowledge of the environment, the location of the camera can be extracted from one or several views. Note that for characteristic environments such as indoors (with windows, doors, ceilings, and angles of walls), location calculation can be quite efficient. The down side is the fact that such positioning is highly relative, in comparison with an absolute positioning such as that achieved based on GNSS.

Luminosity Measurements

As the problem of indoor positioning is a real challenge (see Chapters 9 and 10 for details), many original solutions have been investigated. Among others, the measurement of the level of the surrounding light is an unusual way of applying fingerprinting. This technique relies on the observation, under certain conditions, of the variation of a given parameter (here the light levels). Following a calibration phase, where measurements are carried out throughout the whole place, a database is established. Then, an instantaneous light level measurement allows pattern-matching recognition in the database, and possibly location determination. This general approach can easily be extended to a lot of physical parameters, like temperature or radio power levels, either generated locally (WLAN) or regionally (TV signals for example).

■ 4.3 THE TERRESTRIAL RADIO SYSTEMS

The advent of radio systems, at the beginning of the twentieth century, led to a fantastic era of wireless data transmission. The most well-known of these systems are mobile telecommunications or the GNSS, but there are numerous other radio systems. The main advantages of radio systems compared to optical ones are the following:

- They operate in all weathers, even in cloud, rain or fog[6];
- They can, through the use of appropriate antennas, be used with different radiating patterns, ranging from very directive beams (for example, radar) to almost omnidirectional schemes (for example, GSM);
- By diffraction,[7] they can have a range that is much further than the horizon;
- They can, by diffraction, be received from behind obstacles.

The physical phenomena are identical in principle and very similar in results for optical waves and radio waves, but in the case of radio systems, the wavelengths are better adapted to broadcasting.

Returning to positioning systems, many approaches are then possible, using either the power levels received, the time of flight, or the direction of arrival of the signals.

Amateur Radio Transmissions

The system described below is an extension of the NCDXF[8] and IARU[9] beacon system, and its aim is to provide rough positioning. This system is intended to provide amateur radio with a means of estimating the radio propagation conditions in various frequency bands.

The idea is to implement a transmitting cycle where the power level of the transmission is progressively increased. Let us imagine four possible power levels and the cycle P1, P2, P3, P4, with, as an example, P1 = 1 mW, P2 = 10 mW, P3 = 100 mW, and P4 = 1 W. The rough distance (accuracy is not being sought) can then be determined by evaluating the power level received. This is based on the fact that the power level decreases as the inverse of the square of the distance when propagating in free space. Of course, many phenomena can disturb this relation, but this can be acceptable as a first approximation (and remains quite efficient when the propagation is effectively carried out in free space with no buildings, hills, or obstacles).

Let us now imagine that you receive P2 but not P1. This means that you are at a distance somewhere between the range corresponding to P1 and that between P1 and P2; that is, with an uncertainty equal to the square root of 10 (typically 3). This is used

---

[6]Although heavy snow or very high humidity can cause some trouble at some frequency bands.

[7]Diffraction refers to changes in the directions and intensities of waves when passing an obstacle or through an aperture.

[8]Northern California DX Foundation, Inc.

[9]International Amateur Radio Union.

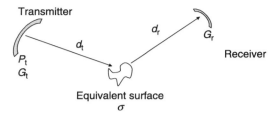

FIGURE 4.4   The bi-static radar principle (P_t is the transmitted power, G_t and G_r are the transmitted and received antenna gains, d_t and d_r are the distances of the transmitter and receiver to the equivalent surface, respectively).

to evaluate the quality of the radio link and not for positioning purposes in the case of the NCDXF and IARU systems, but this extension is very easy to implement.[10]

Radar

The principle of radar (radio detection and ranging) was established at the beginning of the twentieth century. The typical measurements are identical to those carried out by the modern GNSS, namely time measurements in order to provide distance, and frequency[11] measurements in order to provide velocity. Although there are many different types of radar, let us just deal with a simple implementation that should allow us to understand the basic principles.

A transmitter transmits characteristic waves at a given power level. The wave travels in free space and is reflected by any kind of target (a plane or a missile in the first applications) in various directions in space. The function that characterizes the way the target reradiates the incident wave is specified by the so-called "radar effective area" of the target. This corresponds roughly to the directions and intensity of the radiated reflected power. As shown in Fig. 4.4, a receiver can be placed in any location and can measure the power level, time shift, and Doppler shift of the reflected signal at the receiver's end. Knowing the main characteristics of the transmitted signal, the distance and velocity of the target can be computed.

When the transmitter and the receiver are not located in the same place, the radar is called "bi-static," as opposed to the "mono-static." We will now deal with this more common latter configuration. Many technical points are quite easy to address, for instance the time reference. A typical waveform is given in Fig. 4.5: the signal is a repetition of pulses that are equally time spaced. Each pulse is made up of a microwave carrier frequency in order to use the good propagation characteristics of high frequencies and properties linked to the geometry of the antenna. By measuring the time needed for the transmitted wave to be received back at the receiver (considered to be in the same location as the transmitter), one can easily obtain twice

---

[10]Note that this technique is used for indoor WLAN positioning in many cases by considering the received power level throughout a building (in this case taking into account a complex propagation situation). See Chapter 9 for details.

[11]Doppler shift frequency.

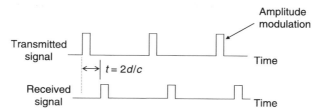

FIGURE 4.5 Distance measurement principle.

the distance from the radar to the target. Of course, the signal received is much lower than the transmitted one. The propagation equation[12] shows that the power decrease is proportional to the fourth power of the distance, meaning that this power decrease is very rapid with distance. Typical distance attenuations of more than 150 dB are usual.

The received frequency can be shifted by the value of the Doppler frequency if the target (or radar) is moving. Thus, by measuring the shift between the transmitted frequency and the received one, the radar can calculate the radial projection of the relative velocity of the target with respect to the radar. It is interesting to note that modern GNSS use exactly the same approach in order to define both the location and velocity of a typical receiver. The time determination is quite different, but the principle is the same: time measurements to deduce distance. As a matter of fact, in the case of the mono-static radar, things are simplified by the co-location of the transmitter and the receiver (which is not really the case with GNSS). Both time and frequency generators are available for comparison of the received signal. In such a case, it is possible to provide a very accurate clock and very accurate frequency shift measurements. Distance and velocity measurements of radars are thus of excellent quality.

Another radar technique, using non-modulated pulses of very short duration, is called the ultra wide band (UWB) radar. The name is associated with the fact that very short pulses have a very large equivalent frequency spectrum. However, one has to remember that the UWB radar is based on a time principle rather than a frequency one. The very short pulse allows a very accurate distance measurement as long as the receiver is able to detect the reflected pulse. One of the main advantages is that when considering the frequency spectrum, the bandwidth is so large that effects related to certain frequency bands can be ignored if they affect only a limited part of the bandwidth. Thus, attenuation when traveling through walls is less than for a narrowband system and multipath can be reduced due to the time pulsed approach (the width of the pulse gives the maximum non-visible multipath-induced delay).

For UWB wireless local area network approaches, although the name is identical, the principle is quite different, in that the time reference is not the same as for radar. This means that in order to define flight times, one will require additional infrastructure for synchronization purposes (see Chapter 9 for details).

---

[12]The propagation equation is given by $P_r = P_t G^2 \lambda^2 \sigma / (4\pi)^3 d^4 L$, where P_t is the transmitted power, P_r the received power level, G the antenna gain, λ the wavelength, σ the radar equivalent surface, d the distance separating the radar from the target, and L all the possible loss along the path.

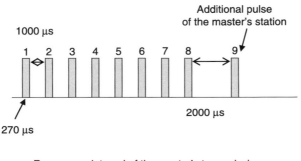

FIGURE 4.6 Principle of the eLoran modulation scheme.

The LORAN and Decca Systems

LORAN (long range navigation) is a terrestrial navigation system using low-frequency signals. The current version, called eLoran,[13] uses the 90–110 kHz band. It is a typical hyperbolic system, because it is based on the time difference between the reception of two signals from two radio stations. If one knows the exact location of both transmitters and they are synchronized, then the time difference allows the calculation of possible locations on a hyperbola whose focal points are the two stations. Thus, the receiver knows, from the first time difference that it lies on a hyperbola: this is of course not enough to define a fix and a second measurement is required. This is achieved through the use of a third station, synchronized with the other two. A new time difference is measured, taking one of the preceding two stations as the reference. This second measurement gives a new hyperbola and the location is determined by the intersection of the two hyperbolae obtained from the two time differences (refer to Fig. 4.13 for details).

In eLoran, the principle is to consider one of the three stations as the master, the other two being the slaves. The implementation considers chains composed of one master station and two or more slave stations. Each LORAN chain uses a unique repetition interval of a sequence of pulses (a typical transmission scheme is given Fig. 4.6), defined in microseconds, identical for all the stations of a given chain.

The recurrence interval of the master's transmission is between 40,000 and 100,000 µs and usually defines the chain.[14] The carrier frequency signal of some

---

[13]The LORAN-A was the first implementation of LORAN, with less accurate positioning than LORAN-C and using the 1750–1950 kHz band. LORAN-B was a phase comparison of the LORAN-A. Refer to Chapter 2 for a basic introduction to LORAN.

[14]For example, the LORAN 6731 station is the one whose repetition interval is 67,310 µs and corresponds to the station at Souston, in the south of France.

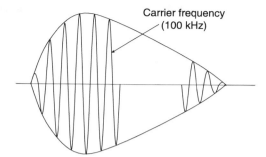

FIGURE 4.7 Typical LORAN pulse shape.

pulses is in the opposite phase compared to the carrier frequency signal of the other pulses; this characterizes the various stations and allows the automatic identification of the station at the receiver. Figure 4.7 gives a typical form of such pulses.

At this stage, two remarks should be made:

- At the frequency of LORAN, waves are split into the ground wave (following the Earth's curvature) and the sky wave (reflected by the ionosphere). The attenuation of the ground wave is stronger, but receivers can differentiate it from the sky wave through the arrival time. The propagation model used for eLoran waves is quite good when the signal is traveling over the sea, and less accurate when traveling over land. Furthermore, as the propagation is supposed to follow the Earth's surface, signal disturbances may occur under certain conditions (especially at night where the ground wave vanishes and the sky wave propagates further). The ionosphere's reaction to sunrise and sunset are also a major concern. In addition, the influence of inland structures, such as cities, is not modeled in eLoran propagation, and accuracy can decrease in such environments.

- When looking at a specific LORAN chart, plotting the various hyperbolae (the time difference is the parameter), as given in Fig. 4.6, one can consider the same approach as that developed for GNSS when considering the dilution of precision (DOP) (see Chapter 8 for details on DOP). It appears that, depending on the location, the intersection of hyperbolae is not likely to provide the same accuracy, taking into account the uncertainties of the measurements. The best approach is to provide the receiver with two perpendicular hyperbolae at the receiver's location. When a LORAN chain is composed of more than three stations, the receiver can choose the best set of stations to optimize the resulting positioning accuracy, or alternatively use the redundancy thus available in order to solve the over-determined problem.

The LORAN coverage of the world is given in Fig. 4.8.

The last feature of eLoran stations to be considered is the implementation of a data channel. This is used both for transmitting LORAN messages and differential GPS messages through the Eurofix method. Eurofix is an integrated navigation and communication system that was proposed and developed by the Delft University of

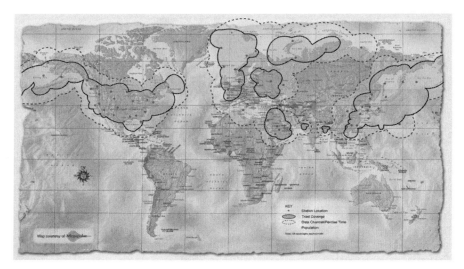

FIGURE 4.8   The worldwide LORAN-C coverage. (*Source*: Megapulse.)

Technology. It is a long-range broadcast system that uses eLoran signals as the carrier. Channel 1, for instance, is devoted to broadcasting differential GPS data, while eLoran navigation capabilities are preserved.

The Decca system, developed in the United Kingdom, and which was operational prior to the end of World War II, was the first hyperbolic radio system. It worked in the 70–128 kHz band and exhibited a 300-mile range at sea. Note that other hyperbolic long-range systems also existed, such as Pulse/8 or Hyperfix. Please refer to Chapter 1, Section 1.5, for a brief, description of older local-coverage hyperbolic systems.

ILS, MLS, VOR, and DME

In some specific cases, such as civil aviation, the need rapidly arose for dedicated systems allowing guidance and landing approach assistance to planes. Terrestrial positioning systems were not accurate or reliable enough. The main systems developed were therefore as follows:

- The VHF Omni-directional radio Range (VOR);
- The Distance Measuring Equipment (DME);
- The Instrument Landing System (ILS); and
- The Microwave Landing System (MLS).

The VOR system is a rotating radio lighthouse. Its transmitted frequency is in the range of 108–118 kHz and the signal is modulated by two 30 Hz signals whose phase difference gives the azimuth with reference to a characteristic direction, which is usually the magnetic north. If this direction is changed to the direction of the VOR station, then the phase difference is zero while the plane stays on the route of the

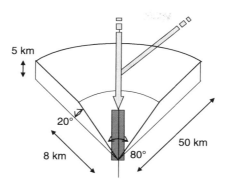

FIGURE 4.9 The MLS concept.

transmitter. With three such VOR stations it is theoretically possible to calculate a location, but VOR is usually used for alignment purposes.

In order to define the distance of the plane from the station, the DME system is used. It works in the 962–1213 MHz band and works around a plane interrogating the DME station. The station answers all the interrogations; and the plane should find the response that corresponds to its request. Note that each transmitter is characterized by a specific pulse sequence.

The VOR and DME systems are coupled in order to provide a location of the plane in polar coordinates. The positioning accuracy is typically a few hundred meters. This accuracy is not enough for a landing phase approach. Thus, the ILS was developed. It defines a light slope rectilinear trajectory for landing by way of the intersection of two surfaces. It then requires two radio lighthouses: the first defines an alignment on the runway (the "localizer") and the second is used for the descent (the "glide-slope"). The system is completed with two or three vertically radiating radio markers that play the role of spot locations in front of the runway. The runway alignment radio lighthouse uses a frequency in the 108–112 kHz band and the descent one in the 328–335 MHz band.

All these systems of so called "goniometry"[15] are limited in accuracy, because the frequencies that are used are not free from multipath disturbances. Nevertheless, the MLS also uses the rotating radio lighthouse principle but at microwave frequencies, in the 5 GHz band. The running mode is always a radio lighthouse one, but the radiating pattern is rather narrow (1–3°) and the free space is successively scanned from right to left and back again. Knowing the scanning pattern, the receiver can determine its angular location by analyzing the time delay separating two successive beams of the MLS. The angles are thus provided in a continuous mode, unlike ILS. The landing path is then evaluated in comparison to a predefined optimal path that can have any desired form (and not necessarily a rectilinear one; see Fig. 4.9 for illustration). This component is not widely deployed because of the number of sites to be equipped and the corresponding costs.[16]

---

[15]The fact of measuring angles.

[16]And the advent of satellite-based navigation systems . . .

Mobile Telecommunication Networks

The GSM Networks

In order to forward communications, the GSM network needs to have access to a database that keeps track of the mobile locations. Indeed, the location is simply the identification of the base station providing the greatest power level to the receiver. Of course, as base stations are rather large installations, their locations are known and they are not movable. Thus, the so-called Cell-Id (see Fig. 4.10), for identification of the telecommunication cell where the terminal is, is embedded in the network. Furthermore, it is implemented for all the deployed networks, as it is compulsory. Therefore, localization features exist in all large-area networks. The advantage of this positioning is that it gives a location in all places where the network is available, that is, where the reception is possible, including indoors. The main disadvantage is the low level of accuracy it provides — the typical size of the result is the cell itself. So, it goes from 100 m in densely populated urban areas to 20 or 30 km in rural zones. Another difficulty where the density of base stations is high is that the cell associated with the mobile is the one whose reception is the most powerful, that is, not necessarily the one that is the nearest. Thus, the accuracy can easily drop to a few hundred meters, even though there are base stations in the vicinity of the mobile terminal.

This method is quite comparable to radio signal strength (RSS) measurements as it is based on power level estimation, but with only one base station. The main advantage of this approach is that it works with only one base station in radio visibility.

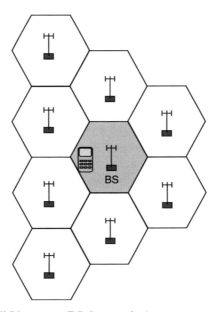

FIGURE 4.10 The Cell-Id concept (BS, base station).

As it is difficult to deal with the very different accuracy figures resulting from the network deployment, Cell-Id positioning has not been used for years, except for telecommunication purposes. Following the proposed FCC regulation consisting of locating every call, the interest in telecommunication-related positioning has grown, with a better accuracy than that allowed by Cell-Id. However, some telecom operators at first, and now almost all operators, have in their catalog of solutions something based on Cell-Id for location-based services (finding a place, estimating a route, and so on).

The next step, before time-based solutions, is to come back to a very old method: measuring angles (see Fig. 4.11). Indeed, in order to increase the capacity of a base station, operators have chosen to develop specific antennas that have the ability to determine the absolute direction of arrival of a signal, also called the "angle of arrival" (AOA), relative to the antenna plane. This enables the channel used by a user in direction, say D1, to be used within the same base station by another user whose direction relative to the antenna of the base is D2, sufficiently different from D1. Thus, once again, this feature has been designed for telecommunications purposes.

Of course, as in the old days, measuring the angle from two or three bases can be used to calculate a location. This is also very comparable to what was done by the sailors when measuring angles from landmarks in order to plot their location. In the case of wide-area telecommunication networks, the idea is to carry out such measurements from three base stations. Assuming the accuracy of the angle measurement is around 1°, and the range around 1 km, one obtains an accuracy of position of about 100 m. The accuracy is therefore not very good and this approach is absolutely not applicable if there is no direct signal (line of sight); that is, urban or indoor positioning is not really intended to be obtained with AOA.

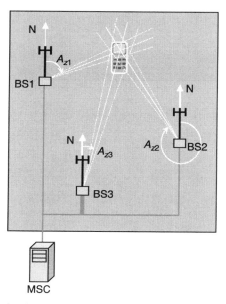

FIGURE 4.11   Angle of arrival principle (MSC, mobile switching center).

The main disadvantage of this technique is that the antennas required are very complex and can only be implemented at the base station end. Furthermore, there is the need for the definition of a reference frame within which all the angles are calculated. Thus, if different base stations have to be used together, this will require a precise orientation common to all three bases. Note also that even if the direction of arrival is calculated in three dimensions, that is, with in fact two angles, the way in which it has been imagined to be used is only in a two-dimensional positioning manner, considering only one angle for the AOA value. No real implementation of such an AOA positioning system is known, certainly because of the many constraints required in order to make the positioning possible and the complexity of the antennas and their deployment.

So, power level measurements are not likely to provide good enough accuracy and an AOA method is too complex and really not acceptable for urban areas. Quite logically, solutions implementing time measurements have been considered. Different possibilities are open to us, such as direct time measurements or difference of times measurements. The main problem of time-based methods in telecommunication networks is that requirements, in terms of time precision, are once again not similar for telecommunication purposes and for positioning purposes. Telecommunication exchanges are based on protocols of transmission that include a synchronization feature, usually through specific heading data sent prior to the real data transmission, in order to define an identical "starting time" for both the transmitter and the receiver. For positioning purposes, one needs, as discussed in previous chapters, very good synchronization, because the resulting localization is directly linked to it. Nevertheless, some methods have been proposed, as shown in Figs 4.12 and 4.13.

The basic idea of "time of arrival" (TOA) is to make direct time measurements between the mobile terminal to be located and various base stations. For similar

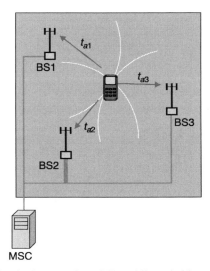

FIGURE 4.12 Time of arrival approach (MSC, mobile switching center).

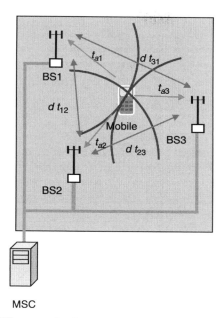

FIGURE 4.13   Time difference of arrival approach (MSC, mobile switching center).

reasons to those applicable to GNSS systems, there is the need for three different measurements in order to calculate a two-dimensional position. As for GNSS, there is the need to know the bias of synchronization for each base station to a reference time (such as the GPS time), as just 10 ns of bias will directly lead to 3 m of error. As the base stations are in a network, it has been found to be easier to implement the time measurements at the base station end. Thus, the mobile sends data and the bases carry out measurements, use the synchronization bias, and finally calculate the mobile location. This location can then be sent back to the terminal, upon request.

Owing to the poor time accuracy in telecommunication networks, the resulting accuracy is around 100 m, in the best cases. Indeed, as the base stations are usually scattered all over the place, direct radio visibility is far from being usual. When multipath occurs, and multipath occurs very often in telecommunication networks, the accuracy drops dramatically to a few hundreds of meters.

A way of minimizing synchronization bias is to use differences of arrival times; this gives good results when the biases are of similar values for the two bases, taking into account the difference.

As with GNSS, considering differences of distances rather than the distances themselves leads to carrying out the intersection of hyperbolae rather than the intersection of circles for this two-dimensional problem. From a system of three base stations, one can obtain three equations when considering the time measurements, and only two when considering the differences. Theoretically, the two systems give the same solution, but practically, Time Difference of Arrival (TDOA) allows less accurate time management. There is currently only one such commercially available system that includes a new method based on multiple measurements and refinement of

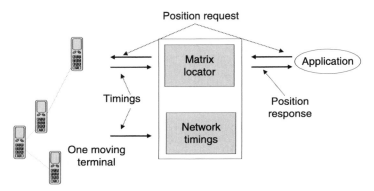

FIGURE 4.14 The Matrix positioning approach.

the positioning when many mobile terminals are in the vicinity of the one we want to locate. This system is called Matrix and is proposed by Cambridge Positioning System (see Fig. 4.14).

This technique could also be used indoors, but due to the same limitations as for TOA and AOA, the results are not very good.[17] An advantage of this technique compared to TOA is the possibility, at no expense, to have it implemented directly at the terminal end; this brings it closer to GNSS approaches.

To be complete on the GSM-related positioning matters, other theoretical techniques should be mentioned. The first is the combination of the Cell-Id with the so-called "timing advance." In fact, within GSM networks, the problem of the potential collision between two transmissions is a major concern. The GSM is based on a time division multiple access (TDMA) scheme, so it means that each transmission is allocated a time slot within a frame of eight time slots. The need for synchronization is then obvious in order to avoid simultaneous transmissions. The way synchronization is achieved through the wireless network is that each terminal transmits with a fixed delay of three time slots (compared to the first time slot received from the base station). Unfortunately, as mobile terminals can be found at quite different distances from the base station, transmissions may still overlap while the propagation delay from the base to the terminals is not taken into account. This collision problem is commonly dealt with by arranging guard times. However, as the maximum radius of a cell has been set to around 35 km, the corresponding guard time needed to avoid transmission overlaps (collisions) would be much too large to be handled with no further refinements, mainly because it would involve a drastic reduction in the network's capacities. Thus, the idea is to advance the transmission of any given mobile terminal (compared to the transmission time corresponding to three time slots after the receipt of the first time slot from the base station transmission) by the amount of its distance to the base station. This approach is called the timing advance and is once again required by the network. In order to provide the terminal with a timing advance value, it is necessary that the base station permanently measures the "round trip time" (more or less the propagation from base station to

---

[17]However, it greatly depends on the application.

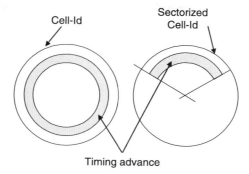

Cell-Id

Sectorized
Cell-Id

Timing advance

FIGURE 4.15 The Cell-Id + timing advance technique: (left) alone and (right) with sectorized antennas.

terminal and return). Then, the terminal transmission is "advanced" from the classical three time slot delay by this amount in order to reduce the required guard time and thus improve the global communication capabilities.

For positioning purposes, this timing advance can of course be used to give a first idea of the distance the terminal is from the base station. It is not really very accurate because its resolution is typically a bit length, hence 3.6 μs. However, this is far better than the cell identification method. The graphical representation of this combined Cell-Id plus timing advance approach is given in Fig. 4.15. The diagram on the left-hand side shows that the resulting position of the terminal is reduced compared with the Cell-Id technique. The graph on the right-hand side is the one obtained when considering, in addition, a sectorized approach where sectorized antennas are used. Using such antennas is quite normal in GSM networks, once again in order to improve the network's capabilities.

Many of the above listed techniques can either be initiated and location calculated at the terminal end or at the base station end.

The UMTS Networks

All the above techniques could have been implemented with GSM. The current mobile telecommunication system is based on UMTS.[18] New names have been used like OTDOA for "observed time difference of arrival," but no real differences exist in comparison with the above-mentioned techniques.[19] The new feature in UMTS, compared to GSM, is that positioning was thought of at the beginning of the standardization.[20] Thus, the possibility to have positioning implemented is taken into account in

---

[18]Universal Mobile Telecommunication System. Note that the "Global" term means total coverage of the Earth for GNSS and only local and not 100% for GSM, as Universal seems to be comparable to the GNSS meaning of Global.

[19]The only difference concerns the way measurements are carried out. Please refer to further readings for details on the methods implemented in the UMTS networks.

[20]"Positioning" here is considered as with GNSS (i.e., in terms of precise location of the terminal), and no longer in terms of "telecommunication-like" positioning (i.e., approximation of the location of the terminal).

the protocols, and also the fact that this positioning could be achieved by means of different techniques, that is, UMTS, but also GNSS or even WLAN. Specific localization data are planned to be included in the protocols.

WPAN, WLAN, and WMAN

The positioning systems based on wireless local area networks are designed for indoor purposes and are dealt with in Chapter 9.

Use of Radio Signals of Various Sources

As already suggested, every radio signal can be used for positioning purposes. If propagation time measurements are used, then the main constraint is time synchronization, and efforts must be carried out to try to overcome this problem. TV signals, for example, have been used in such a way. The system, called LuxTrace,[21] is divided into three parts:

- A mobile terminal, which can be a mobile phone equipped with a TV tuner including a TV measurement module that receives TV signals and calculates pseudo ranges;
- A location server to calculate the position of the mobile terminal;
- A regional monitor unit that measures certain clock characteristics of TV signals and sends time correction data to the location server.

A communication channel is required between the TV measurement module and the location server and between the regional monitor unit and the location server. TV signal range is typically 50–100 km.

Results conducted indoors reported a median position error of less than 50 m, while the 67th and 95th percentile values were 58 and 95 m, respectively. Outdoor results (with line-of-sight to TV transmitters) reported a median position error of less than 5 m, while the 67th and 95th percentile values were 4.9 and 13.6 m, respectively.

■ 4.4 THE SATELLITE RADIO SYSTEMS

The Argos System

The Argos system, like the COSPAS-SARSAT and the DORIS systems, is based on Doppler measurement positioning, as illustrated in Fig. 4.16. One has to remember that the first observations of Sputnik by the members of the Department of Applied Physics of the Johns Hopkins Laboratory concerned the Doppler shift of the transmitted signal.[22] Also remember that the first navigation satellite systems, such as TRANSIT, PARUS, and TSIKADA, were also based on Doppler shift. In the

---

[21]Proposed by Rosum Corporation.
[22]Refer to Chapter 1 for details.

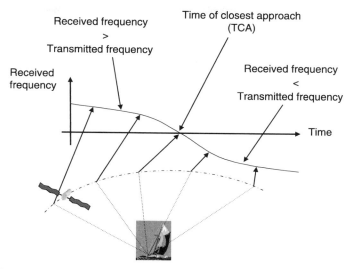

FIGURE 4.16 The Doppler-based positioning technique, I.

present case, the situation is quite different; the transmission is achieved from the Argos beacon and the satellite is the receiving element (see Fig. 4.16 for illustration).

The Doppler shift of the received frequency is zero when the satellite passes the closest point to the transmitter. In such conditions, the beacon is located on a circle of unknown radius, perpendicular to the satellite's orbit. The intersection between this circle and the Earth's surface gives a possible line of locations, as illustrated in Fig. 4.17. Note that this line is perpendicular to the satellite's Earth track at the closest point.

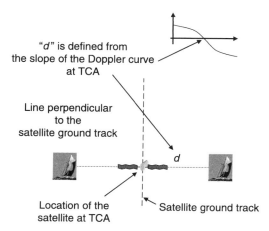

FIGURE 4.17 The Doppler-based positioning technique, II.

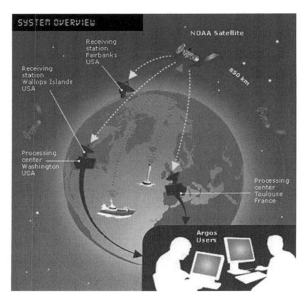

FIGURE 4.18   Overview of the Argos system. (© CLS.)

To be able to provide a more accurate location of the beacon, the slope of the typical Doppler versus time curve is used. Indeed, this slope allows the determination of the angle between the satellite's orbit and the beacon's location. Thus, from both the closest location of the satellite and the slope of the Doppler curve, one is able to provide the user with two points (one on each side of the Earth track of the satellite, perpendicular to the track). The only way to eliminate one point is to wait for another satellite, with another Earth track.

Argos is the result of cooperation between France (CNES[23]) and the United States (NOAA[24]). Different beacons exist that allow various missions to be carried out, from following the migrations of animals to surveillance of polar ice. The smallest beacons can weigh as little as 20 g. The transmitted signal has a frequency of 401.65 MHz and each beacon is allocated a unique identification number. Figure 4.18 shows an overview of the Argos system.

The satellites have a polar orbit with a visibility of about 5000 km in diameter (the altitude of the orbit is between 830 and 870 km). The Earth is scanned a few times each day. Each time a satellite crosses over a global receiving station (there are currently two such stations), it downloads the data collected from Argos beacons. Other regional stations are also included in the distribution process in order to reduce the latency of the system. Finally, there are six processing centers, located in Toulouse (France), Washington DC (United States), Lima (Peru), Tokyo (Japan), Jakarta (Indonesia), and Melbourne (Australia). Their goal

---

[23]Centre National d'Etudes Spatiales (the French Space Agency).
[24]National Oceanic and Atmospheric Administration.

FIGURE 4.19   Overview of the COSPAS-SARSAT system. (© CNES/ill./D. Ducros 2002.)

is to process the raw data from the receiving station in order to make them available to users (with a typical latency of less than 20 min). The accuracy of positioning is typically 300 m.

The COSPAS-SARSAT System

COSPAS-SARSAT is also a Doppler measurement locating system (see Fig. 4.19) aimed at providing assistance to mobile units in distress. Two frequencies are used, leading to different positioning accuracies: 406 MHz (with an accuracy of 2 km) and 121.5 MHz (with an accuracy of 13 km). Each beacon (Fig. 4.20) is typically flown over 24 times a day. More than 10,000 people have been saved since 1982 in the maritime, aerial, and terrestrial domains. The COSPAS (COsmicheskaya Sistema Poiska Avarinykh Sudov, a satellite system for searching for boats in distress) satellites have a quasi-polar orbit at an altitude of 1000 km, and SARSAT (Search and Rescue Satellite-Aided Tracking) satellites orbit at 850 km (also with a quasi-polar orbit).

This program is a joint initiative between the United States (NASA[25]), France (CNES), Canada (DND[26]), and Russia (MORFLOT[27]) and started at the end of the 1970s. The nominal constellation is composed of four satellites, and about 40 ground stations are in operation in 20 associated countries.

---

[25]National Aeronautics and Space Administration.
[26]Department of National Defense.
[27]Ministry of Merchant Marine.

FIGURE 4.20 Examples of COSPAS-SARSAT beacons. (© CNES/P. Jalby 2006.)

The principle of operation is based on the following parts:

- Reception and first processing of signals by the satellites;
- Data transmission to ground stations;
- Identification of people or equipment in distress;
- Processing of data by Mission Control Centers.

Figure 4.20 shows some available COSPAS-SARSAT beacons.

The Galileo program is aimed at providing increased performance to the COSPAS-SARSAT system, first, because of the better positioning accuracy provided, and second, because of the descending radio link from the satellites back to the user, which can also provide a sort of acknowledgment of the distress message.

DORIS

With DORIS (Doppler Orbitography and Radiolocation Integrated by Satellite), the problem of high accuracy is formulated. Satellite orbits are of prime importance when considering the centimeter accuracy of positioning. However, the orbitography (the knowledge of a satellite's orbit) is a difficult problem as it involves complex forces (the Earth's attraction is the most important but not the only one). So, the DORIS project was initiated and included within the Topex-Poseidon satellite. The main objective is to locate terrestrial fixed points. The satellite receives signals transmitted from terrestrial stations and Doppler measurements are analyzed. Taking into account the main forces that act on the satellites (through modeling), the trajectory of the satellite is calculated with a typical accuracy of 10–20 cm within 24 h, and refined processing has improved accuracy to a few centimeters within four weeks. Measurements can also be used the other way to allow reference points to be placed on the Earth and then follow the evolution of their locations. The ground segment is composed of around 60 stations all over the world.

There are also some other parts to the system:

- The master control center, which stores the data and monitors the on-board instrumentation;

- The master beacons, which allow functional data to be uploaded to the satellite;
- The attitude and orbit control processing center, which makes the DORIS orbits available after accuracy verifications (not in real time).

The QZSS Approach

The Quasi-Zenith Satellite System (QZSS) was originally designed as a third generation of satellite system, following the first Doppler-based systems such as TRANSIT and then the current GNSS. It was a third-generation system because of the embedded communication and broadcasting services, as for the planned COMPASS constellation (see later). QZSS originated in 2002 as a joint project between industry and government, with a first launch planned for 2008 and two subsequent ones in 2009. Unfortunately, it seems that the industry part, that is, the communication and broadcast systems, is not going to be implemented.[28] The navigation-related payload is therefore the only one remaining, with an expected first launch in 2009 and subsequent ones (two) several years later.

The QZSS consists of several satellites orbiting on different high inclined planes in such a way that the ground tracks exhibit a repetitive scheme. Eccentricity and inclination of the orbits are chosen in order to provide an elevation angle of more than 70° over the whole trajectory of the satellites when traveling over Japan. At the time of writing the latest parameters published are about 0.1 for eccentricity and an inclination of orbit of about 45°. For users in Japan, this means that they will receive the signal from QZSS with a high elevation angle (almost in the zenithal direction).[29]

This permanent high elevation angle of at least one of the visible QZSS satellites is a very interesting way to provide a solution for the SBAS approach. In fact, the potential problem with SBAS, whether it be WAAS, EGNOS, or MSAS, is that through the use of geostationary satellites, availability is not guaranteed in all urban canyons or in mountainous regions, as a clear sky to the satellite is still required. In northern Europe for instance, a typical inclination of a geostationary satellite is around 45°. For a pedestrian walking along a city street, the 45° clearance is not certain while on the sidewalk. This problem is typical of Japanese cities, and is also present in many northern and southern countries.

Without going into a lot of detail about a constellation that is not yet fully defined, it is planned that QZSS satellites transmit navigation signals with the same characteristics as current and future GPS L1, L2, and L5 signals, including the GPS future L1C signal. These signals will allow QZSS satellites to be used for ranging purposes.

In addition to the abovementioned navigation signals, the so-called L1-Sub meter class Augmentation with Integrity Function (SAIF) and LEX (experimental signal) signals will be available respectively on L1 and E6. SAIF is designed to provide

---

[28]Japanese government announcement of March 2006.
[29]This feature is at the origin of the name of the system: Quasi-Zenith Satellite System.

classical SBAS corrections in order to enhance GPS performances. The LEX signal is intended to allow experience to be gained on faster bit rate characteristics (2 Kbps messages). Of course, as for all SBAS, the codes and messages are of a format compatible with current GPS ones. Note that there are some differences between the WAAS-like MTSAT (multifunctional transport satellite) Satellite-based Augmentation System (MSAS) and QZSS. In terms of ranging, QZSS intends to provide better geometry and more frequencies; this is bound to help in both the DOP-related matters and the capabilities of the dual-frequency mode to remove iono-sphere propagation errors. In addition, with the plan of QZSS to use modernized GPS signals, the civil codes will exhibit higher chip rates than the L1 C/A code used for MSAS. It appears also that differences exist in the way corrections are generated. For instance, ionosphere grids for QZSS will use more reference points than MSAS, mainly because these points will be available in Japan. As MSAS ionosphere corrections are based on a global modeling that requires a less dense grid, it is also expected to be less accurate.

The Japan Aerospace Exploration Agency (JAXA) is in charge of the research and development of the positioning payload, the ground facilities, the integration of the system, and finally of conducting system validation in collaboration with other Japanese institutes.

Arising from reflections on QZSS, the concept of a Regional Satellite Navigation System (RSNS) has been proposed. It is based on the idea that high elevation angle satellites would be useful in many parts of the world. The simulations carried out by JAXA have shown that with four QZSS and four geostationary satellites it would be possible to achieve more than $60°$ satellite elevation angles for a large area over East Asia and Oceania.

GAGAN

In order to complement the augmentation systems all over the world and to keep coherence and compatibility with GPS, the Department of Defense of the United States is cooperating with India in order to develop a Wide Area Augmentation System (WAAS) over Indian airspace. This is the GPS And Geo Augmented Navigation (GAGAN[30]) Satellite-Based Augmentation System (SBAS).

Interoperability with GPS is of course taken into account, as well as with other SBASs, that is, WAAS, EGNOS, and MSAS. Interoperability with GPS is the foundation of the system, but Galileo is also being considered.

The GAGAN infrastructure is based on Indian Reference Stations (INRES), which have a minimum of two identical GPS receivers/antenna subsystems to receive GPS signals (L1, L2, and L5) and GEO signals (L1 and L5) from all the satellites in view. Eight INRES are planned and have been installed at their respective stations. INRES are located at Delhi, Bangalore, Ahmedabad, Kolkata, Jammu, Portblair, Guwahati, and Trivendrum. Five stations are within airport areas. These stations have been chosen to provide service coverage over Indian airspace.

---

[30]GAGAN means "the sky."

In addition to the INRES, one Indian Master Control Station (INMCS) and one Indian Navigation Land Uplink Station (INLUS) are also planned.

A geostationary navigation payload in C band and L1 and L5 frequencies (L band) will be carried on an Indian geostationary satellite. The Indian payload will fly on GSAT-4, which is scheduled for launch in 2008.

Beidou and COMPASS

Beidou is the Chinese name for the Great Bear constellation.[31] It is also the name of the Chinese satellite-based navigation system. Similar to the Galileo program, it is separated into two phases. Beidou-1 is intended to provide China with a SBAS with three geostationary satellites that were launched in 2000 (October and December for the first two) and 2003 (May for the third).

Beidou is designed to provide two-dimensional positioning and communication. In order to achieve this, there is a need for a bidirectional radio link between the users and the satellites. The service provided by Beidou is called the Radio Determination Satellite Service (RDSS). Unlike classical constellations, the ranging is carried out through the bidirectional link by measuring the time taken for the wave to reach the receiver and then return to the satellite. A payload operating center (POC), through one satellite, sends out a navigation signal and the users respond to this signal through two satellites. The transit time allows two-dimensional positioning (one has to know its altitude for instance). The advantage of such a two-way approach is that it does not require an absolutely ultra-precise time reference at the satellite's end, but just a very stable clock in the short term (for the time of the return path).[32] The disadvantage is that a fine calibration is required, because additional electronic delays are included in the time flight measured. Nevertheless, accuracies in the range of 20–100 m have been announced in the vicinity of fixed ground-based calibration sites. In addition, as the mobile receiver is also transmitting, it is thus detectable, which is clearly a problem for some military applications.

These features allow Beidou to propose the availability of some services. Five such services are planned:

- RDSS positioning of mobile users for navigation and fleet management. The accuracy is around 20–100 m "near" the calibration points. As positioning is achieved through a communication-like system, it can also be used for fleet management purposes.

- Broadcast of accuracy corrections and integrity information for use with GPS and GLONASS. Through the S-band downlink, users can access these corrections while receiving GPS and GLONASS on the classical L1 and L2 channels.

- Two-way text messaging; this is a typical text messaging telecommunication service.

---

[31]The intellectual battle to find astronomical and historical references is a popular game in the GNSS community...
[32]Like radar.

- Dissemination of Chinese atomic time for synchronization. Passive (reception only) and active (transmission and reception) modes are available, with respective accuracies of 100 ns and 20 ns. The GPS and GLONASS constellations are used in order to provide such synchronization.

- Communication of GPS-derived mobile users' positions for civil fleet management. Like a telecommunication terrestrial network used to relay the locations of fleet elements, Beidou can be used for large-area coverage purposes over China.

The downlink from satellites operates in the S-band, from 2.4835 to 2.5 GHz, and the uplink, from users' equipment, operates in the L-band, currently from 1.610 to 1.6265 GHz. In the meantime, the SBAS part uses both L1 and L2 frequencies.

Beidou-2, usually called COMPASS, is the second phase: a fourth worldwide GNSS constellation composed of a total of 35 satellites, 5 of which are geostationary, the remaining 30 being medium Earth-orbiting satellites at an altitude comparable to GPS and Galileo, at about 20,000 km. The two services that will be offered are similar to those offered by GPS and GLONASS: an open service and a secured service (the equivalent of M-Code or PRS). The first available values for the open service give an accuracy to within 10 m and 50 ns for both positioning and timing. An additional 30 satellites operating within the L1 and L2 bands will modify Table 6.8 of Chapter 6 as below in Table 4.1.

The following points are based on the planned schedule of available COMPASS satellites, which is pure speculation, but if this fourth constellation is effectively launched, the number of satellites available for navigation purposes will be incredibly high. In fact, this is almost too high for coverage, availability, or any such purposes. Let us imagine the situation where 40 satellites are in the sky anywhere, representing more than 30 different signals (not taking into account the geostationary ones). One should perhaps be afraid of such an increase ... but it also shows the commercial, political, economic, and strategic issues that all great nations are seeing in the GNSS domain.

TABLE 4.1 Evolution of the number of navigation satellites taking into account the provisional deployment of COMPASS.

Year	GPS	GLONASS	Galileo	COMPASS	Total
1980	5				5
1990	12	14			26
1994	27	16			43
1996	24	22			46
1998	27	13			40
2000	28	11			39
2002	28	8			36
2004	29	11			40
2006	30	14	1 (GIOVE)		45
2008	*30*	*18*	*4*	*4*	*56*
2011	*30*	*24*	*21*	*16*	*91*
2014	*30*	*24*	*30*	*30*	*114*

China is also contributing to Galileo by developing applications and regional monitoring capabilities at a global cost of 200 million Euros. International discussions are now required to finalize the frequencies and the modulation scheme in order that all three constellations, GPS, Galileo, and COMPASS (GLONASS is somewhere out of band in terms of frequency and modulation schemes), work in harmony.

■ 4.5 NON-RADIO-BASED SYSTEMS

Inertial systems include all the techniques that take advantage of the inertial properties of any movement. For instance, if a straight line displacement at constant speed is curved, a force F appears and thus a corresponding acceleration γ; both are linked by the simple formula $F = m\gamma$, where m is the mass of an object. The same appears if the movement is simply accelerated. As acceleration is also the first derivative of velocity, it can also be used in order to define the evolution of velocity over time. A second integration could then be carried out in order to obtain the displacement. The gyroscopic effect is also a result of acceleration, but exhibits different sensitivity errors from accelerometers, thus allowing a combined use in order to reduce measurement errors. By extension, other physical measurement systems have been included in inertial systems, such as barometers and magnetometers. The former could help in defining the floor level indoors and the latter the absolute orientation of the mobile terminal.

One has to note that inertial systems are those that first enabled navigation systems to be installed in automobiles. A few months before the official availability of GPS, the first navigation system was available in a car. To achieve such a goal without GPS,[33] there was the need for autonomous sensors that were able to "follow" the car's movement: inertial sensors. Nowadays, GNSS receivers are widely used in order to make the best possible system, thanks to the very good accuracy provided, but for a while, GPS was mainly used to calibrate the inertial system dynamically. The main sensors used in car navigation are accelerometers, gyroscopes, and odometers. Others can be used indoors or for pedestrian purposes, such as barometers or magnetometers.

It is interesting to note that inertial systems are the modern implementation of dead reckoning.[34] In ancient times, the *lochs*[35] were used in order to allow more continuous navigation than was possible with astronomical positioning. Nowadays, when GNSS signals are not available, that is, in a few environments and especially indoors, there is still a demand for continuous positioning. The basic requirements are to determine the velocity vector, that is, both its amplitude and its direction, in real time. As physical measurements are achieved at a given rate, this introduces errors; current systems can typically achieve 100 Hz. The integration (and thus the sum) of errors lead finally to increasing positioning inaccuracy with time.

---

[33]GPS was declared operational in 1995.

[34]"Dead" means "where positioning is no longer possible with the current technique in use."

[35]See Chapter 2 for details.

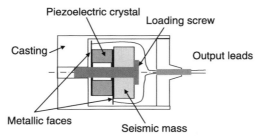

FIGURE 4.21   A piezoelectric accelerometer (modern implementations use nanotechnology).

Inertial personal devices for pedestrians are complex to implement in comparison with those for automobiles, mainly because of the much larger range of physical measurements required and the non-constant attitude of the mobile terminal.[36] For instance, let us imagine a mobile phone equipped with a GNSS receiver and also with an inertial system in order to allow dead reckoning when GNSS signals are no longer available (indoors for instance). As a mobile terminal, the phone is subject to many small but violent movements such as rotation, hand shaking, or even falls. All these movements must be analyzed by the inertial system and should not lead to the accumulation of errors. Unfortunately, as the basic principle is integration, all the errors are added up, and because these kinds of motions are frequent, the resulting error can be significant. The other types of movements that are bound to be difficult for pedestrian mobile terminals are those that are very slow and hesitant. For example, when one is moving from one leg to the other very gently, the resulting signals are almost identical whether there is a real displacement or not. In this case, the errors are not the major concern, but the interpretation of the motion is. Thus, although some remarkable implementations have been achieved, the use of inertial systems for pedestrian mobile terminals has not yet been widely deployed.

Accelerometers

Acceleration is the rate of change of velocity. Accelerometers are used for both measuring vibration and shock (which can be considered as sudden acceleration) and acceleration of bodies. In this latter case, it is possible to use the sensor in order to provide velocity determination (by integrating the signal), positioning (by a double integration), or distance traveled (by successive integrations). Accelerometers are also used to define the attitude of a body, with reference to the horizontal plane, taking the gravity of the Earth (which can be considered as acceleration) as the reference.

The most popular accelerometers use the piezoelectric effect.[37] This consists of a mass attached to the piezoelectric element, as shown in Fig. 4.21. When acceleration occurs, the mass applies a force on the piezoelectric crystal, leading to the appearance

---

[36]Within the car, the horizontal plane should roughly remain unchanged.

[37]The piezoelectric effect was discovered by Pierre and Jacques Curie in 1880. Some crystals, called piezoelectric crystals, when subject to a mechanical constraint, exhibit electric charges of opposite polarity appearing on their sides.

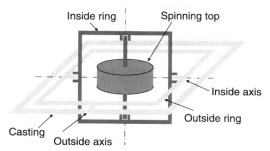

FIGURE 4.22   Principle of a gyroscope.

of a charge across the crystal, thus polarizing its metal faces and producing a voltage. As the electric output signal is a function of force, it is then possible to obtain a measurement of the acceleration.

The main error in accelerometer-based positioning is due to the integration process (the measurement errors add up).

Gyroscopes

The output of a gyroscope is the rate of change of the angle of its axis. This is used in order to measure the variation of the direction of the mobile terminal. A three-dimensional accelerometer could also be used to achieve such a goal, but gyroscopes exhibit slightly different error bias. The use of accelerometers for velocity determination and gyroscopes for angle determination allow both sensors' biases to be compensated.

Although many different technologies are available, let us describe briefly the way they work. A gyroscope is composed of a rotor, rotating at a high speed around its axis, which could have one or two degrees of freedom. The fundamental principle of the operation of a gyroscope is that a moment[38] applied perpendicularly to the rotation axis leads to a displacement that is perpendicular to both the rotation axis and the orientation of the moment. Here again, to be able to define precisely the relative orientation changes in the movement of the mobile, a three-dimensional gyroscope is required. Figure 4.22 shows the principle of a one-axis gyroscope.

Odometers

Although it is possible to use accelerometers in order to obtain the relative distance traveled, this means carrying out a double integration on the measurements. This is certainly not a very efficient process, as errors and biases will accumulate. When possible, in an automobile system for instance, it is preferable to use direct displacement measurement through the use of odometers. In a car, this is simply a sensor that can count the rotation of the wheels and convert the value into a linear distance.

---

[38]A moment is the product of a force applied to the gyroscope and the distance from the force to the center of the gyroscope.

When required, it is also possible to implement a differential approach on the wheels of a given axle in order to define the direction of the displacement.

For indoor purposes, the same idea applies for rolling objects. For pedestrians, it is different, because it is not convenient to use wheels in any way. Once again, accelerometers can be used to count the number of footsteps, and then convert them into a distance.

Magnetometers

As accelerometers, gyroscopes, and odometers are primarily relative sensors, there could be a requirement for absolute ones in order to allow absolute positioning, such as in GNSS fixes. We have seen that with accelerometers designed in order to obtain an inclinometer, it is possible to define the horizontality of the mobile. This is a first approach for an absolute sensor as this is achieved without the need to know any former attitude. Another important parameter is the absolute orientation of the mobile.[39] In applications where the discovery of the environment is required, this feature is a must. For instance, in a museum, the electronic guide should certainly take advantage of the fact that it knows what the visitor is looking at. This is also important when one wants to be oriented when taking a first step. With current GNSS receivers, one needs to start moving before this information is relevant.

Magnetometers are sensitive to the Earth's magnetic field and thus are available all around the world, without any calibration required. The main direction is the magnetic north, which is slightly different from the geographical north (this must be taken into account, at least by staying in the same referential, either magnetic or geographical). The difference is the declination, experimentally discovered by Christopher Columbus during his travels to "India."[40]

Barometers and Altimeters

One of the most important differences between outdoor navigation and indoor navigation is certainly the third dimension. In fact, outdoor navigation is primarily a planar system. Knowledge of the altitude is interesting for general information, but is not really required for most applications. Indoors, the problem is totally different, as knowing the floor level is of primary concern (for emergency aspects for instance, or in order to cope with floor maps that are not identical from one floor to another). In that sense, barometers can be very helpful in allowing quite an accurate determination of the altitude. This can be achieved through the use of so-called micro-altimeters, whose accuracy is typically 1 m on a local and time-limited extension scale. Once this time has elapsed, meteorological fluctuations are bound to occur and cause large bias to the measurement. The idea is then to reset the altimeter when entering the building at a known location (and known absolute altitude relative to any referential), and to use the micro-altimeter to determine the floor level. An accuracy of 1 m is enough to achieve such a goal. With an increased accuracy of a fraction of a

---

[39]Note that GNSS signals do not provide this information, unless in dynamic mode.
[40]See Chapters 1 and 2 for details.

meter, one could also imagine determining whether the mobile lies on the floor or is really handheld, for evaluating if the user is standing up or lying down.

A variation in air pressure can be given by the relation $\Delta P = \rho g Z$, where ρ is air density, g the acceleration due to gravity, and Z the altitude. Calibration at the bottom floor of the building should allow the calculation of the air density ($\rho = P/rT$) under the specific current conditions. Some GPS receivers are currently equipped with micro-barometers and allow floor-level determination without ambiguity.

✅ 4.6 EXERCISES

Exercise 4.1 Describe the main differences between optical and radio positioning systems in terms of

- Accuracy;
- Coverage;
- Sensitivity to environments;
- Main physical principles.

Note that the idea of this exercise it to organize your own representation of different positioning approaches. Feel free not to follow the book's organization.

Exercise 4.2 By analyzing the laser method (as described in the section "Lasers" in this Chapter), the obvious difficulty appears to be the presence of an obstacle in the line of sight or an open door. Can you imagine a way to overcome this drawback? What would be the number of distances required for a two-dimensional positioning? Answer the same question for a three-dimensional positioning?

Exercise 4.3 Let us consider the free space propagation model for the transmission of a radio electric wave as given by the following formula:

$$P_t G_t = \left(\frac{4\pi d}{\lambda}\right)^2 P_r G_r \eta,$$

where P_r and P_t are respectively the received and the transmitted power levels, G_r and G_t the received and transmitted antenna gains, d the distance separating the transmitter and the receiver, λ the wavelength of the signal, and η an efficiency coefficient.

Taking into account an approach similar to that described in the section "Amateur radio transmissions" in this chapter, find the corresponding accuracy of positioning. How many transmitters are required in order to provide an acceptable location? Propose further developments of this technique in order to improve accuracy. Feel free to be innovative.

Exercise 4.4 The measurement of distances with radar can be very accurate. Explain the main reason for this fact (think in terms of time synchronization). If one would like to build up a positioning system with such a technique, what could be the global architecture considered? Would it be interesting (give details) to take into account the Doppler measurements?

Exercise 4.5 Give advantages and drawbacks of the following mobile network techniques for positioning:

- Angle of arrival;
- Time of arrival;

- Time difference of arrival;
- Cell identification;
- Timing advance.

By looking at further reading, specify the actual implementations and give the main reasons why some techniques are not currently used.

Exercise 4.6 What are the two different ways to solve the ambiguity of the two possible location points obtained by the Doppler-based positioning technique (using satellite measurements)? Considering a satellite at 1000 km altitude with a polar orbit, draw the Doppler curve obtained from the signal of a beacon transmitting at 400 MHz, assuming the satellite passes directly over the beacon. Find out from the equations the major sources of error. What is the form of the Doppler curve if the satellite is viewed from the beacon with an angle of 45°? Give numerical values too.

Exercise 4.7 Explain how a position can be obtained from accelerometers and/or gyroscopes. Note that different techniques are in use; choose the one you want and give full details of how it functions.

Exercise 4.8 How is it possible to implement an odometer measurement technique for

- A car;
- A pedestrian;
- An object (a parcel for example).

Exercise 4.9 Describe how to use a micro-barometer of 1 m of accuracy (in altitude) when wanting to determine the floor in a multistorey building. By looking at further reading, give the actual physical and usage limitations of this approach.

■ BIBLIOGRAPHY

Caffery JJ. Wireless location in CDMA cellular radio systems. Kluwer Academic Publishers, 2000. Boston.

Caffery JJ, Stüber GL. Overview of radiolocation in CDMA cellular systems. IEE Communications Magazine; 1998.

Carroll J. GPS + LORAN-C: performance analysis of an integrated tracking system. GPS World 2006;17(7):40−47.

Duffett-Smith P, Rowe R. Comparative A-GPS and 3G-MATRIX testing in a dense urban environment. In: ION GNSS 2006: Proceedings; Forth Worth (TX).

Duffett-Smith PJ, Tarlow B. E-GPS: indoor mobile phone positioning on GSM and W-CDMA. In: ION GNSS 2005: Proceedings; Long Beach (CA).

eLoran definition document. Version 1.0. International Loran Association; 12 January 2007.

El-Sheimy N, Niu X. The promise of MEMS to the navigation community. InsideGNSS 2007;2(2):46−56.

Kaplan ED, Hegarty C. Understanding GPS: principles and applications. 2nd ed. Artech House; 2006.

Kogure S, Sawabe M, Kishimoto M. Status of QZSS navigation system in Japan. In: ION GNSS 19th International Technical Meeting of the Satellite Division: Proceedings; Fort Worth (TX); 2006.

Küpper A. Location based services — fundamentals and operation. England: John Wiley and Sons; 2005.

Ladetto Q, Merminod B. Digital magnetic compass and gyroscope integration for pedestrian navigation. In: GPS/GNSS 2003: Proceedings; Portland, USA.

Martone M, Metzler J. Prime time positioning: using broadcast TV signals to fill GPS acquisition gaps. GPS World 2005;16(9):52–59.

Parkinson BW, Spilker Jr. JJ. Global positioning system: theory and applications. American Institute of Aeronautics and Astronautics; 1996.

Patwari N, Ash JN, Kyperountas S, Hero III AO, Moses RL, Correal NS. Locating the nodes — cooperative localization in wireless sensor networks. IEEE Signal Processing Magazine; July 2005.

Shankar PM. Introduction to wireless systems. John Wiley & Sons; 2002. UK.

SnapTrack. Location technologies for GSM, GPRS and UMTS networks. White Paper; available at: http://www.snaptrack.com.

Toran F, et al. Position via internet: SISNET catches GPS in urban canyons. GPS World 2004;15(4):28–35.

Weinburg H. MEMS sensors are driving the automotive industry. Sensors 2002;19(2):36–41.

Yazdi N, et al. Micromachined inertial sensors. Proc. IEEE 1998;86(8):1640–1659.

Web Links

http://www.cls.fr/welcome_en.html. Nov 14, 2007.

http://www.cospas-sarsat.org. Nov 14, 2007.

CHAPTER 5

GNSS System Descriptions

In Chapter 3, a global description of the main satellite-based positioning systems was given. However, a more complete description is needed for the three main constellations (GPS, GLONASS, and Galileo), as the development of geographical positioning is mainly due to GNSS. Chapters 5, 6, and 7 are intended to provide a more detailed view of the architectures of the systems (Chapter 5), the various signals available (Chapter 6), and the way these signals are actually acquired and processed (Chapter 7). In the present chapter, GNSS systems are described and, in the second part, the basics of positioning are set out with the goal of providing clues for the understanding of the main technical choices (satellites, orbits, signals, and so on).

GNSS systems are quite complex, involving many different components. Of course, the satellites are the most visible part, but they require a heavy ground infrastructure in order to deliver the right signals with the right parameters to the users. The user community has, until now, only had a link to the satellites, and only a downlink from the satellites. This feature means that it is impossible to saturate the system. How these various segments are linked to each other and what are the main principles of GNSS positioning are explained in the first part of this chapter. This leads to a description of Satellite-Based Augmentation System (SBAS) infrastructures as they are designed to improve GNSS performances.

■ 5.1 SYSTEM DESCRIPTION

The three constellations are based on the same global architecture: a user segment that is made up of all possible receivers, a space segment regrouping the satellites, and a ground segment that is composed of monitoring, controlling, and uploading stations. The number and the geographical distribution of these

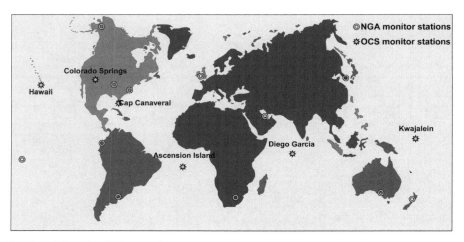

FIGURE 5.1 The GPS ground segment.

stations are important concerns with regard to integrity,[1] which is one of the challenges to be faced.

The Ground Segments

The main functions of the ground segments are to

- Monitor the satellites;
- Estimate the on-board clock state and define the corresponding parameters to be broadcast (with reference to the constellation's master time);
- Define the orbits of each satellite in order to predict the ephemeris data, together with the almanac;
- Determine the attitude and location orders to be sent to the satellites to correct their orbits.

GPS Ground Segment

For the GPS system, the ground segment is composed of a master control station (MCS) located in Colorado Springs and three uploading stations located in Ascension, Diego Garcia, and Kwajalein. In addition, 11 surveillance stations are used to carry out the measurements required for the definition of the data to be uploaded[2]: 5 GPS stations are located in Hawaii, Colorado Springs, Ascension, Diego Garcia, and Kwajalein, and also 6 National Geospatial Agency (NGA) stations since September 2005. Figure 5.1 gives details of the locations of the GPS ground segment.

---

[1]Integrity is the characteristic that allows a user to evaluate the confidence he can attribute to the positioning the receiver is providing (as a crude definition).

[2]Note that these data, absolutely essential to the receiver for location calculation, are described in Section "Reasons for a Navigation Message," Chapter 6.

At the Schriever Air Force base, Colorado, the back-up "master clock" of the United States Naval Observatory is located, which has a stability of less than 1 s in 20 million years.

GLONASS Ground Segment

For the GLONASS system, the ground segment is composed of a system control center (SCC) located in Krasnoznamensk in the Moscow region, which is in charge of satellite control, orbit determination, and time synchronization. In addition there are five telemetry, tracking, and control (TT&C) stations located in the St Petersburg region, Schelkovo in the Moscow region, Yenisseysk, and Komsomolsk-Amur. Synchronization monitoring is centralized at Schelkovo. Figure 5.2 gives details of the geographical distribution of the GLONASS ground segment.

For security and deployment reasons, all the ground segment of GLONASS are located on Russian territory. This helps in monitoring the system but reduces the range of uploading and surveillance stations. A modernization of the ground segment of GLONASS is under way.

Galileo Ground Segment

For the Galileo system, the ground segment will be composed of two control stations located in Europe, five monitoring and uploading stations, and nine stations specific to the uploading of navigation messages. This latter feature allows an increased uploading rate compared to GPS, which will help in improving accuracy by providing more accurate ephemeris data. In addition, between 30 and 40 surveillance stations, distributed throughout the world, are used to follow the satellite signals in order to provide the definition of the data to be uploaded. Figure 5.3 gives details of the geographical distribution of the Galileo ground segment stations.

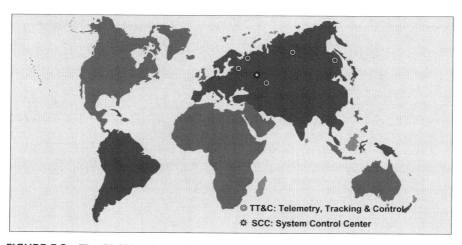

FIGURE 5.2 The GLONASS ground segment.

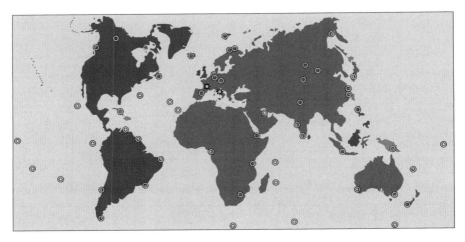

FIGURE 5.3 The Galileo ground segment.

In addition, the ground segment of Galileo is planned to include both regional and local components. The idea is that satellite coverage is limited and that there will be a need, in specific cases, for additional components such as terrestrial transmitters, for instance.

The Space Segments

GPS Space Segment

The nominal constellation is composed of 24 satellites, although there are currently always between 28 and 30 actually in operation. There are six quasi-circular orbits with an inclination of 55° with reference to the equatorial plane, at an altitude of 20,183 km (see Fig. 3.8). The distance traveled by the signal from a satellite to a receiver then varies from around 20,200 km if the satellite is in the zenith, to around 25,600 km when the satellite is at the horizon.[3]

A satellite makes a total Earth circumference in exactly half a sidereal day, or 11 h 57 min 58 s. The tracks of a given satellite on the Earth's surface are therefore almost identical from one day to the next. The difference between the sidereal day's duration and 24 h is due to the fact the Earth is moving around the Sun and then makes one additional rotation per year. Taking this into account leads to a revolution period of the satellites that has repetitive tracks on the Earth's surface.

On each orbital plane, four satellites (nominal) are positioned so as to allow visibility of at least four satellites at any location on Earth and at any time. When reducing the horizon to 15° above the horizon, a minimum of four satellites are still

---

[3]These two values correspond respectively to 67 ms and 82 ms of travel time at the speed of light. This will have a fundamental effect on the satellite orbital modeling in the system. One must take into account this time when evaluating the location of the satellites at transmission time.

visible, allowing three-dimensional positioning. In Paris, between 8 and 12 satellites are usually available when there is a clear sky.

Different generations of satellites — Block I, II, IIA, IIR, and IIF — co-exist. Eleven satellites of Block I were launched between 1978 and 1985. The last one was in use until 1995. The constellation was declared operational once 24 Block II satellites were launched at the end of 1993 (initial operational capability — IOC). Full operational capability (FOC) was declared in March 1995.

The main functions of a satellite are as follows.

- It receives and stores data from the control segment.
- It maintains a very precise time. In order to achieve such a goal, each satellite usually carries four atomic clocks of two different technologies (cesium and rubidium), depending on the generation of the satellite.
- It transmits data to users through the use of two frequencies, L1 and L2, and in the future also L5.
- It controls both its attitude and position.
- It enables a wireless link between satellites in the case of GPS III.

The Block II and IIA satellites transmit the C/A code on L1 and the P(Y) code on both L1 and L2. The first launch occurred on August 14, 1989. A total of 28 Block II/ IIA satellites were constructed and 15 are still orbiting at an altitude of 20,350 km. The initially specified life-time was 7.5 years, well below the real life-time of more than 10 years.

The Block IIR[4] satellites are more powerful and should have a life-time of 10 years. The first such satellite was launched on July 22, 1997. Twenty-one have been bought and 12 are currently in orbit. The same signals are transmitted as for Block II/IIA.

The Block IIR-M[5] satellites (see Fig. 5.4) add three new signals to the existing ones: the L2C civil signal and the M code on both L1 and L2. The M code is the new military code. The first launch occurred on September 25, 2005. A total of eight such satellites have been built and are waiting to be launched. Their estimated life-time is also 10 years.

The Block IIF satellites will add another new civil signal, on L5, designed mainly as a back-up necessary for civil aviation purposes. The first launch is planned for 2009 and a total of between 12 and 19 satellites should be built (nine are already available). Their estimated life-time is 12 years.

GLONASS Space Segment

The nominal constellation is composed of 24 satellites distributed on three orbital planes whose inclination is 64.8° with reference to the equatorial plane. The altitude of the quasi-circular orbits is around 19,100 km, leading to a revolution period of 11 h 15 min 44 s. The orbital planes are 120° apart in longitude. Eight

---

[4]R stands for replenishment.
[5]M stands for modernized.

FIGURE 5.4 A Block IIR-M satellite. (*Source*: Lockheed Martin.)

satellites are regularly spaced in each orbital plane (45° spacing). The orbital planes are also 15° apart in latitude, allowing a complete coverage of the Earth and surrounding space. In 1995, the constellation in orbit was completed and the system was declared operational. In March 1995, it also became available for public use for the long term. Unfortunately, following financial and political difficulties and changes, between 1996 and 1998 the GLONASS constellation was not maintained and the number of operational satellites decreased dramatically to seven in 2002 (see Fig. 5.5 for details).

In addition to the aforementioned financial difficulties, the life-time of the satellites (3 years, compared to 10 years for the GPS satellites) was the major cause of this situation. To keep the system operational, numerous launches had to be made, which led to more financial difficulties. In 2001, directives were approved and undertaken by the government and the President towards the development and the maintenance of GLONASS.[6] The major goals of this program were as follows (source, GSIC/ Federal Space Agency):

- Modernization of the GLONASS program in GLONASS-M;
- Current development of GLONASS-K for improved performances;
- Development of the ground segment;
- Research and development towards international cooperation for GLONASS use.

---

[6]Under the Federal Dedicated Program "Global Navigation System" approved in August 2001 for a 10-year period.

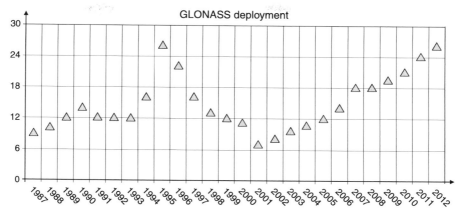

△ Satellite in constellation

FIGURE 5.5 Evolution of the number of GLONASS operational satellites: past, present, and future. (*Source*: GSIC/Federal Space Agency.)

This was achieved progressively by maintaining the constellation to a minimal level and successively adding new satellites, making improvements in the life-time and performances of the GLONASS-M satellites, and the development of new smaller GLONASS-K satellites to deploy a full 24-satellite constellation of both GLONASS-M and GLONASS-K for domestic and international availability. The aim is to have 18 satellites in 2008 and 24 in 2011.[7] The availability of GLONASS-based positioning can be estimated through the availability figures given on the web site of the Federal Space Agency (www://http.GLONASS-ianc.rsa.ru), which shows a typical number of four available satellites in many parts of the world. For example, with a total of 14 GLONASS satellites, at least four are visible from Paris, France, about 80% of the time.

The main feature of the GLONASS-M satellite is the transmission of a civil signal on L2. Note that the first "M" satellites were launched in 2003. The GLONASS-K satellites are planned to be launched in 2008, which will add a third civil signal in L3, designed for safety-of-life applications, where back-up is required.

Galileo Space Segment

The final constellation will be composed of 30 satellites distributed on three orbital planes (medium earth orbits, MEO) at 23,222 km altitude. The inclination of the orbital planes with reference to the equatorial plane will be 56°, allowing a better coverage than GPS for the northern countries of Europe, for instance. Each orbital plane will have 10 satellites (nine operational plus one spare that will be on standby until another breaks down), equally spaced (40°). Such a configuration leads to a revolution time of 14 h 4 min, which gives a 17/10 value in comparison with the total sidereal duration of a day.[8] This feature means that every ten days there will be a repetition rate of ground tracks of Galileo satellites.

---

[7]The current state (March 2007) is 19 available satellites.

FIGURE 5.6 GIOVE-A satellite. (ESA-JL ATELLEYN.)

The development phase is planned to first include the launches of two satellites, GIOVE-A and GIOVE-B,[9] for preliminary validations and tests. GIOVE-A (see Fig. 5.6), built by Surrey Satellite Technology Ltd, was successfully launched on December 28, 2005. Its main goals were to secure the frequency band allocations, to validate in orbit some technological issues such as clocks and signal generators, and to characterize the MEO environment. The frequency plan had to be secured before June 2006 in order to follow the ITU requirements: it was successfully achieved in March 2006. Thus, the initial mission of GIOVE-B, which was built in case the GIOVE-A launch failed, has been modified in order to provide further achievements. The expected life-time of these two satellites is about two years.

The next steps are

- The launch of GIOVE-B, planned at the end of 2007;
- The launch (planned in 2008) of the first four Galileo satellites in order to validate the final signals and to allow a complete operational demonstration — these four satellites will be integrated into the final constellation;
- Following this first validation, the complete constellation (26 remaining satellites) will be progressively deployed from 2009–2012;
- The initial operational capability phase should be reached in 2012.[10]

---

[8]Which is twice the revolution time of a GPS satellite (11 h 57 min 58 s).
[9]GIOVE is the Italian name of the planet Jupiter, whose moons Galileo observed, leading to the first precise longitude calculation. GIOVE also stands for Galileo In-Orbit Validation Element.
[10]Refer to Chapter 3 for current status.

The User (Terminal) Segments

The user segment is composed of a great variety of terminals, from military to mass market ones, and also scientific ones. The major tasks of a receiver are to

- Select the satellites in view;
- Acquire the corresponding signals and evaluate their health;
- Carry out the propagation time measurements;
- Carry out the Doppler shift measurements;
- Calculate the location of the terminal and estimate the user range error;
- Calculate the speed of the terminal; and
- Provide accurate time.

Therefore, users will have at their disposal a single terminal allowing localization, time reference, altitude determination, speed indicator, and so on.

The reader can refer to Chapter 11 for a quick overview of the various receivers currently available. This section is intended to show some current comparable receivers of the three constellations. Figure 5.7 shows a combined GPS/GLONASS receiver and a GIOVE-A receiver.

The NovAtel 15a receiver tracks and decodes GPS L1 and L5, SBAS L1 and L5, and Galileo L1 and E5a. The Aschtech GG24 sensor is a dual constellation (GPS and GLONASS) 24-channel receiver. The Septentrio GeNeRx1 is a combined GPS/Galileo receiver (48 GPS channels and 6 generic Galileo channels) that can track GPS L1, L2, and L5 and Galileo L1, E5a, E5b, and E6 signals.

Note that Javad has always produced multi-frequency multi-constellation products and Javad is already providing research institutes with a chipset called GeNiuSS that integrates all the receiving capabilities of the three constellations into a single chip (see Fig. 5.8). The integration of GPS and Galileo, especially for the L1 common band, is bound to be a standard, but is full integration with GLONASS also the future? Wait and see!

FIGURE 5.7 An Ashtech GPS/GLONASS GG24 sensor (left; *Source*: with used by permission, 2007 Magellan Navigation, Inc.) and a Septentrio GIOVE-A receiver (right; *Source*: Septentrio).

FIGURE 5.8 The Javad TTGYG chip. (*Source*: Javad Navigation Systems.)

The Services Offered

At the early stages of the development of the GPS program, a need arose for two different types of signals:

- The first, to meet civilian needs;
- The second, to provide robustness and potentially higher accuracy (for military users).

We have already seen that this has ultimately been achieved through the use of two frequencies and different codes. This gave rise to two well identified so-called services: the Standard Positioning Service (SPS) and the Precise Positioning Service (PPS).

All GPS users have access to the SPS. This service is subject to voluntary degradations depending on geopolitical or strategic issues decided by the United States. For example, the so-called Selective Availability (SA) was switched on from the beginning of GPS operation until May 1, 2000, in order to decrease accuracy for civilian receivers.[11] With the withdrawal of the SA, the horizontal accuracy improved from typically 100 m to around 15 m. The SPS is based on the Coarse/Acquisition code sequence available only on L1 (1.57542 GHz).

Only authorized users have access to the PPS. Indeed one needs to know the P code, which is such a long sequence (at a higher chip rate than SPS) that any real-time search would fail, at least with current computing capabilities. Thus, one needs to know the sequence and moreover to know the starting point of the sequence

---

[11]Some people believe that this withdrawal of the SA was intended to thwart Galileo and to show that there was no need for a new global constellation. One of the major arguments for developing Galileo was precisely that the GPS system was solely in the hands of the United States...

at each initial GPS week (midnight Saturdays).[12] Furthermore, there is the possibility to have the P code encrypted, then noted P(Y), through the use of a specific unknown W code. This approach is called anti-spoofing (AS).

The GLONASS signals are also organized so that two similar services are available. The radio transmission principles remain, with two frequency bands, L1 and L2, and two types of signals, civil and military. The corresponding GLONASS services are the Standard Precision Service (SPS) and the High Precision Service (HPS).

Galileo has been built around the notion of services. Of course, achievable technical performances and constraints are key drivers, but the architecture of the signals and systems has been based on services. Four such services will be available and, additionally, a support to a search and rescue service will also be provided. They are intended to cover the needs of mass market, professional, and scientific users, but also to answer governmental expectations and safety-of-life requirements in positioning-related fields. These services will be met by using only Galileo signals, although it is not foreseen to have Galileo-only receivers, but multiconstellation receivers. The resulting five services are as follows.

- *The Open Service (OS).* This is composed of corresponding open signals and is free of charge, like the GPS and GLONASS equivalent. The performances should be comparable to those of GPS.

- *The Safety of Life Service (SoL).* This provides integrity in addition to the OS. It means that a user should be warned when the positioning fails to meet certain margins of accuracy. Certain service levels are already available, given by international transportation laws for example. In order to satisfy most local regulations it is also planned to provide a guarantee. A way to achieve this feature would be to have the possibility to activate a coded signature of the signal, allowing a user to be sure of the Galileo incident signal. Galileo's integrity should be available on a world-wide range, but non-European countries could use regional augmentations in order to provide integrity data. The two levels specified so far are intended to deal with the different applications certain to be interested in this service, notably in the transportation field. The first level applies to applications where time factors are highly important, such as aviation in landing phase operations. The second level applies where the timing is less sensitive, such as maritime navigation, but where high continuity and integrity are still required.

- *The Commercial Service (CS).* This is characterized by two additional signals, allows faster data rates and potentially improved accuracy. In addition, the signals are encrypted to allow the service to be subject to a charge. Like the SoL, a service guarantee will be provided. The CS will be operated by the Galileo Operating Concessionaire (GOC), following the general framework of the license agreement concluded with the Galileo Supervisory Authority (GSA).

---

[12] All the GPS codes are well known and quite easily reproducible. Nevertheless, the P code is 37 weeks long and the GPS sequence is limited to one week (it is reset each Saturday midnight to a new value). So, knowing the starting point is crucial; only authorized people have access to this fundamental piece of information.

- *The Public Regulated Service (PRS).* This provides positioning and timing information to users authorized by governments. It requires a high level of performance with both continuity and access control. Two such signals will be delivered. The main potential users are either European structures like the European Police Office (Europol), the European anti fraud Office (OLAF), and civilian security forces. Member states' structures such as national security services, frontier surveillance forces, or criminal repression forces are also involved.
- *The Search And Rescue (SAR) Service.* This is Galileo's contribution to the international COSPAS-SARSAT[13] system. Galileo is intended to intercept emergency beacon signals and relay them to national rescue centers. These emergency signals could be transmitted from a ship, a plane, or even from individuals. The advantage of using Galileo is clearly its high accuracy, allowing positioning, together with timing information. This allows a real-time alert in all conditions.

The details of the technical implementation of all these services are given in the following sections (with regard to frequencies, signals, and so on).

■ 5.2 SUMMARY AND COMPARISON OF THE THREE SYSTEMS

Table 5.1 summarizes the main characteristics of the three systems. Note that the number of GPS satellites is usually maintained between 28 and 30, as compared to the 24 required for nominal operation.

Concerning the frequency bands, note that these are planned configurations, as some bands are not yet fully defined[14] (GPS L5, GLONASS L3, and Galileo E5, E6). Four of the bands of Galileo are L1, E5a, E5b, and E6. The fifth band (L6) is a SAR-specific band and is reserved for this purpose.

■ 5.3 BASICS OF GNSS POSITIONING PARAMETERS

The following discussions are intended to give the reader some explanations of the main parameters of GNSS positioning. The comments are based mainly on the GPS system but apply similarly to GLONASS and Galileo, unless specified. The goal of this section is to "decode" the requirements for positioning and to explain the major past, present, and future developments.

---

[13]COSPAS-SARSAT stands for Cosmicheskaya Systyema Poiska Avariynich Sudov — Search and Rescue Satellite Aided Tracking. See Chapter 4 for details.

[14]One has to wait for the definitive public release of corresponding Interface Control Documents (ICDs).

TABLE 5.1 **Major system parameters for GPS, GLONASS, and Galileo.**

Parameters	GPS	GLONASS	Galileo
Ground Segment			
Master control station	1 in U.S.	1 in Russia	2 in EU
Surveillance stations	11	7	30–40
Space Segment			
Satellites	30 (03/2007)	19 (03/2007)	27 + 3 (2011)
Orbital planes	6	3	3
Inclination	55°	64.8°	56°
Altitude (km)	20,180	19,100	23,222
Ground track repetition	1 day	7 d 23 h 27 min 28 s	10 days
Revolution time	11 h 57 min 58 s	11 h 15 min 44 s	14 h 4 min
Selective availability	Off	No	No
Anti-spoofing	Yes	No	No
User Segment			
Services	2	2	5
Frequency bands	3	3	4 + 1

Position-Related Parameters

Required Number of Satellites

As the requirements for GNSS are quite different, and in particular positioning should be available 24 h a day in all weather conditions, its technical features are largely different from those of TRANSIT or TSIKADA. There is also a desire to be able to deliver in real time the velocity of a mobile receiver and a precise time. The basic principle is to make propagation time measurements of the signal from the satellite to the receiver in order to evaluate the corresponding distance separating them. This means having a very good synchronization between all the satellite transmissions in order to be able to make comparisons. Once achieved, and assuming no measurement error, a single measurement means that the receiver lies somewhere on the surface of a sphere whose radius is precisely the value of the measured time, converted into distance (by simply, at the first stage, multiplying the time by the speed of light).[15] Clearly, a single measurement made from a space signal is far from sufficient for positioning. Thus, others are required. Let us imagine a second satellite is in radio electric visibility and detected by the same receiver, making two simultaneous measurements (T1 from satellite number one, and T2 from satellite number two). Then, as detailed in Fig. 5.9, the possible receiver locations are located at the

---

[15]One can easily "feel" that corrections are going to be required to reduce the inaccuracy of such a multiplication, especially for propagation in atmospheric layers where the electromagnetic wave does not travel at the speed of light.

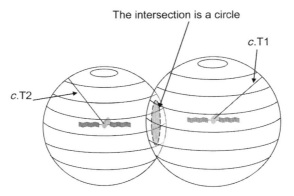

FIGURE 5.9 Intersection of two sphere surfaces.

intersections of two spherical surfaces having respectively radius c.T1 and c.T2, c being the speed of light.

Such an intersection is indeed a circle, whose radius depends on the respective locations of the receiver and the satellites. As the resulting circle has a large radius, there is the need for a third measurement, T3, from a third satellite. The intersection of the three spheres' surfaces is now represented by two points as described in Fig. 5.10. The important point to understand at this stage is that the location one is looking for physically exists and the problem is not a purely mathematical one. The receiver is physically at a real location that gives T1, T2, and T3 time measurements from satellites 1, 2, and 3. If it happens that the mathematical resolution does not converge, it simply means that errors have occurred, either in the physical propagation path (multipath for example) or within the modeling of one or several parameters of the global problem (see later developments for a full understanding of these parameters).

In fact, there is the need for a fourth measurement, T4, to obtain a full three-dimensional location: this is the general case. However, the resulting two points of

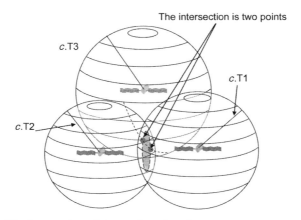

FIGURE 5.10 The intersection of three spheres' surfaces.

the intersection of the three spheres' surfaces present specificity: they are located one above and the other below the plane containing the three satellites, which are the centers of the spheres. Terrestrial positioning is therefore a little bit special in the sense that "visible" satellites are all located above the horizon.[16] This means that the mentioned plane cannot intersect the Earth, so the intersection lying above this plane is an impossible solution (remember this is true only for terrestrial positioning).

Time Synchronization

Thus, from a mathematical point of view, there is generally the need for four measurements, but only three time measurements in the case of terrestrial positioning. Unfortunately, this is not the end of the story, as making time measurements requires very good synchronization between the various players, satellites and receiver, in order to be accurate. The measurement of the flight time of a signal from a satellite to the receiver requires two variables: the time the signal was transmitted (the time at which it leaves the satellite) and the bias between the satellite clock and that of the receiver. The GNSS considered that managing the sending time is the role of the ground controlling segment of the constellations. Thus, this point is in fact part of the definition of the system. However, the so-called clock bias of the receiver is a real difficulty to overcome, as there is the need for very precise local synchronization.[17] Telecommunication systems have this same difficulty in order to recognize the messages sent by a transmitter, but not with the same accuracy, as the order of a fraction of a microsecond is enough. In telecommunications, synchronization is commonly achieved through the use of specific bits sent over the transmission channel, before the transmission of real data. This is possible because the time synchronization thus achieved is of the order of the channel bandwidth, around 100 ns (typically 10 MHz) or 1 μs (typically 1 MHz). For positioning purposes, the same technique would lead to a bandwidth that was clearly not available[18] for 1 ns synchronization,

In a global positioning system, one has the good fortune that the receiver clock bias is identical for all measurements, T1 through T4, for all the satellite signals, as long as these measurements are made at the same time. In such a case, which is typical for modern receivers, which normally have 12–20 parallel channels for simultaneous measurements, the clock bias in fact includes all the common bias to all the signals. Thus, there is a new "unknown" of the positioning problem, which is essentially to determine the three space variables defining any location: x, y, and z in a Cartesian referential. Thus, the solution vector of the GNSS positioning is made up of four coordinates: x, y, z, and t. Before going ahead with how to solve this problem, let us come back to the most important difficulty of positioning: errors. For this chapter, we shall deal only with the effect of errors on the measurements used in the navigation solution (and not the physical effects or the eventual corrections).

---

[16]In this case, the "visible horizon" is identical to the "radio electric horizon."

[17]One has to remember that 1 ns is equivalent to 30 cm, or 1 m to 3.3 ns. The synchronization needs to achieve such a level of accuracy.

[18]The radio-electric bandwidth is a rare resource and needs to be carefully used. International bodies, such as the UIT (Union of International Telecommunications), are in charge of such matters.

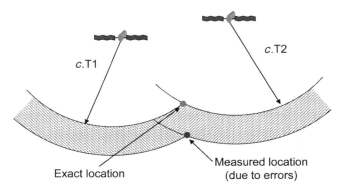

FIGURE 5.11 A real problem considering errors.

As the various measurements are all affected by different errors, the real problem of sphere intersection is slightly different from the one described previously. Figure 5.11 shows the real problem. Both the clock bias of the receiver and the modeling and propagation errors of each measurement are included. The clock bias is identical for all satellite measurements, as other errors are specific to each signal.

Because of the clock bias error added to the measurement, the distance obtained from the time measurement is called a Pseudo Range.

Returning to positioning, there is the need to find a way to extract the bias clock variable from the new set of equations. Indeed, as the clock bias is associated with the receiver, a new measurement from a new satellite will not introduce a new unknown from a spatial point of view, thus leading to a complete set of equations. This leads to the fact that instead of three (respectively four) measurements relating to the purely geometrical aspects for a terrestrial (respectively general) positioning, there is the need for one more measurement, say four (respectively five). For a terrestrial positioning the set of equations to be solved is as follows:

$$\rho_1 = \sqrt{(x_r - x_1)^2 + (y_r - y_1)^2 + (z_r - z_1)^2} + ct_r \tag{5.1a}$$

$$\rho_2 = \sqrt{(x_r - x_2)^2 + (y_r - y_2)^2 + (z_r - z_2)^2} + ct_r \tag{5.1b}$$

$$\rho_3 = \sqrt{(x_r - x_3)^2 + (y_r - y_3)^2 + (z_r - z_3)^2} + ct_r \tag{5.1c}$$

$$\rho_4 = \sqrt{(x_r - x_4)^2 + (y_r - y_4)^2 + (z_r - z_4)^2} + ct_r, \tag{5.1d}$$

where (x_i, y_i, z_i) are the coordinates of the satellites used for measurements, ρ_i are the pseudo ranges obtained from these measurements and (x_r, y_r, z_r, t_r) is the solution vector composed of the three spatial coordinates and the time clock bias.

Satellite Orbits and Location Calculations

Many of the system's characteristics can be classified from the very basic equations given above. The first is that there is a need for the receiver to know the satellites' locations. Second, the accuracy of the pseudo range is fundamental. If the pseudo range is wrong, then the resulting location will be wrong. Finally, the physical constants used by the system are also very important. For example, the speed of light, introduced in the above system of equations, must be 299,792,458 m/s for GPS (and not 3×10^8 m/s).[19]

The satellite's location is provided by the satellite through the so-called ephemeris, just as in the early days of navigation where there was the need for Sun location ephemeris for latitude determination. Here, ephemeris has to be a little bit more accurate, because of the increased accuracy needed. A new question arises from this discussion — do we need to take into account the displacement of the satellite while it is transmitting its signal to the receivers? In other words, has the time of propagation between the satellites and the receiver to be taken into account? To give part of the answer to this question, let us just remember some elementary facts concerning the three major constellations, existing or future.

The orbits of the satellites have been chosen in order to provide a full coverage of the Earth. This is therefore a compromise between the number of satellites, their height (involving the power to be transmitted and then the life-time of the satellites), and the number of satellites visible from any terrestrial place. Remember that the GLONASS satellites are at an altitude of 19,100 km, those of GPS at 20,200 km, and those of Galileo at 23,200 km.

In such conditions, it appears that the velocity of the satellites is around 3675 m/s for Galileo, 3870 m/s for GPS, and 3950 m/s for GLONASS.[20] The distances separating the satellites from the receiver once again depend on the constellation. Considering the nearest and the farthest satellites for each constellation, one obtains 19,100 km and 24,680 km for GLONASS, 20,200 km and 25,820 km for GPS, and 23,222 km and 28,920 km for Galileo. These values lead directly to the propagation time (considering the speed of light to be the speed of the satellite signals, which is good enough for this limited demonstration) of 64 ms and 82 ms for GLONASS signals, 67 ms and 86 ms for GPS, and 77 ms and 96 ms for Galileo. It is thus now possible to extract the distances traveled by the satellite during the time of transmission. This gives 252 m (respectively 325 m) for GLONASS, 260 m (respectively 333 m) for GPS, and 285 m (respectively 355 m) for the smaller distance (respectively larger distance). It can easily be seen that it is not possible to neglect this time of flight for the calculation of satellite location, without having a direct effect on positioning accuracy. Indeed, to achieve a positioning accurate to a few meters, the accuracy of the satellite's location must be much smaller than the above values (typically a few meters). Thus, the location needed is actually the location where the satellite was at the instant it transmitted the signal that the receiver received at the time of measurement, that is, a few tens of milliseconds before.

The orbital modeling implemented in GNSS is to provide the receiver with parameters that enable the calculation of the location of the satellites at each time.

Signal-Related Parameters

Once the choices of the orbits and of the ephemeris data have been made, there is still the need for signals. What were the reasons for the choices of the frequencies and codes used?

---

[19]This is of great influence due to the fact that real "t" is typically of the order of a fraction of a millisecond, leading to high values of ct.

[20]All the values have been calculated based on a value of 6400 km for the radius of the Earth.

Choice of Frequencies

The choice of frequencies is a complex combination of many technical and non-technical parameters. Of course, there is a need for a frequency (or frequencies) whose propagation capabilities through the terrestrial atmosphere are of good quality. Essentially this involves frequencies above 1 GHz. The date the system was designed is also very significant as it stipulates the technology used for both transmitter and receiver. The GPS program started in 1964 with the first development phases and the choice of final frequencies had to be made. At that time, the L-band was chosen. Then, to be able to operate the system throughout the whole world, there was also the need for reserved frequencies. This means that no other devices are allowed to use these frequencies, allowing very low power levels without fear of interference. In the case of GPS, with regard to the C/A code, a very small interference is likely to prevent the detection of the signal (a few decibels of degradation over the signal-to-noise ratio is enough). The reservation of such frequencies was achieved after a very long and demanding process through the World Radio communication Conference (WRC), held every three years or so under the coordination of the ITU.

Choice of Codes and Modulations

Once the frequency band has been chosen,[21] the signal structure must be defined. In a satellite-based positioning system, there is the need for three components: the identification of the satellite, the transmission of data required for the computation of the location (typically ephemeris data), and finally a means to achieve the physical measurement of the time delay of the transmission. Different choices are available in order to fulfil these requirements, but physical limitations have to be respected, such as the bandwidth allowed.[22] Thus, the choice was made to use a Pseudo Random Noise code[23] (PRN) for both identification and time measurement, and a very low rate of data transmission (typically 50 Hz) for so-called navigation data (ephemeris, time synchronization values of the satellites, clock correction values, propagation modeling correction factors, and so on). In such a way, the signal spectrum remains within the allocated bandwidth and positioning is possible. Because the civilian aspect of GPS had been planned for years, the idea of having two different codes emerged: the first, C/A, should allow a less accurate positioning than the second one, P, which was reserved for authorized people only. There arose the need for a "security" aspect; therefore the P code must not be easily available to non-authorized people.

This problem of robust codes arose in many applications, and GPS took advantage of the works of M. Gold (carried out in 1967 and 1968), who inventoried all the possible combinations of feedback shift registers (see Chapter 6 for more detail on the codes) and finally chose codes based on two 10-bit shift registers for C/A and

---

[21]This frequency is called the carrier frequency because its role is mainly to carry the information that is going to be the modulating data. Thus, high frequencies are chosen for wireless applications because of propagation capabilities.

[22]The bandwidths allocated to GPS were 24 MHz in L1 and 22 MHz in L2. The future L5 will have 28 MHz allocated.

[23]PRN codes are specific codes that look like random combinations of 0 and 1, but of course they are not at all random. This is the reason it is possible to obtain them at the receivers' end.

codes based on four 12-bit shift registers for P. The last point to be defined with these codes was the cadency, that is, the rate or the frequency used to generate these codes. C/A has a total length of 1023 bits, while P has a total length of more than 2.35×10^{14} bits. The main idea was to allow rapid acquisition, identification, and measurement, even if the accuracy is not at its best for C/A, and for an improved accuracy and interference immunity for P, even if it is necessary to use the C/A signal for the first pre-acquisition phase. Thus, the rate of C/A was chosen to be 1.023 MHz, leading to a 1 ms repeating code, and 10.23 MHz for the P code (which has a complete duration of 266 days). This much longer feature of the P code helps provide immunity to interference to a much higher degree than for the C/A code. It can easily be understood that this feature is of uppermost importance for military applications. Furthermore, the fact that the P code is ten times faster that the C/A one has also led to an improvement in the accuracy of the physical time measurement.

Another point of interest is the number of frequencies that are going to be used for a given system. For both GPS and GLONASS, the precursors, two frequencies were used. The main reason for this was the fact that the main error source is the delay in the ionosphere. This high atmosphere layer is composed of ionized particles that have the direct effect of slowing down the information transmitted. Instead of propagating at the speed of light, it travels a little more slowly. This effect has to be taken into account when considering the translation from time measurement into pseudo range. The way this can be carried out is by modeling the ionosphere's thickness and the proportions of the ionized particles. Unfortunately, this modeling is very complex — the thickness and the concentration of ionized particles depend on the effect of the Sun for instance. Thus, it appears to be dependent on season, on the Sun's activity, on the temperature, and of course on the actual path of the wave (thus it depends on the relative location of the satellite and the receiver on Earth), and so on. The modeling takes all these aspects into account, but the remaining error is still the largest part of the global error of the signal. Fortunately, the physics that lies behind the behavior of the ionosphere is non-linear (the ionosphere is a dispersive[24] medium). Thus, having access to two measurements made at two different frequencies can help in totally removing this perturbing effect. As the reader can establish for himself, the P code is available on L1 and L2, unless the C/A code is only available on L1. P-code receivers thus have the ability to overcome the main error, but C/A receivers cannot. This was a deliberate choice of the U.S. and Soviet Union programs in-order to limit the availability of this dual frequency to authorized people only. A civil C/A code could have been sent on L2 also, but has not been. The only impact for the receiver is to deal with two frequencies instead of only one. Current electronics cannot deal with L1 and L2 using the same radio frequency front-end.[25] In the case of a civil transmission on L2, the need for a more

---

[24]A dispersive medium is one in which various frequencies exhibit various behaviors. This is the case when the mathematical expression of the phenomenon observed is not a linear function of the parameter being considered. In the current case of propagation delay while passing through the ionosphere, the equation exhibits the presence of the frequency to the square.

[25]This RF Front-End is the part of the receiver that deals with the incoming signals. Current and future architectures are described in Chapter 7.

complex electronic part in the receiver would have been required, increasing the price of this civilian equipment. Note that this very interesting feature is now being considered by both GPS and GLONASS providing L2C civil signals (today for a limited number of satellites).

Velocity Calculation: Doppler Shift Measurement

The reader should now understand the way the different choices were made towards the definition of the GPS system,[26] as well as their advantages and disadvantages. This approach has some other impacts on the performances and capabilities of the output of a receiver: time is also available (which can be accepted quite easily as a consequence of the description made previously). Moreover, the velocity of the receiver, as a three-dimensional vector, can also be given with good accuracy. Indeed, in the early days of satellite transmission, that is, Sputnik, the problem of the Doppler shift of the signal while the satellite was moving had arisen. The observations made by George C. Weiffenbach and William H. Guier, which led to the TRANSIT system, were based on the Doppler shift measurement, and the TRANSIT system itself was based on Doppler shifts. For GPS, a need was identified to be able to have the speed of the receiver available. This is achieved, once again, through Doppler shift analysis. In fact, when the receivers' hardware (see Chapter 7 for details) tries to "find" the signal of a given satellite, it has to take into account two components of the Doppler shift: the motion of the satellite and that of the receiver. The Doppler results from the relative displacement of both. It is compulsory to take the Doppler shift into account in order to achieve the correlation process,[27] which allows identification and time measurement.

The Doppler is then precisely (see Chapter 7 for figures) evaluated within the first stages of the receiver. This is achieved through the measurement of the frequency shift of the local oscillator. The interesting feature here is that this measurement, associated with a given satellite, is independent of the time measurement carried out for the pseudo range determination. Thus, as in a radar system, time and frequency measurements are linked to each other for calculation, but independent in the physical measurements. Following a similar approach to that for location finding, it is possible to obtain the equations from four such frequency measurements, from four different satellites (see Chapter 8, section "Calculation of velocity " for details). The solution of such a system is then a vector in which the first three coordinates are the three space-based coordinates of the velocity of the receiver (that is, v_x, v_y and v_z) and the fourth is the derivative (relative to time) of the clock bias, that is, the receiver clock drift.[28]

---

[26]The main differences with GLONASS and Galileo are also given in Chapters 3 and 6.

[27]The correlation is the mechanism that compares the received code, specific to a given satellite, to a local replica of this code in the receiver. When the two codes match in time, then the correlation is maximal. When the two codes are slightly shifted backward or forward, the correlation is almost zero.

[28]Like the clock bias, this calculated value is an intermediate variable in the computations. Nevertheless, the first GPS receivers were made with a single channel of reception, as compared to the usually 12 to 20 parallel channels today. For these first receivers, there was the need to "follow" the clock bias in order to calculate the receiver position (because in such a case, the clock bias changes at each measurement, making a new unknown in each equation, then requiring another input). The receiver clock drift allowed such calculations to be carried out.

Navigation Message

From the above discussion, it is obvious that modeling the orbits of satellites is required in order to define their location or time synchronization. Furthermore, physical propagation models are also required to adjust the time to distance conversion of pseudo ranges. As the corresponding parameters are not stationary, that is, are not constant with time, there is a need for data transmission to the receiver. At the time of the design and development of GPS and GLONASS, telecommunication capabilities worldwide were not available and the choice made was to transmit these data through a navigation message, added to the code of a satellite before modulation of the carrier frequency. The update rate of the data was not very high (the orbital parameters are valid for a few hours) and in order to reduce the bandwidth, the choice of a very low data rate message (50 bits per second for GPS) was retained. For Galileo, a new approach could have been envisaged, like the one proposed by Assisted-GNSS, where the navigation message content is sent to the receiver through the use of modern high speed communication networks, thus reducing the dependence of the system on this very slow message. Nevertheless, a similar approach to GPS has been implemented. Interoperability is probably one cause, as the fact that a receiver should always be able to compute a location on a "stand-alone" basis, that is, without any external or additional data to that from the GNSS signals.

Relativistic Effects

Two main effects are considered in GNSS: the gravitational field affecting the satellites (quite different from the Earth's gravitational field) and the speed of displacement of the satellite clocks (with respect to the same clocks on the Earth's surface). The difference in the gravitational field leads to a clock running 43 μs faster for the satellite, compared to the same clock on the Earth's surface, and the speed of the satellite results in a clock that is 9 μs slower. The total for the satellite clock is then 34 μs faster, per day.

The combination of these two effects leads to a clock rate offset of 4.45×10^{-10} ($\Delta f/f$) faster for the satellite clock. Rather than having to take this into account at the receiver end, it was decided to deal with relativistic effects at the satellite end, by having the frequency of the central oscillator shifted in such a way that the behavior is equivalent to that which would have been observed with no relativity. Thus, a central frequency of 10.22999999543 MHz is used instead of 10.23 MHz.

For GLONASS, the same problem led to the nominal value of the frequency, as observed at the satellite, to be biased from 5.0 MHz by the relative value $\Delta f/f = -4.36 \times 10^{-10}$. This gives a central frequency of 4.99999999782 MHz (the value is given for the nominal orbital height 19,100 km).

Other relativistic effects are also considered, such as the Sagnac effect (due to the Earth's rotation), or the space–time curvature due to the Earth's gravitational field.

Modernization

A consideration of the modernization signals follows on from the preceding discussion. As the community of mass market users has increased, and the Galileo program has become a reality, the need for real improvements in the system have arisen. First, and most importantly in terms of impact on accuracy, is the need to make two frequencies available. The first step of the modernization of GPS was precisely to introduce an L2C signal, a civil code on L2. Developments of GPS electronics made the total cost of a typical car navigation system approximately U.S.$500. Within this total cost, the GPS receiver represents less than $10 today, the rest being accounted for in the packaging, the color screen, the software included that allows interesting features, and, of course, the cartography. Improving accuracy will, if no further improvements are carried out, certainly raise the price by some fraction of the electronic parts, say typically a few dollars. This increase is minimal and would not affect sales of the products.

With regard to mass market applications, having two frequencies would certainly be enough. The modernization of signals is nevertheless going to provide a third civil signal in the L5 (1176 MHz) band. Different reasons have played a role in this decision. The main one is that L2 is not an Aeronautical Radio Navigation Service (ARNS) band; thus, for aviation, dual-frequency receivers require another ARNS band (that is, L5). The two other driving concepts are integrity[29] and interoperability,[30] and have been widely supported by the Galileo program. Integrity is a very important feature for some applications where security is of great concern, for instance, in civil aviation. In such a case, there is the need for an alert procedure, telling the receiver that the location calculated could present a weakness (or even be wrong). In addition, for the GPS system, a relative redundancy is compulsory in order to provide the "service."[31] A third civil signal, in an ARNS band, is thus of tremendous importance to achieve such demanding application integrity values. The second concept, interoperability, requires coordination between program managers, but also coordination in the frequency bands used. Navigation bands are not that large but there is still enough room for individual bands. For instance, the GLONASS bands do not overlap with the GPS and Galileo ones (except the future L3). In order to avoid interference, which is of major concern for navigation signals, the decision was taken that no Galileo signal would lie in the L2 band.[32] Thus, if one decides to stay with L1 and L2 frequency bands for each constellation, then there will be only one overlapping, namely L1. In such a case, in order to reduce the error of ionosphere propagation, mass market receivers would need to have a three-frequency-band capability. Thus, having a third frequency for GPS, as it is for Galileo, is a very good compromise for both integrity and interoperability. Of course, one could envisage using L1 and L5/E5a for a dual-frequency receiver, although the L5/E5a band includes fewer signals than the L2 or E6 ones.

---

[29] Integrity can be seen as a means, embedded in the system, to give acknowledgement to the users that some part of the system is not working properly.

[30] Interoperability is the ability of future receivers to use both constellations to provide an "embedded" positioning, that is, a location obtained by using satellites from both GPS and Galileo constellations. Because of different bands, interoperability with GLONASS is more complicated, although possible.

[31] For the first time with Galileo this notion of "services" has been used to define the primary needs and then the signals.

[32] This point has even led to a petition from users asking GPS and Galileo program leaders to find a compromise to use the same bands, for the benefit of the users' communities.

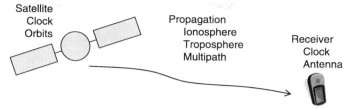

FIGURE 5.12 A diagrammatic representation of a satellite-to-receiver radio link.

Of course, modernization is not limited to adding signals; a profound remodeling of these signals has been undertaken, as described in Chapters 3 and 6.

■ 5.4 INTRODUCTION TO ERROR SOURCES

As soon as a physical measurement is carried out, the precision of this measurement should be evaluated. In GNSS, some errors are likely to degrade the signals. Let us consider the diagrammatic representation of a satellite to receiver link as given in Fig. 5.12.

Error sources can be split into three categories, depending on the physical location where they take place:[33]

- Errors due to satellite based uncertainties;
- Errors due to signal propagation; and
- Errors due to receiver-based uncertainties.

At the satellite end, both clock synchronization biases, with reference to constellation time and satellite location accuracy, are certain to generate final positioning errors. At the receiver end, clock bias and also the location of the center of phase of the antenna are sources of positioning inaccuracy. In addition, the signal does not propagate at a constant speed because of variations induced by the successive layers of the atmosphere. Additional delays can also be induced by multipath. These are physical-based potential errors, but voluntary errors were also added in the early days of GPS (Selective Availability). The main difficulties of GNSS positioning are signal acquisition and tracking, as well as error mitigation. A description of error sources is given in Chapter 6.

■ 5.5 CONCEPTS OF DIFFERENTIAL APPROACHES

It appears that, under certain conditions, there is a need to carry out differences in order to allow higher accuracy. The real problem of GNSS signals lies in the various error sources that are present in the propagation time measurements. In order to reduce

---

[33]Note that there are of course other ways of representing error sources.

the impact of these errors, one approach consists in carrying out differences between measurements. This allows, in certain cases, the removal of quantities that are identical in different measurements, thus leading to a significant improvement. Achieving such differences is known by the generic term "differential."

Depending on the goal one wants to achieve, there are many differential approaches. In Chapter 3, the differential method using a single frequency implemented in WAAS, EGNOS, and MSAS was described. For centimeter-accuracy positioning, the so-called phase measurement technique is required.[34] Obviously, these two differential approaches are quite different from each other, other than the fact that they both calculate differences.

In order to propose a classification of differential GNSS techniques, let us discriminate four criteria based on three major characteristics, namely the position, the frequency, and the time. The four proposed classes of differential techniques are as follows:

- *Receiver position differential*, where the difference of receiver positions is carried out.

- *Satellite position differential*, where the difference of satellite positions is carried out. In this category, methods applying corrections at the pseudo range level, such as SBAS for instance, are also considered.

- *Frequency differential*, where the difference of signal propagation time is carried out for different frequencies. Note that this is mainly used to remove ionosphere-related delays.

- *Time differential*, where the difference of signal propagation time is carried out at different instants. Note that the main use is currently to remove the ambiguity of carrier-phase-related measurements.

Differences can be weighted differences, where coefficients are applied, as for instance in the case of calculating the frequency differential.[35] In order to choose one or more methods, it is necessary to understand the principles behind each method. There is sometimes the need to carry out differences of position at the same instant in order to remove time-related parameters (receiver or satellite clock biases for example), and sometimes the need to carry out differences at two different instants. Phase measurement is a good example of a combined differential approach.

Table 5.2 presents an attempt to summarize the differential methods in comparison with the stand-alone receiver mode where the receiver achieves the positioning without any external aid.[36] It is then clear that there is no phase measurement stand-alone mode because of the phase ambiguity that requires external aid. The

---

[34]The carrier phase measurement techniques are often referred to as "relative" techniques in order to differentiate them from "differential" code-phase-based techniques.

[35]The difference is carried out with respect to the frequency ratio to the power of two. The formula is finally $[\rho1 - \gamma\rho2]/[1 - \gamma]$ where γ is $[L1/L2]^2$ and $\rho1$ and $\rho2$ the respective pseudo ranges at L1 and L2.

[36]Note that this classification does not make reference to the way the external aid is achieved. For instance, some local differential networks are available for positioning with accuracy to the centimeter using a phase differential approach in real time through the use of a single receiver plus a wireless data link to a differential service provider. In the present classification, such an approach is considered as being a differential method, and not a stand-alone method.

TABLE 5.2 **Current (2007) differential methods.**

	Code Measurements	Phase Measurements
1. Stand-alone receiver	GNSS A-GNSS Pseudolites	
2. Receiver position differential	GNSS w LAAS	GNSS SD w (3,4,5) Pseudolites SD w (3,4)
3. Satellite position differential	GNSS w SBAS GNSS w GBAS	GNSS DD w (2,4,5) Pseudolites DD w (2,4)
4. Time differential		GNSS TD w (2,3,5) Pseudolites TD w (2,3)
5. Frequency differential		GNSS dual-freq. w (2,3,4)

SD, DD, TD, Simple, double, triple differences; "w" indicates that the method is used in conjunction with the methods in brackets. SBAS, Satellite-Based Augmentation System; GBAS, Ground-Based Augmentation System; LAAS, Local Area Augmentation System.

three global systems considered are the GNSS, the Assisted-GNSS (A-GNSS), and the pseudolites (refer to Chapter 10 for details).

In Table 5.2 the most commonly currently used positioning methods are presented: the stand-alone, either with GNSS, A-GNSS, or pseudolites, where the receiver handles incoming signals only from the satellites. The only current possibility is to carry out code measurements because there is no solution for ambiguity resolution for phase measurements in a stand-alone mode. The next step for code measurements is to implement the receiver position differential. Indeed, the idea is simply to put a reference receiver at a known location to output a "common bias error vector," which will be removed from the resulting navigation solution of the mobile receiver. This differential requires the availability of a fixed receiver in the vicinity of the mobile one — this is the basic differential approach implemented on a large scale with LAAS. This is clearly related to the locations of receivers. Of course, this differential method can be implemented through local facilities (it is sufficient to place a fixed receiver somewhere), leading to row 2 of Table 5.2, second column. For code measurements, although other approaches are possible (see Table 5.3), this is the only method implemented. For the phase measurements, the concept is based on three successive differences (SD, DD, and TD), which respectively implement a receiver differential technique, a satellite differential technique where two satellites, at the same instant, are used, and a time differential method where a given satellite is followed while moving. The frequency differential method must also be implemented in order to reduce ionosphere propagation error effects and to allow centimeter accuracy. Thus, the current phase measurement technique uses four differential methods together. This approach can also be implemented locally with pseudolites. The main difference is then the fact that dual frequency is not required as the reduced area covered does not involve ionosphere propagation.

Besides these implemented methods, developing SD and DD techniques with code measurements in order to totally remove the clock biases of the receivers and satellites could be envisaged, in some specific cases such as when initializing an

TABLE 5.3 Current (2007) potential differential methods.

	Code Measurements	Phase Measurements
1. Stand-alone receiver	**GNSS** **A-GNSS** **Pseudolites**	
2. Receiver position differential	**GNSS w LAAS** GNSS SD w (3) Pseudolites SD w (3)	**GNSS SD w (3,4,5)** **Pseudolites SD w (3,4)**
3. Satellite position differential	**GNSS w SBAS** **GNSS w GBAS** GNSS DD w (2) Pseudolites DD w (2)	**GNSS DD w (2,4,5)** **Pseudolites DD w (2,4)**
4. Time differential		**GNSS TD w (2,3,5)** **Pseudolites TD w (2,3)**
5. Frequency differential		**GNSS dual-freq. w (2,3,4)**

Methods that appear in Table 5.2 are shown in bold.

initial position in a dynamic positioning using carrier phase measurements. The TD is of no real interest because the removal of ambiguity concerning the code is more easily carried out by calculation (refer to Chapter 8). Table 5.3 adds these possible approaches to Table 5.2. The reason they are not implemented is due to their increased complexity with respect to potential gain. Of course, the gain will be real, because the clock biases would be thoroughly removed, but remains too small in comparison to the other sources.

Table 5.4 provides the foreseeable new features that will be available in the coming years with the modernization of GPS, the renewal of GLONASS, and the advent of Galileo. This does not take into consideration the corresponding complexity or cost of receivers. The possibility of achieving code dual frequency is a substantial improvement for mass market civilian users. This has to be balanced by the fact that corresponding applications have to be found, because, with current receiver architectures, the cost of the GNSS positioning functionality would be about twice the cost of a mono-frequency system (one has to implement two radio front-ends). Note that the interest in dual frequency, from a technical point of view, is both in removing the ionosphere errors and in the fact that it is possible to implement this latter correction in an SD or DD method.

The next point that could be of interest would be to implement a differential method taking advantage of the fact that there are three different constellations. Thus, one could have imagined using different signals or satellites from different constellations in order to achieve a new differential approach. Although the interoperability concept consists effectively in the possibility of computing a position using satellites from different constellations (see Chapter 8 for details of the computations), a differential method is not achievable simply because the various signals are transmitted from different satellites, and thus any potential interest vanishes.[37] Remember that for the ionosphere differential approach, the compulsory requirement is to receive

---

[37]It would be quite different if all the satellites from all the constellations were planned to transmit all the signals. This is perhaps the next step in international cooperation for satellite-based navigation signals.

TABLE 5.4 Additional features provided by future constellations.

	Code Measurements	Phase Measurements
1. Stand-alone receiver	**GNSS** **A-GNSS** **Pseudolites**	
2. Receiver position differential	**GNSS w LAAS** **GNSS SD w (3,5)** **Pseudolites SD w (3,5)**	**GNSS SD w (3,4,5)** **Pseudolites SD w (3,4)**
3. Satellite position differential	**GNSS w SBAS** **GNSS w GBAS** **GNSS DD w (2,5)** **Pseudolites DD w (2,5)**	**GNSS DD w (2,4,5)** **Pseudolites DD w (2,4)**
4. Time differential		**GNSS TD w (2,3,5)** **Pseudolites TD w (2,3)**
5. Frequency differential	GNSS L1–L2 GNSS L1–L3 GNSS L1–L5 GNSS L1–E5 etc. . . .	**GNSS dual-freq. w (2,3,4)**

Methods that appear in Table 5.3 are shown in bold.

two signals at two frequencies issuing from the same (located) transmitter; this is the same for all the differential methods discussed above.

■ 5.6 SBAS SYSTEM DESCRIPTION (WAAS AND EGNOS)

A major problem lies in the fact that there is no alert message originally planned to be transmitted by any means in the case where a navigation satellite breaks down or sends incorrect data. Thus, SBAS have been designed to overcome this difficulty and therefore to allow applications where safety of life is concerned (note the transport domain was the original target).

The various SBAS, either already available or under development, follow a similar principle, where a network of receiving stations is distributed over the geographical region to be covered. These stations carry out measurements from navigation satellites and transmit these measurements to a control center that evaluates the accuracy of the positioning service provided by the navigation systems. Such a network allows two different features: differential corrections and integrity. The first feature consists simply in sending differential corrections to receivers through a transmission channel. The clever idea was to use a code similar to that of the navigation satellite (similar to the C/A code) so any receiver could access these data. The second feature helps a receiver to use or not use a given navigation satellite in its navigation solution depending on the reliability of the sending signals. This is evaluated using the control center's calculations.

The infrastructure of the European system EGNOS (European Geostationary Navigation Overlay Service) is as follows:

- About 30 receiving stations called RIMS (ranging and integrity monitoring stations), mainly deployed in the European area.

- Four control and computation centers called MCC (master control centers), where the differential and integrity messages are generated. These centers are located in Europe and are able to cope with the potential breakdown of one center.
- Six transmitting stations called NLES (navigation land earth stations) that allow transmission of ground computed differential and integrity messages to the geostationary satellites.
- A center for performance evaluation called PACF (performance assessment and check-out facility), located in Toulouse, France.
- A qualification center called ASQF (application-specific qualification facility), located in Spain.

The infrastructure of the United States system WAAS (Wide Area Augmentation System) is as follows:

- About 25 receiving stations called WRS (wide area reference station), distributed over the United States.
- Two control and computation centers called WMS (wide area master stations) located respectively on the East and West coasts of the United States.
- A few transmitting stations called GUS (ground uplink stations) that allow transmission to the geostationary WAAS satellites.

In addition, the systems include geostationary satellites. There are three such satellites for EGNOS (Inmarsat 3 AOR-E, Inmarsat 3 IOR, and Artemis) and four such satellites for WAAS (Inmarsat POR, Inmarsat AOR-W, Telesat Anik F1R, and Panamsat Galaxy 15).

A similar SBAS system is also planned for Galileo, with comparable features. The main difference will be that the transmission from satellites to receivers will be based on the Galileo satellites rather than on geostationary satellites. This system is called ERIS (External Regional Integrity Service).

✓ 5.7 EXERCISES

Exercise 5.1 What are the respective tasks of the three segments of a GNSS system?

Exercise 5.2 Describe the various services provided by GPS and GLONASS. How are these services implemented?

Exercise 5.3 What is the philosophy of Galileo? How have the proposed services been implemented in terms of frequencies? Why?

Exercise 5.4 How important is the fact there is only a downlink from satellites to receivers? Note that all three constellations are similar (GPS, GLONASS, and Galileo) on this point. Note also that the foreseen COMPASS plans to implement a different concept. Give comments on this latter point.

Exercise 5.5 When is it absolutely necessary to have five satellites in order to compute a three-dimensional positioning?

Exercise 5.6 Class the effect of time synchronization on the following parameters:

- Time of flight measurements;
- Computation of satellite location;
- Effect of receiver motion.

And conclude concerning the synchronization necessary to reach a few meters of accuracy (do not take into account propagation-related errors in this exercise).

Exercise 5.7 Synthesize the compromises that have to be carried out between

- Codes (length, rates, and so on);
- Frequencies;
- Modulations;
- Navigation message (rates, data, and so on).

Use this synthesis to explain the current choices for both modernized GPS and for Galileo.

Exercise 5.8 When is the use of the frequency differential method crucial? Answer the same question for the satellite position differential method? Can you explain the reason why the satellite position differential method is almost never used with single-frequency receivers?

Exercise 5.9 What could be the most interesting feature of a time differential method for a single-frequency receiver?

Exercise 5.10 What are the main characteristics of the ground segment of EGNOS or WAAS systems? Develop your answer by doing further reading.

■ BIBLIOGRAPHY

Braasch MS. Autocorrelation sidelobe considerations in the characterization of multipath errors. IEEE Transactions on Aerospace and Electronic Systems January 1997;30(1).

Brunner FK, Gu M. An improved model for the dual frequency ionospheric correction of GPS observations. Manuscripta Geodaetica 1991;16:205–214.

Chop J. et al. Local corrections, disparate uses: cooperation spawns natinal differential GPS. GPS World 2002;13.

Collins JP. Assessment and development of a tropospheric delay model for aircraft users of the "Global Positioning System." M.Sc.E. thesis, Department of Geodesy and Geomatics Engineering Technical Report No. 203; University of New Brunswick; Fredericton: New Brunswick; Canada. 1999. Available at: http://gge.unb.ca/Pubs/TR203.pdf. Nov 14, 2007.

Elgered G. Tropospheric radio path delay. In: Jansse MA, editor. Atmospheric remote sensing by microwave radiometry. New York: Wiley; 1993.

EGNOS (The European Geostationary Navigation Overlay Service). A cornerstone of Galileo. ESA SP-1303. Available at: http://www.egnos-pro.esa.int/Publications/GNSS%202001/SBAS_integrity.pdf.

Galileo open service – signal in space interface control document. (OS SIS ICD) Draft 0; European Space Agency/Galileo Joint Undertaking; 23 May 2006.

Global navigation satellite system. GLONASS; Interface Control Document; ICD02_e; Moscow; 2002.

Hudnut KW, Titus B. GPS L1 civil signal modernization (L1C). The Interagency GPS Executive Board; July 2004.

Kaplan ED, Hegarty C. Understanding GPS: principles and applications. 2nd ed. Artech House; 2006. Norwood, MA, USA.

Kelly JM, Braasch MS. Mitigation of GPS multipath via exploitation of signal dynamics. In: ION Annual Meeting: Proceedings; 1999.

Lachapelle G, Bruton A, Henriksen J, Cannon ME, McMillan C. Evaluation of high performance multipath reduction technologies for precise DGPS shipborne positioning. The Hydrographic Journal 1996;82:11–17.

Langley RB. The GPS error budget. GPS World, 1997;8(3):51–56.

Lavrakas J, Shank C. Inside GPS: the master control station. GPS World Magazine, n°9 (September 1994);46–54.

Lisano ME, Carpenter JR, Gomez S. Navigation, attitude determination, and multipath analysis: results from the STS-77 GPS attitude and navigation experiment (GANE). Navigation 1999;46(3):175–192.

Local area augmentation system. LAAS; September 2004. FAA web site; http://gps.faa.gov.

NAVSTAR GPS space segment/Navigation user interfaces. ICD-GPS 200; Revision D; IRN-200-D-001; public release version. ARINC Engineering Services; LLC; March 7, 2006.

NAVSTAR GPS Space segment/User Segment L5 Interfaces. IS-GPS-705; IRN-705-003. ARINC Engineering Services; LLC; September 22, 2005.

NAVSTAR GPS Space segment/User Segment L1C Interfaces. Draft IS-GPS-800. ARINC Engineering Services; LLC; April 19, 2006.

Ober PB. SBAS integrity concept—towards SBAS validation. European Organisation for the Safety of Air Navigation, Eurocontrol; May 2001.

Parkinson BW, Spilker Jr. JJ. Global positioning system: theory and applications. American Institute of Aeronautics and Astronautics; 1996.

Raymond D, Henry B, Micheal W. WAAS geostationary communication segment (GCS) requirements analysis. In: IEEE Position, Location and Navigation Symposium (PLANS): Proceedings: April 15–18, 2002.

Specification for the wide area augmentation system. U.S. Department of Transportation; Federal Aviation Administration; FAA-E-2892b; August 13, 2001.

System safety architecture Description for the WAAS. Prepared for the FAA by Raytheon Company; October 10, 2001.

Van Nee R. Multipath effects on GPS code phase measurements. In: *ION GPS-9 1*: Proceedings; 1991;915–924.

Van Nee R. Multipath and multi-transmitter interference in spread-spectrum communication and navigation systems. Ph.D. dissertation; Delft University of Technology; Delft; The Netherlands; 1995.

Web Links

http://www.esa.int/esaNA/egnos.html. Nov 14, 2007.

http://www.egnos-pro.esa.int/publications.html. Nov 14, 2007.

http://igscb.jpl.nasa.gov. Nov 14, 2007.

http://gps.faa.gov. Nov 14, 2007.

http://tycho.usno.navy.mil/gps.html. Nov 14, 2007.

http://www.Colorado.EDU/geography/gcraft/notes/gps/gps.html. Nov 14, 2007.

http://www.navcen.uscg.gov/gps. Nov 14, 2007.

http://www.ngs.noaa.gov/. Nov 14, 2007.

6

GNSS Navigation Signals: Description and Details

The interest in having different constellations arises from the fact that they complement each other rather than compete with each other, although the recent developments of GPS have been driven by the advent of Galileo. The goal of this chapter is to provide a technical comparison of constellations in order to highlight the synergies that will eventually be used in future receivers. User communities will benefit from this situation and no-one believes that there will be constellation-specific receivers, but rather multi-constellation receivers (at least for GPS and Galileo). The dramatic increase in both the number of satellites and signals available in a few years will certainly be the trigger for the development of performances and applications.

When dealing with positioning and navigation, satellite-based systems appear as kings in the whole palette of possible solutions. Thus, the Global Navigation Satellite Systems (GNSS) are the subject of a few specific chapters here. This chapter will deal with GNSS signals in space, then Chapter 7 will look at acquisition techniques, before Chapter 8, which will deal with calculation techniques. The main idea of these chapters is to carry out a comparison between the current and near-future constellations: namely GPS, GLONASS, and Galileo. The other satellite-based systems for positioning were described in Chapter 4. The descriptions are not exhaustive, but rather are oriented towards explanations in order to allow the reader to understand the main issues.

■ 6.1 NAVIGATION SIGNAL STRUCTURES AND MODULATIONS FOR GPS, GLONASS, AND GALILEO

In order to achieve all the proposed services, there is of course the need for signals. The number of signals in space is growing rapidly. For the GPS and

GLONASS constellations there is the advent of new civil signals on a second frequency, plus also a third frequency civil signal. Launches do take time, and a typical user will have to wait a few years in order to really appreciate the difference. The GLONASS program, launching typically three satellites per year, with improved life-times, is generating a rapid availability of new signals. For instance, the L2 civil signal has been available since 2003 and now there are 11 satellites transmitting this signal.[1] No satellites are yet transmitting the L3 civil signals for GLONASS.[2] GPS launches are limited to the bare essentials and the availability of L2C civil signals is currently limited to three satellites.[3] This makes quite a difference when compared with Galileo, which will provide all its new signals at once during the 2011–2013 period. The availability of Galileo signals will be progressive due to satellite deployment, but improved coverage and availability will occur over a short period of two or three years.

From the users' point of view, the new capabilities allowed by these new signals will lead to potential new applications that will be developed as soon as the signals are really available in space, although it will take several years to be set up. Thus, this section is devoted to the description of navigation signals, both already available and those foreseen.

Structures and Modulations for GPS and GLONASS

All the satellites of the current systems are transmitting two frequencies (L1 and L2) and two codes (C/A and P). A few satellites are transmitting an L2 civil signal, both for GPS and GLONASS, and some GPS satellites are also transmitting a new military code, the M-Code, on both L1 and L2. In the near future, a third frequency will be available for both constellations, L5 for GPS and L3 for GLONASS. In addition, following an interoperability agreement between the United States and the European Union, a new L1C civil signal is envisaged for GPS. They are represented in Fig. 6.1 (GPS) and in Fig. 6.2 (GLONASS).

Note that there is a fundamental difference between GPS and GLONASS regarding the spectrum, which is induced by the addressing techniques used for identification of the satellites. The basic difficulty, as with all radio communication systems, is to find the best way to allow multiple satellites to access the same receiver without interference; the corresponding techniques are called "multiple accesses." There are three main techniques available: the Frequency Division Multiple Access (FDMA), where specific frequencies are allocated to transmitters in order to be identified; the Time Division Multiple Access (TDMA), where individual time slots allow identification; and finally the Code Division Multiple Access (CDMA), where codes are assigned to each transmitter.[4]

---

[1]On March 27, 2007.
[2]On March 27, 2007.
[3]On March 27, 2007.
[4]Note that some systems are using a combination of these techniques. The main advantage of CDMA is that the whole bandwidth allocated can be used by every user, as long as they are recognized by their specific codes. This allows high performance in terms of interference mitigation. Furthermore, when the number of users is reduced, it also allows excellent noise behavior.

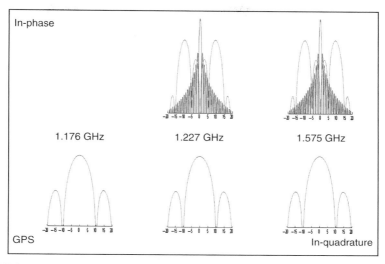

FIGURE 6.1 The GPS signal spectrum.

For GPS, the technique used is the CDMA, each satellite being characterized by its code, and all satellites transmit on the same frequency. In the case of GLONASS, the technique is the FDMA, where each satellite is characterized by its frequency, and all satellites have the same code.

It has already been mentioned that two frequency bands, and soon more, are available. The GPS radio frequency link 1 and 2 (L1 and L2), as with the GLONASS Link 1 and 2 (G1 and G2), are at a high frequency in order to provide the systems with good propagation performances. The future L5 (GPS) and L3 (GLONASS) are also in the same L-band, between 1 and 2 GHz. Onto these so-called

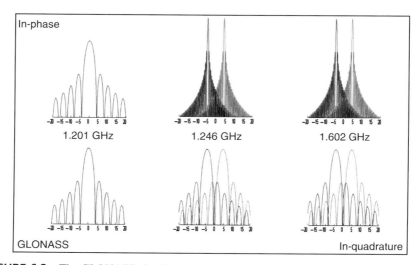

FIGURE 6.2 The GLONASS signal spectrum.

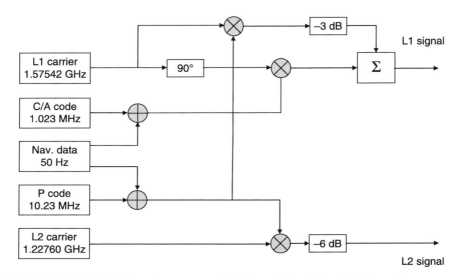

FIGURE 6.3 The GPS signal structure on L1(C/A and P codes) and L2 (P code).

carrier frequencies, "data" are added through multiplications — this process is called "modulation." Various types of modulations exist and provide an optimum data rate according to bandwidth. For GPS, as all satellites use the same frequency (at both L1 and L2), there is a need for some identifying codes, and a coarse code and a precise code have been designed for each satellite. The C/A code (coarse acquisition) is available on L1, and the P code (precise) is available on both L1 and L2.

The modulations used for both L1 and L2 of GPS are Binary Phase Shift Keying (BPSK). This means that for each piece of data to be transmitted, the phase of the sinusoidal carrier is either kept continuous (in this case the binary state of the data remains unchanged) or is shifted by half a period, or 180° (in this case the new piece of data changes). The codes are indeed repeating sequences that will allow satellite identification in the case of GPS. However, we have already mentioned the fact that the receiver also requires data concerning the health of the satellites, their orbital parameters or the synchronization of their clocks. All these data are included in the so-called "navigation message" sent by all satellites. In such a way, there is no direct link between the ground segment and the user segment. In order not to impact too much on the resulting spectrum, the data rate of this message has been set to 50 bits per second.[5] The main drawback of this rather slow data rate is the time required for a receiver to receive the entire message. Figures 6.3 and 6.4 give the simplified representations of the structures of GPS and GLONASS signals (only L1 C/A and L2 C/A and P).

When considering the resulting signal, including the navigation data, the code, and the BPSK modulation, it is important to keep in mind the relative durations of each element: a navigation bit lasts 20 ms, but a code chip lasts less than 1 μs

---

[5]Note that current thinking is either to reduce or to increase this data rate. As will be presented later, some signals do not even include navigation data.

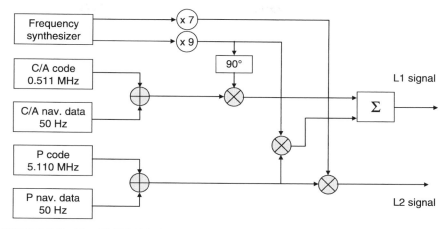

FIGURE 6.4 The GLONASS signal structure on L1(C/A and P codes) and L2 (P code).

(in the case of the C/A code), with a carrier cycle duration of 670 ps. Figure 6.5 illustrates this point.

The details of these modulations are outside the scope of this book. Please refer to further readings for deeper analysis. Note just that the oscillator frequency driving the codes for GPS is 10.23 MHz compared to the 5.11 MHz for GLONASS. This involves spectrum occupancy for GPS that is twice the size that of GLONASS, for one satellite. Regarding the total spectrum, this is compensated by the fact that GPS satellites all use the same bandwidth, as GLONASS satellites are shifted from each other due to the FDMA technique.

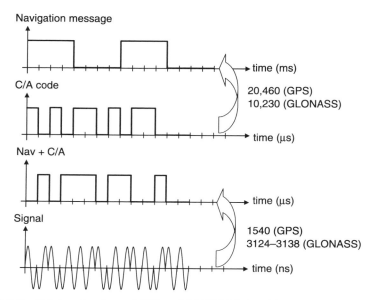

FIGURE 6.5 Relative durations of GPS and GLONASS signal items.

Structure and Modulations for Galileo

Regarding apparent complexity of Galileo signals, we decided to devote a separate paragraph to Galileo. Galileo is based on the same principles, although in a more complex form (note that future GPS signals will also be a little bit more complex than current ones).

The signals of Galileo have been designed in accordance with associated services and the corresponding requirements in terms of availability, power levels, performances, and so on. Thus, the service signals cannot be seen as a sum of individual contributions, but rather as a combination of frequency items designed on purpose. Let us review these arrangements relating to services before entering the discussion. Figure 6.6 gives a graphical representation of the total band allocation devoted to Galileo signals and services.

Open Service (OS)

The three OS frequencies are L1, E5a, and E5b, allowing ionosphere propagation correction. Each frequency is composed of two telemetric codes, in-phase and in-quadrature phase. Navigation data are added only to the in-phase signal, leaving the second signal without navigation data in order to allow long integration, very useful when operating in weak receiving conditions, such as under heavy foliage or indoors. This second signal is called the "pilot signal" or "pilot tone." These signals are transmitted on the three frequencies and operations are planned using one, two, or even three frequencies.

Commercial Service (CS)

In addition to the OS frequencies, the commercial service will use the E6 band (1278.25 MHz), also composed of an in-phase signal incorporating navigation data and an in-quadrature phase pilot signal. The signals in the E6 band will be encrypted for restricted access.

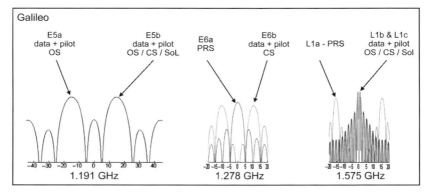

FIGURE 6.6 The Galileo navigation signal allocation: frequencies and services. (*Source*: GJU.)

Safety of Life Service (SoL)

The SoL frequencies are identical to the ones allocated to the OS, but Galileo will provide specific integrity features together with back-up service provision in case of poor conditions (in case a frequency no longer operates, for example).

Public Regulated Service (PRS)

The signals are separated from the OS frequencies and modulations in order to be able to still deliver this service in case of high interference conditions. This means that codes will be much longer, probably encrypted, and exhibiting a larger bandwidth in order to provide a much more reliable continuity of location service. The Interface Control Document concerning the PRS has not yet been released.

Search and Rescue Service (SAR)

The search and rescue "repeater" of some Galileo satellites can detect alert messages sent by COSPAS-SARSAT beacons in the 406–406.1MHz band. This information is transmitted to terrestrial COSPAS-SARSAT stations through the use of a specific L6 downlink (1.544–1.545 GHz). Then, the COSPAT-SARSAT system takes over for recovery operations. In addition, the Galileo satellites concerned send a feedback message to the initial sender of the alert message in order to tell him the message is being processed. A few (not all) Galileo satellites will be SAR equipped.

Signals Spectrum

The 10 signals of Galileo are transmitted in the 1164–1215 MHz band for E5a and E5b, 1260–1300 MHz for E6 and 1559–1591 MHz for L1 (often referred to as E2–L1–E1 because the initial GPS L1 band was only 24 MHz wide). In addition, the L6 downlink for SAR lies in the 1544–1545 MHz band: this is the 11th signal. Figure 6.7 provides a simplified view of the signals (L6 not included).

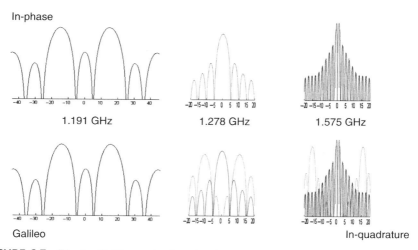

FIGURE 6.7   The ten Galileo navigation signals.

TABLE 6.1 Primary Galileo navigation signal parameters.

Signal	Channel	Modulation Type	Chip Rate (Mcps)	Symbol Rate (sps)	User Minimum Received Power[a] (dBW)
E5	E5a data	AltBOC(15,10)	10.23	50	−155
	E5a pilot			NA[b]	
	E5b data			250	−155
	E5b pilot			NA	
E6	E6-B data	BPSK(5)	5.115	1000	−155
	E6-C pilot			NA	
E1	E1-B data	BOC(1,1)	1.023	250	−157
	E1-C pilot			NA	

[a]For a satellite above $10°$ elevation and based on an ideally matched and isotropic 0 dBi antenna, and lossless atmosphere.
[b]NA, Not Applicable.
Source: GAL OS SIS ICD/D.0.

The data channels are represented in-phase and the pilot channels are given in in-quadrature phase. For all the ten signals, the multiple access technique used is CDMA. Long codes and short codes are designed to offer a wide range of performances and possible optimization in various environments. For instance, short codes allow a quick acquisition time, even if they are more sensitive to interference or in weak signal conditions. Long codes allow accurate correlation and robustness under weak conditions, but acquisition is more difficult. The combination of both codes has been achieved in Galileo (like modernized signals of GPS).

Concerning the modulations, for the three OS associated signals for which a draft version of the ICD was released in May 2006, Table 6.1 gives the considered choice (subject to change in a final version of the ICD). Besides the classical BPSK, both the BOC and a modified version of the BOC, the AltBOC have been selected. This latter allows a clever combination of the two E5a and E5b signals into a constant envelop signal that can be injected directly on a wide bandwidth channel. This signal can then easily be processed at the receiver's end.

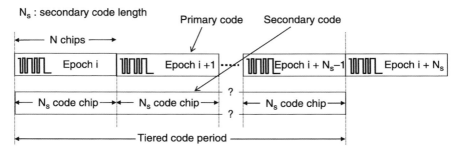

FIGURE 6.8 The codes interlink concept. (*Source*: GAL OS SIS ICD/D.0.)

TABLE 6.2 Code lengths assignment.

Channel	Code Length (ms)	Code Length (chips)	
		Primary	Secondary
E5a-I	20	10,230	20
E5a-Q	100	10,230	100
E5b-I	4	10,230	4
E5b-Q	100	10,230	100
E6a	—	—	—
E6b-I	1	5,115	—
E6b-Q	100	5,115	100
E1-B	4	4,092	—
E1-C	100	4,092	25

Source: GAL OS SIS ICD/D.0.

Note that an error detection and correction technique (Forward Error Correction, FEC) is implemented for the navigation message: each bit is encoded by a number of symbols (two in Galileo).

In addition, the PRS service should implement a BOC(10, 5) scheme with a chip rate of 5.115 Mega chips per second (Mcps). The corresponding symbol rate should be at 250 symbols per second (sps).

The codes (long and short) are designed to be embedded following the diagrammatic representation of Fig. 6.8.

Table 6.2 gives the detail of the code length considered in relation to frequency bands. A complete description of the codes is outside the scope of this book, but one has to note that all codes are once again fully available in the corresponding ICD, as with GPS and GLONASS.

The mathematical expressions of the Galileo signals are unnecessarily complicated for our purpose. Please refer to the corresponding ICDs for full details.

■ 6.2 SOME EXPLANATIONS OF THE CONCEPTS AND DETAILS OF THE CODES

It is of course fundamental to identify satellites and to get the constellation data into the receiver, but we must remember that the basic principle of satellite positioning is trilateration. Thus there should be a mechanism, within the signal, allowing basic propagation time measurements. This is also achieved through the use of codes, sometimes called telemetric codes, because they also have this functionality. The main idea is to make the correlation between the expected code and a local replica of it. This involves knowing the code one is seeking, and this is possible as all the codes are well assigned to the satellites and perfectly known.[6] Let us assume that

---

[6]This is applicable to GPS and Galileo, as there is only one code for GLONASS. The assignation table is available in the ICDs for GPS, but has not yet been fully released for Galileo (November 2007).

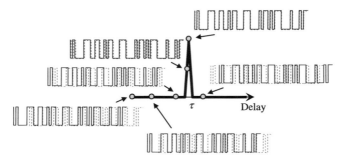

FIGURE 6.9   A typical GPS correlation function.

the receiver is seeking satellite "one" of code C1. The sequence being known, the receiver tries successively to correlate the incoming C1i ("i" for incoming) from the satellite with the local C1l ("l" for local). Of course, the receiver needs to shift the C1l in time in order to obtain the best coincidence with C1i. Once coincidence is obtained, that is, the correlation is at its maximum, the time shift of C1l gives the propagation time needed. A typical form of the GPS correlation function is given in Fig. 6.9. The dotted line code is C1i and the black one C1l, which is progressively shifted in order to reach the correlation peak. The corresponding time τ is the propagation time delay. Further details on correlations are given in Chapter 8, together with location calculations.

The correlation function of Galileo codes is a little bit different in form (as described in Chapter 7), but the main idea is identical.

The next paragraphs are devoted to explanations of codes, frequencies, navigation messages, and multiple access scheme choices and modulations.

Reasons for Different Codes

Codes are thus mainly required to carry out time measurements. For GPS and Galileo, they are also used for identification purposes, but not for GLONASS, where the frequency is used to identify the satellite. In the case of GPS and Galileo, as all satellites use the same frequency, there is the need to avoid misunderstanding of a satellite's identifier. Thus, the codes should be of better "quality" than within GLONASS. A code is mainly characterized by the shape of the correlation function and the shape of the cross-correlation functions. The first should present an abrupt peak in order to clearly identify the delay, and the second one, obtained when trying to correlate two different codes, should exhibit a nearly zero response curve. Thus, only very specific codes optimally provide such characteristics. A complete study, carried out in the 1960s by Gold,[7] has been used to reserve 36 C/A codes (and 37 P codes) for the GPS satellites.

---

[7]Gold drew up the tables of all the combinations of the linear feedback shift register of various lengths. He also showed that combinations of such codes could provide much better correlation and cross-correlation performances. Thus GPS used Gold codes.

The data rate of the GPS C/A codes is 1.023 MHz, while that for GLONASS is 0.511 MHz. This means that chips[8] are output at this frequency. The size of the code (its length) is given by the size of the shift registers used to implement it. In the case of GPS, these is a 1023 chip length for the C/A codes, and 511 chips for the GLONASS C/A codes. The Galileo codes are of similar shapes, although sometimes longer. This clearly means that the signal is a continuous repetition of a code sequence of 1 ms or so.[9]

Two different codes were initially used in both GPS and GLONASS: the C/A code just described, and the P code. The P code is a much longer sequence at a higher data rate (typically ten times higher than the C/A codes, that is, 10.23 MHz for GPS and 5.11 MHz for GLONASS) for both GPS and GLONASS. There are two different reasons for these very long codes. The first concerns both constellations: the P code, although fully described in the respective ICDs, is supposed to be reserved for authorized people only. How could this be achieved if the codes are known? "Simply" by having the code so long that it would be almost impossible to find the current running chip without the precise knowledge of a given starting sequence (at a given time). For example, the GPS P code is 266 days long at a chip rate of 10.23 MHz. The number of chips is greater than 2.35×10^{14}! Furthermore, the sequence is periodically reset at midnight every Saturday. The problem is that, when not authorized, one does not know the starting point within the whole sequence: this is the way protection is achieved, although codes are known. In addition, the P code can be encrypted, leading to the Y code. This feature allows additional protection.

The second reason for long codes is linked to the correlation, cross-correlation, and immunity from interference domains. It concerns GPS, GLONASS, and Galileo, but we shall use GPS and Galileo for the explanation because of the CDMA scheme implemented. In fact, when using CDMA, the main idea is to allow each satellite to use the complete bandwidth in order to allow more efficient and powerful modulations. This is obtained by so-called "spread spectrum" techniques, consisting of using the coding, at the transmission's end, to spread the signal over the whole bandwidth. The CDMA therefore presents a great advantage in terms of immunity to interference, because the correlation function achieved at the receiver end will mathematically "de-spread" the coded signal (the one from the satellite), but also spread the interference one. Thus, the interferer will be naturally rejected. Figure 6.10 gives a picture of this mechanism.

The problem is that each signal can be considered as noise before receiver correlation. If too many signals are incoming, then the correlation peak might no longer be observable. The "processing gain," as indicated in Fig. 6.10, corresponds to the correlation gain provided by the association between the code and the correlation function. This gain greatly depends on the length of the code.[10] Very long codes provide a very high processing gain that involves a high noise resistance (military designed codes should, of course, exhibit such a feature).

---

[8]The term "chip" is used instead of "bit" to show that it does not carry any pieces of information.

[9]Note that 1 ms or so is less than the 64–96 ms needed for the signal to reach the receiver from the satellites. So, the propagation time is made up of many code lengths. This aspect will be dealt with in Chapter 8 when explaining position calculations.

[10]Typically $10 \times \log N$ where N is the length of the code.

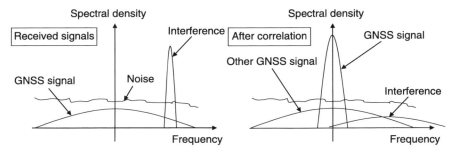

FIGURE 6.10   The natural immunity of the spread spectrum to interference.

Reasons for Different Frequencies

The first GPS satellites of the Block IIR-M were launched in September 2005 and are implementing L1 C/A, L1 P, L2 P, as for all the satellites, but also the L2 C, L1 M, and L2 M signals. From the original three signals, satellites are now transmitting six signals (three in each frequency band), two of which are available for civilian users. Meanwhile, GLONASS-M satellites, launched from 2003, add an L2 C signal to the previously existing L1 C/A, L1 P, and L2 P. The GLONASS-K, whose first launches are planned in 2008, will add a third civil frequency L3. Nevertheless, dual-frequency receivers already exist. For instance, surveyors, who have nothing to do with military applications, have been using such receivers for years — this was possible through the use of L1 and L2 carrier phase measurements (and not code phase measurements). So, dual-frequency civilian receivers are not based on code measurements. The first Galileo experimental satellite, GIOVE-A, is already transmitting three frequencies, although not simultaneously, and the future definitive satellites will transmit ten signals in three frequency bands: E5, E6, and L1.

As already mentioned several times, the main advantage of multifrequency signals arises from the possibility to remove the ionosphere-related propagation time error, because the induced delay is inversely proportional to the square of frequency. It is then possible to obtain a so-called "ionosphere-free" set of equations. The idea is to carry out propagation time measurements from the satellite to the receiver for two (or more) frequencies. The fundamental point here is that the signals are perfectly synchronized at the satellite's end. Thus, at the receiver's end, the difference in arrival time can be measured. The first assumption made is that both signals follow the same path. The second assumption is that, regarding propagation, it is only the ionosphere contribution that depends on frequency. Thus, the difference, with regard to the frequency, gives the ionosphere delay. The mathematical expressions are given by

$$\Delta t = \frac{40.3}{cf^2} \int N \, ds \rightarrow \begin{cases} \rho_1 = \rho + \dfrac{40.3}{cf_1^2} \int N \, ds \\[2mm] \rho_2 = \rho + \dfrac{40.3}{cf_2^2} \int N \, ds \end{cases} \tag{6.1}$$

$$\rho = \frac{\rho_2 - \gamma \rho_1}{1 - \gamma}, \qquad \text{where } \gamma = \left(\frac{f_1}{f_2}\right)^2 \tag{6.2}$$

and where Δt is the additional delay due to ionosphere propagation conditions, c is the speed of light, and N characterizes the number of ionized particles that causes the delay. $\int N ds$ is the integration of N along the signal path "ds". Thus the pseudo ranges for frequencies 1 and 2, ρ_1 and ρ_2, are given with respect to the "real" pseudo range and the respective frequencies f_1 and f_2. It is then possible to calculate the iono-sphere-free pseudo range given by the last equation. From these equations appears the fact that the value of γ is of importance. For instance, a value near to one is not really acceptable as the ionosphere-free pseudo range will then be very sensitive to measure-ment errors (because of the $(1 - \gamma)$ coefficient, near to zero).

Thus, the expected increase of frequency bands, including for civilian users, would be a real advantage for positioning accuracy, because the ionosphere-induced delay is the most important item in the global error budget. Nevertheless, one should choose the right couple of frequencies: Galileo E5a and E5b, for instance, are not really a very good choice.

Another interest in multifrequency signals lies in the availability of multiple channels for single-frequency positioning. This allows positioning on L2 when L1 is no longer available, either because of a system breakdown or intentional pertur-bations. Again, the initial choice of providing this feature only for military appli-cations was clearly stated. However, as the number of such channels increases, the global system's reliability is also increased (from the positioning capability point of view).

As a conclusion to this paragraph, let us come back to the remark concerning the fact that dual-frequency receivers have been available for years for civilian users. This is true when considering phase measurements, as in that case the receiver is not dealing with codes but with the carrier frequency. This technique has been developed in order to allow high accuracy positioning, that is, centimeter or even millimeter range accuracy, while not having access to the P codes. With this level of accuracy one must get rid of ionosphere-induced delays,[11] and a dual-frequency approach is usually required.[12]

Reasons for a Navigation Message

Data required in order to calculate a position are of different types: measure-ments, typically the pseudo ranges, but also data such as satellite locations at time of transmission, clock synchronization between satellites, relative locations of given satellites, and so on. Once all these data are gathered, the calculation can be carried out. Concerning measurements, Chapter 8 will describe the way a receiver obtains them. For all the other data, concerning the state and behavior of the space segment, the satellites send out a navigation message (through a permanent link to the ground segment). It includes all the data available for orbit calculation, satellite clock corrections, health of signals, and so on. This is transmitted through the

---

[11]Not only ionosphere delays.

[12]Note that this remark holds only for positioning using outdoor propagation. Refer to Chapter 10 for description of phase indoor positioning based on pseudolites: there is no need for dual frequency in this specific case.

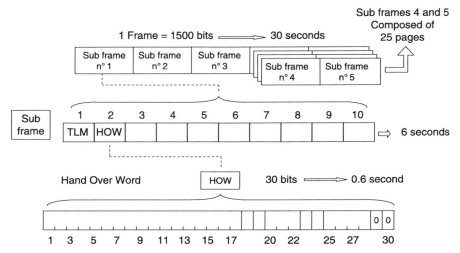

FIGURE 6.11 The GPS navigation message structure.

already mentioned navigation message, at a data rate of 50 bits/second for GPS and GLONASS. Furthermore, it is included on all signals, except for so-called "pilot tones."

Regarding GPS, Fig. 6.11 shows the various elements included. It is composed of 25 pages (or frames) of 1500 bits each (5 sub-frames of 300 bits). The three first sub-frames are identical for all 25 pages, only the last two sub-frames being different from one page to the next. The first three sub-frames are specific to the satellite that the receiver is currently decoding, while the last two sub-frames are relative to the whole constellation. Within these three sub-frames, satellite clock correction parameters, as well as ephemeris[13] data are included. In such a way, the receiver needs 30 s to obtain the entire set of required parameters in order to carry out position calculations, once the satellite is acquired.[14]

The last two sub-frames include data that allow the optimization of the ionosphere propagation correction, time transfer data from GPS time to UTC[15] and an almanac for all the satellites. Almanacs are indeed raw ephemeris, required by the receiver in order to evaluate the satellites it has to start searching for when switched on. In fact, the receiver makes the assumption that its local time is right and that it is in a similar place to where it was the last time it was switched off. Then, using the stored

---

[13]Ephemeris data allow the receiver to make a precise calculation of the satellite location versus time. This enables the calculation of the location of the satellite at the time of transmission, using the same data for all the receivers, wherever they are located and whatever the propagation time from the satellite.

[14]Thus the reader can understand the typical Time To First Fix (TTFF) values of around 45 s for non-aided receivers in a warm start configuration (meaning ephemeris stored within the receiver are too old to be considered for calculations). The few seconds required to acquire enough satellites plus the 30 s devoted to the navigation data retrieval plus finally the time needed for position calculations.

[15]UTC stands for Universal Time Coordinated.

almanac, if any, it can define the constellation representation and start searching for "probable" satellites in view. Of course, if either the almanac is not available or the above assumptions are not true, then a complete first search algorithm would be required. This feature is intended to reduce the time needed for the first fix. Please refer to Chapter 10 for recent approaches implemented in order to further decrease this time (Assisted-GNSS).

The case of GLONASS is roughly the same as GPS, but the navigation message is of course different. The navigation message includes immediate and non-immediate data. The immediate data are related to the satellite currently transmitting data and are as following:

- Time marks;
- Synchronization difference between satellite clock and GLONASS time;
- Correction on the satellite frequency with regards to the theoretical one (due to the FDMA scheme addressed in GLONASS);
- The satellite ephemeris.

The non-immediate data include the almanac data:

- Health of the satellite;
- Raw clock corrections to the GLONASS time;
- Orbit parameters for all the satellites of the constellation (orbit almanac)
- The GLONASS time correction relative to UTC (SU).

The navigation message comprises a succession of super-frames. A super-frame lasts 2.5 min and is made up of 5 frames of 30 s. Each frame is composed of 15 strings of 2 s. The immediate data are fully transmitted in every frame (see Table 6.3 for details).

The navigation message of Galileo is split into three different messages: the F/NAV (Freely Accessible Navigation message), the I/NAV (Integrity Navigation message), and the C/NAV (Commercial Navigation message). The latter has not yet been released. The F/NAV is transmitted for the OS on E5a-I, and I/NAV is transmitted for the OS, CS, and SoL on E5b-I and E1-B. The navigation message is transmitted on each data channel as a sequence of frames. A frame is composed of several sub-frames, and a sub-frame is composed of several pages. As an example, the F/NAV message structure (see Table 6.4) is a frame that lasts 600 s, is composed of 12 sub-frames of 50 s each, a sub-frame being composed of 5 pages of 10 s each. Within a given sub-frame, the first page is devoted to satellite clock correction, ionosphere data, validation status, and some other parameters that allow the receiver to carry out a quick location calculation. The second, third, and fourth pages are devoted to the ephemeris data, together with the GPS-to-Galileo synchronization data. This feature is very important when thinking of the multiconstellation positioning concepts where the calculations are carried out with, say, two GPS satellites and two Galileo satellites. These constellation time synchronization data allow such calculations without the need to introduce a new unknown variable (the constellation time bias). With GLONASS, this can only be indirectly achieved by obtaining

TABLE 6.3 The GLONASS navigation message structure.

Frame Number	String Numbers	
I	1–3	Immediate data for transmitting satellite
	4–15	Non-immediate data (almanac) for five satellites
II	1–3	Immediate data for transmitting satellite
	4–15	Non-immediate data (almanac) for five satellites
III	1–3	Immediate data for transmitting satellite
	4–15	Non-immediate data (almanac) for five satellites
IV	1–3	Immediate data for transmitting satellite
	4–15	Non-immediate data (almanac) for five satellites
V	1–3	Immediate data for transmitting satellite
	4–15	Non-immediate data (almanac) for five satellites

the GLONASS time bias relative to the UTC. The fifth page is the almanac for the rest of the constellation and changes while the 12-page sub-frames are being transmitted.

Possible Choices for Multiple Access and Modulations Schemes

GPS and Galileo on the one hand and GLONASS on the other have made different choices concerning the multiple access scheme used (CDMA for GPS and Galileo and FDMA for GLONASS), and have made various choices of modulation schemes (BPSK for both GPS and GLONASS for the early L1 and L2 signals, and BPSK, BOC, and AltBOC schemes for Galileo[16]). Although such choices have to integrate a lot of constraints (economic, technical, strategic, and so on), it is a pity for the users' communities that the same choice has not been made concerning the multiple access technique. The main reason concerns the receivers; a GLONASS receiver should have quite different hardware than a GPS or Galileo receiver. Current GPS/GLONASS constellation receivers are in fact the integration of two different receivers into a single case. Of course, this still allows integrated calculations where signals of both constellations can be used together, but this integration is not possible within the hardware. Discussion between the United States and the European Union have taken this very important point into account in order to define "compatible" signals, that is, signals compatible at the receiver end.

TDMA is not a good choice, because the transmission of satellite signals must be continuous to provide permanent ranging capabilities. So, the two remaining possible choices were FDMA and CDMA. Note that the current most advanced telecommunication systems are implementing, in one way or another, CDMA-based schemes.

---

[16]Note that the GPS M-Code also implements a BOC modulation. It is also foreseen to use a BOC(1,1) modulation for the future L1C signal.

TABLE 6.4 The Galileo F/NAV navigation message structure.[a]

	Page Type	Page Content
Sub-frame 1	1	SVID, clock correction, SISA, ionospheric correction, BGD, signal health status, GST and data validity status
	2	Ephemeris (1/3) and GST
	3	Ephemeris (2/3) and GST
	4	Ephemeris (3/3), GST–UTC conversion, GST–GPS conversion and TOW
	5	Almanac for satellite k and almanac for satellite $k + 1$ part 1
Sub-frame 2	1	SVID, clock correction, SISA, ionospheric correction, BGD, signal health status, GST and data validity status
	2	Ephemeris (1/3) and GST
	3	Ephemeris (2/3) and GST
	4	Ephemeris (3/3), GST–UTC conversion, GST–GPS conversion and TOW
	5	Almanac for satellite $k + 1$ part 2 and almanac for satellite $(k + 2)$
Sub-frame 3	1	SVID, clock correction, SISA, ionospheric correction, BGD, signal health status, GST and data validity status
	2	Ephemeris (1/3) and GST
	3	Ephemeris (2/3) and GST
	4	Ephemeris (3/3), GST–UTC conversion, GST–GPS conversion and TOW
	5	Almanac for satellite $k + 3$ and almanac for satellite $k + 4$ part 1
⋮	⋮	⋮

[a]Note that this is subject to change.
SVID, Space Vehicle IDentificator; SISA, Signal In Space Accuracy; BGD, Broadcast Group Delay; GST, Galileo System Time; UTC, Universal Time Coordinated; TOW, Time of Week.

Chapter 7 gives some elements concerning receiver architectures, but the mid-term future seems to be software receivers. Most of the processing is done by the software (signal processing), leaving the hardware to carry out the first conversion from physical signals on the antenna to the high rate digital sampling. In such a case, provided that this conversion shows sufficiently good performance, the above-mentioned limitation from the multiple access schemes will no longer really apply. Nevertheless, the agreement between the United States and the European Union concerning at least L1 is a good approach in the short term.

In order to share the same band, especially in L1, modulations have shifted from classical BPSK (which is still preserved) to modern Binary Offset Carrier (BOC) modulations. The corresponding spectrum of an L1 centered BOC is shifted to each side of the central frequency, allowing many different signals on the same central frequency. Thus, as described in the following sections, the L1 band will soon be occupied by GPS L1 C/A, GPS L1 P, GPS L1 M, Galileo OS, and Galileo PRS signals, all together. It is also planned to add a new GPS L1C signal (comparable to the OS Galileo one).

■ 6.3 MATHEMATICAL FORMULATION OF THE SIGNALS

Deriving from all the preceding discussions, it is possible to give the mathematical expressions of the various signals.

S1 (for L1 C/A and P codes) and S2 (for L2 P code) GPS are given by

$$S_1 = A_p(P \oplus D)\cos(2\pi L_1 t + \phi) + A_c(C/A \oplus D)\sin(2\pi L_1 t + \phi) \qquad (6.3)$$

$$S_2 = B_p(P \oplus D)\cos(2\pi L_2 t + \phi), \qquad (6.4)$$

where L1 is 1.57542 MHz, L2 is 1.2267 MHz, D is the navigation message data bits, P is the P-code chips, C/A the C/A code chips, A_p and A_c the respective amplitudes of the P and C/A codes on L1, and B_p the amplitude of P-code on L2. Note that the relative phases of P and C/A codes on both frequencies are identical because of the embedded synchronization, at the satellite end, of the various signal generators. The respective amplitudes are specified at -128.5 dBm[17] for C/A, -131.5 dBm for P on L1, and -131.5 dBm for P on L2. These figures are summarized in later sections. Note also that all the frequencies used to generate the GPS signals are derived from a single frequency ultra-stable oscillator (the atomic clocks) tuned at 10.23 MHz.

For GLONASS, the form of the signals is slightly different because of its use of the FDMA scheme. Conserving the same notations as for the GPS signals, the mathematical expressions for L1 and L2 signals are as follows:

$$S_{K1} = PD_p\cos(2\pi f_{K1}t + \phi) + (C/A)D_{C/A}\sin(2\pi f_{K1}t + \phi) \qquad (6.5)$$

$$S_{K2} = PD_p\cos(2\pi f_{K2}t + \phi). \qquad (6.6)$$

The C/A and P codes are the same for all satellites and, because an FDMA scheme is used, various frequencies are used for each satellite. D_p and $D_{C/A}$ are the navigation data added on the P and the C/A codes respectively. The nominal allocated frequencies, for each satellite K, in L1 and L2 are defined in Chapter 3, in the section on "GLONASS."

Note that K is provided in the non-immediate data in the navigation message, for all satellites. The power levels intended to be available on Earth assuming a 3 dBi gain receiving antenna, are -131 dBm on L1 and -137 dBm on L2.

The equivalent mathematical expressions for Galileo signals are too complex for our present purpose and can be found in the ICD.[18] We will just give some representation of the E5, E6, and L1 signals.

---

[17] A dBm is a unit of power. It is expressed in a logarithm scale with reference to one milliwatt. -130 dBm stands for 10^{-13} mW, which is quite small!

[18] The same applies to GPS L5, L2C, and M.

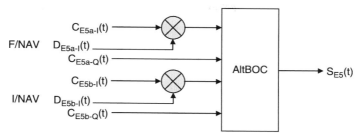

FIGURE 6.12 Galileo E5 signal components.

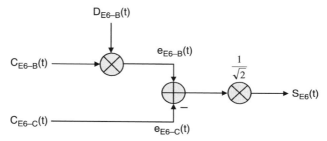

FIGURE 6.13 Galileo E6 signal components.

The E5 signal (see Fig. 6.12) includes the following components and is obtained through an AltBOC modulation:

- e_{E5a-I} from the F/NAV navigation data stream D_{E5a-I} modulated with the unencrypted ranging code C_{E5a-I};
- e_{E5a-Q} (pilot channel) from the unencrypted ranging code C_{E5a-Q};
- e_{E5b-I} from the I/NAV navigation data stream DE_{5b-I} modulated with the unencrypted ranging code CY;
- e_{E5b-Q} (pilot channel) from the unencrypted ranging code C_{E5b-Q}.

The E6 signal (see Fig. 6.13) includes the following components and is obtained through a BPSK modulation:

- e_{E6-B} from the C/NAV navigation data stream D_{E6-B} modulated with the ranging code C_{E6-B};
- e_{E6-C} (pilot channel) from the ranging code C_{E6-C}.

The E1 signal (see Fig. 6.14) includes the following components and is obtained through a BOC modulation:

- e_{E1-B} from the I/NAV navigation data stream D_{E1-B}, modulated with the ranging code C_{E1-B} and the sub-carrier sc_{E1-B};
- e_{E1-C} (pilot channel) from the ranging code C_{E1-C} modulated with the sub-carrier sc_{E1-C}.

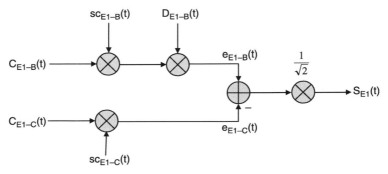

FIGURE 6.14 Galileo E1 signal components.

■ 6.4 SUMMARY AND COMPARISON OF THE THREE SYSTEMS

This section aims at providing the reader with a summary of figures and tables. Care must be taken when looking at these figures: they are only summaries of whole systems that are quite complex. Please do not consider that definitive conclusions can be drawn from these tables. Further analyses are required, depending on one's goals.

Reasons for Compatibility of Frequencies and Receivers

Please refer to Fig. 3.19 for a diagrammatic representation of all the future satellite-based navigation signals for GPS, GLONASS, and Galileo.

Only L1 is a full GPS and Galileo band, where L2 is GPS only, E5b is Galileo only, as is E6. E5a and L5 are common to GPS and Galileo. E5, and L5, as with L1, are Aeronautical Radio Navigation Service (ARNS) bands. Their main advantage results from the severe restriction for other systems to transmit spurious signals in these bands. L2 and E6 are not ARNS bands. The larger bandwidths of E5 and L5 are an advantage for tracking accuracy. Shared bands allow cost reduction when implementing multi-constellation receivers, but reduce the bandwidth allocated to each individual signal, leading to lower performances for each constellation taken independently.

Recap Tables

The following tables (Tables 6.5 and 6.6) are intended to present a summary of the various specifications of each constellation, showing the redundancies and complementarities of the constellations. Many elements are not yet fully defined, especially for future signals (L5 GPS, L3 GLONASS, PRS Galileo, and so on); thus, the reading of these tables must be undertaken carefully.

TABLE 6.5 **Major system parameters for GPS, GLONASS, and Galileo concerning the signals.**

	GPS	GLONASS	Galileo
Signals			
Satellite identification	Own code	Own frequency	Own code
Multiple access technique	CDMA	FDMA	CDMA
Frequencies/Bandwidth (MHz)			
Galileo E5a-I			1176.45/27.795
Galileo E5a-Q			1176.45/27.795
GPS L5 C-I	1176.45/27.795		
GPS L5 C-Q	1176.45/27.795		
GLONASS L3 C/A		1198–1203/14	
GLONASS L3 P		1198–1203/14	
Galileo E5b-I			1207.14/23.205
Galileo E5b-Q			1207.14/23.205
GPS L2 CM	1227.6/22		
GPS L2 CL	1227.6/22		
GPS L2 P	1227.6/22		
GPS L2 M	1227.6/22		
GLONASS L2 C/A		1243–1248/16	
GLONASS L2 P		1243–1248/16	
Galileo E6a			1278.75/40
Galileo E6b-I			1278.75/40
Galileo E6b-Q			1278.75/40
GPS L1 C/A	1575.42/24		
GPS L1 C	1575.42/24		
GPS L1 P	1575.42/24		
GPS L1 M	1575.42/24		
Galileo L1A			1575.42/32
Galileo L1B			1575.42/32
Galileo L1C			1575.42/32
GLONASS L1 C/A		1598–1605/18	
GLONASS L1 P		1598–1605/18	
Modulation			
Galileo E5a-I			AltBOC(15,10)
Galileo E5a-Q			AltBOC(15,10)
GPS L5 C-I	BPSK(10)		
GPS L5 C-Q	BPSK(10)		
GLONASS L3 C/A		BPSK(5.11)	
GLONASS L3 P		BPSK(5.11)	
Galileo E5b-I			AltBOC(15,10)
Galileo E5b-Q			AltBOC(15,10)
GPS L2 CM	BPSK(0.5115)		
GPS L2 CL	BPSK(0.5115)		

(Continued)

TABLE 6.5 *Continued*

	GPS	GLONASS	Galileo
GPS L2 P	BPSK(10)		
GPS L2 M	BOC(10,5)		
GLONASS L2 C/A		BPSK(0.511)	
GLONASS L2 P		BPSK(5.11)	
Galileo E6a			BOCcos(10,5)
Galileo E6b-I			BPSK(5)
Galileo E6b-Q			BPSK(5)
GPS L1 C/A	BPSK(1)		
GPS L1 C	BOC(1,1)		
GPS L1 P	BPSK(10)		
GPS L1 M	BOC(10,5)		
Galileo L1A			BOCcos(15,2.5)
Galileo L1B			MBOC(1,1)
Galileo L1C			MBOC(1,1)
GLONASS L1 C/A		BPSK(0.511)	
GLONASS L1 P		BPSK(5.11)	
	Power levels[a] (dBm)		
Galileo E5a-I			−125
Galileo E5a-Q			−125
GPS L5 C-I	−127.9		
GPS L5 C-Q	−127.9		
GLONASS L3 C/A		NA	
GLONASS L3 P		NA	
Galileo E5b-I			−125
Galileo E5b-Q			−125
GPS L2 CM	−130		
GPS L2 CL	−130		
GPS L2 P	−131.5		
GPS L2 M	NA		
GLONASS L2 C/A		−131	
GLONASS L2 P		−131	
Galileo E6a			−125
Galileo E6b-I			−125
Galileo E6b-Q			−125
GPS L1 C/A	−128.5		
GPS L1 C	−125.5		
GPS L1 P	−131.5		
GPS L1 M	NA		
Galileo L1A			−127
Galileo L1B			−127
Galileo L1C			−127
GLONASS L1 C/A		−137	
GLONASS L1 P		−137	

[a]For GPS, 5° elevation and 3 dBi linearly polarized antenna; for GLONASS, 5° elevation and 3 dBi linearly polarized antenna; for Galileo, 10° elevation and 0 dBi isotropic antenna.
NA, Not available.

TABLE 6.6 Major system parameters for GPS, GLONASS, and Galileo concerning the codes and navigation data.

	Code Length Chip Rate Secondary Code		
	GPS	GLONASS	Galileo
Galileo E5a-I			10,230/10.23/Y
Galileo E5a-Q			10,230/10.23/Y
GPS L5 C-I	10,230/10.23/Y		
GPS L5 C-Q	10,230/10.23/Y		
GLONASS L3 C/A		NA/4.096/N	
GLONASS L3 P		NA/4.096/N	
Galileo E5b-I			10,230/10.23/Y
Galileo E5b-Q			10,230/10.23/Y
GPS L2 CM	10,230/0.5115/N		
GPS L2 CL	1.5s/0.5115/N		
GPS L2 P	7 days/10.23/N		
GPS L2 M	NA/5.115/NA		
GLONASS L2 C/A		511/0.511/N	
GLONASS L2 P		1 s/5.115/N	
Galileo E6a			NA/5.115/NA
Galileo E6b-I			NA/5.115/NA
Galileo E6b-Q			NA/5.115/NA
GPS L1 C/A	1023/1.023/N		
GPS L1 C	10230/NA/N A		
GPS L1 P	7 days/10.23/N		
GPS L1 M	NA/5.115/NA		
Galileo L1A			NA/2.5575/NA
Galileo L1B			4096/1.023/N
Galileo L1C			8192/1.023/Y
GLONASS L1 C/A		511/0.511/N	
GLONASS L1 P		1 s/5.115/N	

	Navigation Data (sps)		
	GPS	GLONASS	Galileo
Galileo E5a-I			50
Galileo E5a-Q			PILOT
GPS L5 C-I	100		
GPS L5 C-Q	PILOT		
GLONASS L3 C/A		NA	
GLONASS L3 P		NA	
Galileo E5b-I			250
Galileo E5b-Q			PILOT
GPS L2 CM	50		
GPS L2 CL	PILOT		
GPS L2 P	50 or N		
GPS L2 M	NA		
GLONASS L2 C/A		50	

(*Continued*)

TABLE 6.6 *Continued*

	Code Length Chip Rate Secondary Code		
	GPS	GLONASS	Galileo
GLONASS L2 P		50	
Galileo E6a			110
Galileo E6b-I			1000
Galileo E6b-Q			PILOT
GPS L1 C/A	50		
GPS L1 C	25		
GPS L1 P	50		
GPS L1 M	NA		
Galileo L1A			110
Galileo L1B			250
Galileo L1C			PILOT
GLONASS L1 C/A		50	
GLONASS L1 P		50	

NA, not available; Y, yes; N, no; sps, symbols per second.

Some values are subject to further definition or change. "Y" stands for Yes, "N" for No, and "NA" for either Not Applicable or Not Available, depending on the lines. "PILOT" stands for Pilot tones that do not include navigation data.

■ 6.5 DEVELOPMENTS

The number of satellites and the number of associated signals available in space over the years are interesting features for the users. Tables 6.7 and 6.8 provide these

TABLE 6.7 **Number of navigation satellites over the years.**

Year	GPS	GLONASS	Galileo	Total
1980	5			5
1990	12	14		26
1994	27	16		43
1996	24	22		46
1998	27	13		40
2000	28	11		39
2002	28	8		36
2004	29	11		40
2006	30	14	1 (GIOVE)	45
2008	**30**	**18**	**4**	**52**
2011	**30**	**24**	**21**	**75**
2014	**30**	**24**	**30**	**84**

Numbers in bold type indicate predicted values.

TABLE 6.8 **Number of navigation signals over the years.**

Year	GPS	GLONASS	Galileo	Total
1980	3			3
1990	3	3		6
1994	3	3		6
1996	3	3		6
1998	3	3		6
2000	3	3		6
2002	3	3		6
2004	3	4 (L2C)		7
2006	6 (L2C, M)	4	3 (GIOVE-A)	13
2008	**6**	**6 (L3)**	**4 (GIOVE-B)**	**16**
2011	**8 (L5)**	**6**	**10**	**24**
2014	**9 (L1C)**	**6**	**10**	**25**

Numbers in bold type indicated predicted data.

figures. Anticipated deployments are given in the first columns relating to the three constellations. The early TRANSIT satellites, like Parus and Tsikada, are not included in the tables. Note also that the Satellite Based Augmentation Systems (SBAS) are not included, although some such satellites do enable ranging as well.

In Table 6.8, the number of signals for each constellation is the total number of transmitted signals, irrespective of the fact they are civil, military, or reserved. In parentheses are the names of the newly added codes. Care must be taken when analyzing this table. For instance, there were three signals at the early stage of GPS, only one being civil. But civilian receivers using phase measurements on L1 and L2 were available. Table 6.8 has counted only one signal in this case.

■ 6.6 ERROR SOURCES

As soon as a physical measurement is carried out, the accuracy of this measurement should be evaluated. In the case of positioning, this is also true, because these measurements are the foundations of the location calculation. Furthermore, the various descriptions presented so far have dealt with a "perfect" signal and theoretical assumptions that no errors degrade the signals. This is of course untrue. Let us consider the diagrammatic representation of a satellite-to-receiver link as given in Fig. 5.12.

The error sources can be split into three categories, depending on where they take place[19]: errors due to satellite-based uncertainties, errors due to signal propagation, and errors due to receiver-based uncertainties. On the satellite side, both clock synchronization biases (with reference to constellation time) and satellite location accuracy are sure to generate final positioning errors. On the receiver side, clock bias and also antenna center of phase location are sources of

---

[19]Note that there are of course other ways to represent error sources.

positioning inaccuracy. Finally, a very strong assumption has been made until now — that signal propagation is achieved at a constant speed, namely the speed of light. This is clearly not true when the signal is traveling through the atmosphere and this aspect has to be taken into account. Also in the propagation domain, the problem of multi-reflected signal path, called multipath, is certain to increase the actual length of the path from satellite to receiver, leading once again to resulting positioning inaccuracy.

In addition to these physical errors, a voluntary noise can be introduced, in some constellations, by the system management. This was the case for the GPS system with its so-called Selective Availability (SA). The main idea was to generate intentional error sources in both the time synchronization data of the satellites and the ephemeris data. In such a way, the calculations carried out at the receiver end were affected by this intentional error. GPS signals were affected by the SA until May 1, 2000, when the United States decided to withdraw it.[20]

Impact of an Error in Pseudo Ranges

All the abovementioned errors lead to an incorrect estimation of the corresponding pseudo ranges. The calculation method will be described in Chapter 8, but it is quite easy to understand that any pseudo range error leads to a positioning error.[21] There are also three possible classifications:

- **Synchronization Errors.** These occur either at the satellite end or at the receiver end, and cause the transmitting time and the receiving time not to be considered in the same time reference frame.

- **Propagation Errors.** These cause the signal flight time from the satellite to the receiver to appear different from what it actually is. The two effects concerned are the atmosphere propagation physics and the multipath. In the first case, the information transmitted is slowed down while crossing some layers, which should be taken into account. In the second case, the signal is lengthened. Note that both effects lead to an increased value of the pseudo range.

- **Location Errors.** These either relate to the satellite or to the receiver. When looking at the position calculation, the locations of satellites are required data. The receiver knows these locations through the use of ephemeris data sent by the navigation message. The accuracy of these satellite location calculations has a direct influence. There is also a problem concerning the real physical location of the receiver, which is calculated through positioning. This sounds strange, as its location is precisely what is being looked for. Indeed, when dealing with high precision positioning, that is, centimeter accuracy, one should think of the real significance of this centimeter. In particular, the size of the receiver is much larger than a centimeter, so where is the location

---

[20]See Chapter 3 for further comments.

[21]Although the link between one or a few pseudo ranges' errors and positioning error is not direct.

point to be considered? The answer is a specific point on the receiving antenna: the center of phase. Thus, one has to be aware of the real center of phase location of the antenna in order to use the positioning result.[22]

In addition, one should include all the noise-like errors such as internal receiver noise, thermal noise, and so on.

Time Synchronization Related Errors

The problem of synchronization has to be clearly distinguished from those of precision or stability. The satellite-based time, driven by atomic clocks for each constellation, is very accurate. One could have imagined some receivers also to be equipped with atomic clocks. The time flow would have been very precise too. However, synchronization is related to the fact that all the system components deal with a common timescale; this is of primary importance when carrying out time measurements in the GNSS fields. One has to keep in mind the fact that 1 ns is equivalent to 30 cm.[23] It is then necessary to know the satellites' time drift and the synchronization of the receiver must be very . . . accurate!

The problems of satellite synchronization and receiver synchronization are very different, as are the techniques implemented to deal with these problems. At the satellite end, we have already mentioned that atomic clocks are used, but in order to achieve synchronization, the ground segment predicts both the satellite's position and clock at future times and estimates the accuracy of these predictions. Specific computations are carried out (Kalman filtering) in order to improve the accuracy of this extracted data, used to update the navigation message. In such a way, the ground segment monitors the satellites' clocks and uploads the corresponding parameters to each satellite. Thus, the synchronization of satellites' clocks to GPS time is obtained from the navigation message. The problem of receiver synchronization is even more crucial as it must be carried out on a frequent time base, the receiver not being equipped with atomic clocks.

The solution to this fundamental question is achieved in a clever way: the bias of the receiver's clock is considered as being an unknown variable in the positioning problem of the receiver. Indeed, the solution vector of three-dimensional positioning is made up of three spatial variables, typically X, Y, and Z, or longitude, latitude, and altitude, plus one temporal variable, the receiver's clock bias with respect to the GPS time. Of course, an additional unknown means that an additional measurement will be required,[24] but this also allows the constraints on the receiver clock to be loosened. This feature was required to open the way to affordable mass market receivers.

---

[22]Note that the same remark applies to the satellite antenna. The satellite location obtained from the ephemeris data is bound to be the satellite antenna center of phase location.

[23]Even with 10^{-14} relative time accuracy, this means that the potential time bias per day is around 9 ns. This has to be taken into account.

[24]Please refer to Chapter 8 for solution details.

Propagation-Related Errors

While traveling from the satellite to the receiver, the signal has to cross various layers of the atmosphere. This crossing induces two effects: the slowing down of the signal and the deviation of its path. The second effect is not considered in the following. The first effect, however, is fundamental and must be eliminated, either by dual-frequency analysis[25] (for ionosphere-related effects) or by appropriate modeling.

Ionosphere Propagation

The ionosphere is an atmospheric layer that lies typically between 50 and 1,000 km of altitude. This is a dispersive medium,[26] ionized through the action of the Sun's radiation. The corresponding thermal agitation of the ionosphere depends also on the latitude of the observation point (it is more agitated at the poles and at the equator) and on the Sun's activity (pseudo period of 11 years). Furthermore, the time of the year, seasons, and time of day also have a high impact on both the height of the ionosphere layer and the density of its ionized particles. These physical parameters have a direct influence on propagation delays. The corresponding pseudo range errors vary then from 0 to 15 m at zenithal incidence, and as much as 45 m for low incidence signals. This means in particular that the delays are highly different from one satellite to another, given their relative locations with respect to the receiver's location.

The solutions to this problem adopted to date are either to use dual-frequency receivers, as stated in the section "Reasons for Different Frequencies," which are rather expensive, or to implement specific modeling. In this latter case, residual errors are bound to remain. The navigation message is also in charge of providing propagation parameters associated with the Klobuchar model for GPS, GLONASS, and also Galileo.

For local differential positioning,[27] one has to distinguish the techniques where the distance between both receivers is less than 10 km where it can be considered that the path from the satellite to the receivers is almost identical,[28] and the case where the distance is greater. In this latter situation, dual-frequency receivers are required.

Troposphere Propagation

The troposphere is the lowest layer of the atmosphere, and is directly in contact with the Earth's surface. Its height varies from 7 to 14 km, depending on the observation point. This is a non-dispersive medium for frequencies under 30 GHz. The induced effect on pseudo range of this layer varies from 2 m if the satellite is at the zenith to 30 m for a 5° elevation satellite. This delay[29] depends on temperature, pressure, humidity, and elevation of satellite.

---

[25]Refer to Section 6.2, subsection "Reasons for Different Frequencies," for details.

[26]Dispersive means that the behavior depends on the frequency of the signal.

[27]Refer to Chapter 5 for a description of the differential approaches.

[28]This value obviously depends on real conditions: this figure is given as a typical value.

[29]As usual, delays (time) and pseudo ranges (distance) are indiscriminately used to qualify the same quantities.

The troposphere delay is mainly affected by the altitude of the receiver and is usually considered as being the combination of a dry component and a wet component, the dry one being about 90% of the total. The modeling of this latter component is quite good. The wet component depends on local atmospheric conditions and varies very quickly. Local measurements could help in achieving better accuracy, but are rather difficult to implement. Thus, models are usually also used for this second component.

The differential technique could be used for receivers close to each other when it is possible to consider equivalent meteorological conditions. Nevertheless, the user should check for very close altitude values of both receivers in order to achieve acceptable differential elimination of the troposphere delays.

The modeling currently used for GPS, GLONASS, and Galileo is based on Saastamoinen and Hopfield works and such parameters are included at the receiver level. Other models exist and can be used for specific purposes, such as the exponential model.

Multipath

In the propagation domain, it is usual to distinguish between two configurations when considering the radio link between a transmitter and a receiver: the Line of Sight (LOS) configuration where there is a direct geometrical non-obstructed path, and the Non Line of Sight (NLOS) configuration, where no such direct path exists. When considering a NLOS situation in satellite-based navigation, it appears that as the receiver is fundamentally based on the hypothesis that the signal it is receiving is LOS, the resulting positioning will be erroneous. However, even when considering LOS configuration, there may also be a reflected path that is, for example, obtained by reflection of the satellite signal on metallic surfaces such as building facades. Figure 6.15 shows such reflected signals.

In such a case, the incident signal on the receiving antenna will in fact be the time combination of all the incoming signals, namely the LOS one but also the reflected ones. As the receiver has not been designed to have any knowledge of its environment, it assumes that the signal received is LOS. The real problem is in fact the time resolution of the receiver. If it is able to discriminate between two signals delayed by an amount of time dt, then the interference effect is reduced to reflected signals that reach the receiving antenna less than dt after the LOS did. Unfortunately, the discrimination capabilities of a code correlation based receiver are not infinite. A typical value is given by the correlation spacing in Early-Late- based correlations.[30] This leads to values in the order of $10-15$ m of remaining errors for so-called narrow correlation and $100-150$m for one chip spaced correlators. Specific correlation configurations have been developed by almost all the receiver manufacturers to reduce multipath impact on the resulting positioning. Figure 6.16 shows the typical form of multipath performance representation for various correlators. It gives the pseudo range error (in meters) on the y-axis, in terms of reflected delay, as compared to LOS, on the x-axis (in chips).

---

[30]Refer to Chapter 7 for details on the correlation architecture of receivers.

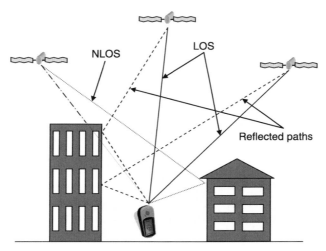

FIGURE 6.15 A diagrammatic representation of a satellite-to-receiver radio link.

It can be seen that performances of narrow correlators are much better than those of wide correlators, but also that other techniques such as edge correlators are even better. These techniques are typical signal processing approaches, as compared to "hardware" solutions consisting in reducing the amount of reflected signal received by the antenna. These latter methods can only be carried out when the environment

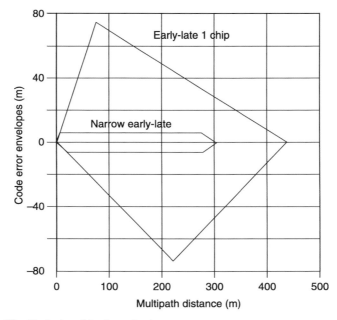

FIGURE 6.16 Typical multipath evaluation curve.

FIGURE 6.17 A choke-ring antenna. (*Source*: Used with permission, © 2007 Magellan Navigation, Inc.)

is known or specific. This is for example the case for ground-based reflection that can be mitigated using so-called Choke-Ring antenna designed in order to trap the waves reflected by the ground (see Fig. 6.17). A long-term analysis of the so-called residuals can also help in eliminating the multipath induced error; this approach, clearly not possible for real-time application, is reserved for specific scientific purposes.

The case of phase-based receivers is even more complicated, because direct and reflected signals are added to the antenna. The resulting signal exhibits a phase that can be quite different from the LOS one. Such effects have to be carefully analyzed when using phase measurements in multipath environments.

Location-Related Errors

Within the geometrical problem of trilateration, it is important, when trying to estimate the receiver's position, to know the satellite's position at time of signal transmission (that is, when the signal that reaches the receiver left the satellite).[31] The requirement is then to know, with the best possible accuracy, the position of the satellite. Unfortunately, it moves, rather quickly, on its orbit.[32] Furthermore, the satellite's orbit is determined by many effects that have to be calculated for a first estimation of the satellite's trajectory, in order to upload the ephemeris data in the navigation message. Measurements, usually achieved through the use of laser telemetry, allow an adjustment of the models. The various aspects to be considered are typically the Earth's, Moon's, and Sun's gravitational effects, but also the solar radiation and solar and lunar tidal effects.

---

[31] A more complete discussion on this subject is given in Chapter 5.
[32] This is easier in the case of terrestrial fixed transmitters such as pseudolites for instance.

TABLE 6.9 **Estimation of error sources on pseudo range.**[a]

	GPS	
	PPS	SPS
Synchronization errors		
Satellite (ephemeris)	3.0	3.0
Receiver (residual noise)	0.2	1.5
Propagation errors		
Ionosphere	2.3	4.9–9.8
Troposphere	2.0	2.0
Multipath	1.2	2.5
Location error		
Satellite location (ephemeris)	4.2	4.2
Other errors		
Ground/spatial/user (sum)	2.9	2.9

[a]All the values, given in meters, are typical one sigma values.
PPS, Precise Positioning Service; SPS, Standard Positioning Service.

The second error related to location applies to the location calculated by the receiver. Instinctively, when dealing with typical accuracy values of around a few meters, the question of knowing the exact position of the calculated point does not arise: it is assumed to be "the receiver's antenna" that is usually a few centimeters tall. It is even probable that a car driver does not even ask himself the question and assumes that this is the center of his car.[33] This question should be considered when dealing with high accuracy positioning. The physical point where signals are "collected" is the center of phase of the antenna. This immaterial point is very difficult to precisely define and also depends on the incident angle of arrival of the signal: thus, this point is not necessarily geometrically fixed. As this is not the physical center of the antenna, perfect calibration of the antenna is therefore required. When centimeter accuracy is needed, this is fundamental.

Estimation of Error Budget

The ultimate goal of the GNSS community is to reduce the various error effects on the receiver's positioning. A first analysis must give indications of the respective importance of each individual contribution in the global error budget. The assumption made here is that the budget on an individual pseudo range is directly linked to the corresponding receiver location (which is not completely true but is nevertheless an acceptable hypothesis). Table 6.9 gives an idea of respective values associated with different sources on a single pseudo range measurement. In addition to the

---

[33]It is also interesting to analyze the human perception of the accuracy: when the street number the driver is looking for is in his visual field and the receiver tells him he has arrived, the driver usually considers the system as "incredibly accurate."

errors described above, there is the need to draw a new line to incorporate all the non-predictable sources at the ground, spatial, and user segments.

From this first analysis, it appears that two items are the greatest contributors: ionosphere propagation modeling and ephemeris data, both for satellite clock synchronization and orbit determination.

These are domains where great efforts have been made in recent years. Although the solution to errors due to the ionosphere remains the dual-frequency approach, mass market receivers are bound to remain L1 only for some years. Thus, another solution had to be found in order to reduce the resulting impact of the ionosphere. Identically, the orbit determination of GPS and GLONASS satellites asks for a wider monitoring of the satellites than currently achieved by the ground segments. The way such requirements have been implemented is presented in the next section and relates to the Satellite-Based Augmentation Systems (SBAS): EGNOS, WAAS, MSAS, and some others.

SBAS Contribution to Error Mitigation

In this area, the navigation community has found some interesting solutions, one being satellite-based augmentation systems. For instance, the European Geostationary Navigation Overlay Service, EGNOS, is intended to provide the GPS and GLONASS systems with correction values for propagation errors as well as both orbits and clock bias of satellites. EGNOS is also a system that transmits differential corrections for pseudo ranges. It can moreover be used for ranging, as can any other GPS satellite. Finally, it provides a binary integrity indicator: "use/don't use."

Please refer to Chapters 3 and 5 for details concerning EGNOS and their worldwide equivalent, the WAAS in northern America, the MSAS in Japan, and GAGAN in India. Just remember that to achieve such a goal requires a large deployment of terrestrial ground stations in order to be able to develop new atmosphere models (or to improve the parameter values used in current standard models), and to improve the clock bias and orbit determination of each individual satellite. The corresponding correction parameters are then transmitted to the receivers, through a specific navigation message.

Dealing with the most important terms within the total budget error, SBAS are of great interest to single-frequency user communities. After acquiring and decoding the SBAS data, the multipath-related errors still remain, as well as those associated with the receiver.

▪ 6.7 TIME REFERENCE SYSTEMS

There are different scales related to time. This arises from the historical fact that when very stable atomic clocks were built, synchronization needs arose with respect to astronomical time, which had been used for centuries. The two corresponding timescales are the TAI (International Atomic Time) and the Universal Time Coordinated (UTC).

TAI was established by the Bureau International des Poids et Mesures (BIPM), a section of the Bureau International de l'Heure (BIH) in Paris and is the weighted mean value of more than 200 atomic clocks distributed throughout the world.[34] TAI is a continuous timescale that has a fundamental problem in practice: the Earth's rotation with respect to the Sun is slowing down by a variable amount that is currently about 1 s per year. Thus TAI could become inconveniently out of synchronization with the solar day. This problem has been overcome by introducing UTC, which runs at the same rate as TAI but is incremented by one second jumps ("leap seconds") when necessary. UTC is thus obtained by periodically adding or subtracting one second from TAI in order to build up a reference time that follows the Earth's rotation.

In addition to these times, the constellations have developed their own system times: GPS time (GPST), the GLONASS System Time (GLONASSST), and the Galileo System Time (GST). The GPST is generated from all the atomic clocks of the system, including those in the satellites. Nevertheless, this timescale is not totally free but is subjugated in the long term to UTC (USNO).[35] A regular monitoring of the GPS time with the UTC (USNO) allows a very tight link over the long term. For historical reasons, GPST has been shifted by 19 s from IAT and by 13 s from UTC (USNO).

The GLONASS System Time (GLONASSST) is generated on a similar basis and is related to UT(CIS).[36] UTC (CIS) is maintained by the Main Metrological Center of the Russian Time and Frequency Service (VNIIFTRI) at Mendeleevo in the Moscow region. When UTC is incremented or decremented, the GLONASSST is incremented or decremented too. Therefore, there is no integer-second difference between GLONASSST and UTC. However, there is a constant offset of 3 h between GLONASSST and UTC(CIS) due to features specific to GLONASS monitoring.[37]

The Galileo System Time (GST), in contrast to GPST, is produced only with terrestrial clocks available in the two redundant Precise Time Facilities (PTF) of Galileo. GST is optimized in order to achieve a very good short-term stability required for the functions associated with navigation (short-term means typically less than a day). The reason the space segment has not been considered has two aspects: the internal satellite clocks' frequency drift and the inherent measurement error associated with satellite clocks. Both aspects would have had the effect of decreasing GST accuracy. Thus, for metrological aspects, comparisons with clocks external to Galileo will be used.

---

[34]The fundamental unit of TAI is the "second," defined as "the duration of 9,192,631,770 periods of the radiation corresponding to the transition between two hyperfine levels of the ground state of the cesium 133 atom."

[35]UTC (USNO) is the UTC part, which is obtained through the use of all the United States Naval Observatory terrestrial clocks.

[36]Also referred to as UTC(SU).

[37]GlonassST = UTC + 3 h 0 min.

✅ 6.8 EXERCISES

Exercise 6.1 Describe the link between the receiver's immunity to noise and the code and modulation choices. Conclude briefly with a comparison of noise behavior for both C/A and P codes of GPS.

Exercise 6.2 GPS and GLONASS signals are different in the multiple access scheme implemented: CDMA for GPS and FDMA for GLONASS. Comment on the sensitivity of both constellations regarding intentional and non-intentional jamming. Is there a difference between civil and military signals concerning this matter? Please give details.

Exercise 6.3 The data contained in the navigation message are required in order to compute a location. Find some reasons why 50 Hz was chosen for GPS for C/A and P codes. Can you also comment on the discussions carried out when defining the future L1C signal data rate: the possible choices were 25, 50, or 100 Hz.

Exercise 6.4 What is the interest of primary and secondary codes?

Exercise 6.5 Give advantages and drawbacks of short and long codes.

Exercise 6.6 Pilot tones are dataless signals: explain how this absence of a navigation message is potentially a help in finding the signals in weak reception conditions.

Exercise 6.7 Detail the basics of a correlation process. Considering a 1 ms code length with 1023 chips (GPS C/A), give the typical form of the correlation function for the total duration needed for the signal to travel from the satellite to the receiver (consider for example a typical duration of 70 ms). What are the consequences of this form on the computation of a location?

Exercise 6.8 The interest of two frequencies relies clearly on ionosphere propagation error mitigation. What would be your choice, among all the signals that will be available in a few years, for these two frequencies? Please take into account the following aspects to strengthen your answer:

- Ease of making receivers;
- Efficiency of ionosphere error removal;
- Interoperability of constellations (if two or more constellations are considered).

Exercise 6.9 Regarding the various spectrums of GPS, GLONASS, and Galileo and taking into account the SBAS signals to come, what would be the best choice of frequency and bandwidth for a single-frequency receiver? Comment on the proposed system.

Exercise 6.10 After further reading, set out the state of the relative importance of the error sources and give mitigation approaches that could be followed for each type of error. From this analysis, propose your own improved system. Compare with the current developments of the GPS and Galileo constellations.

Exercise 6.11 What is the effect of the absence of a direct path from a satellite to the receiver? What could be the solutions to solve this problem?

Exercise 6.12 Explain why a narrow spacing correlation helps in decreasing the effect of a multipath on the resulting accuracy of a pseudo range measurement. Why is there a limitation to the reduction of this spacing?

■ BIBLIOGRAPHY

Betz JW, Blanco MA, Cahn CR, Dafesh PA, Hegarty CJ, Hudnut KW, Kasemsri V, Keegan R, Kovach K, Lenahan LS, Ma HH, Rushanan JJ, Sklar D, Stansell TA, Wang CC, Yi SK. Description of the L1C signal. In: IONGNSS 2006: Proceedings; Fort Worth (TX); September 2006.

Erhard P, Armengou-Miret E. Status and description of Galileo signals structure and frequency plan. European Space Agency Technical Note; April 2004.

Gold R. Optimal binary sequences for spread spectrum multiplexing. IEEE Transactions on Information Theory; October 1967; p 619–621.

Global positioning system standard positioning service. Performance Standard October; Assistant Secretary Of Defense For Command, Control, Communications, and Intelligence; 2001. Available at: http://www.navcen.uscg.gov/gps/geninfo/2001SPS PerformanceStandardFINAL.pdf. Nov 14, 2007.

GPS SPS signal specification. Annexes A–C; 1995. Available at: http://www.navcen.uscg.gov/pubs/gps/sigspec/gpsspsa.pdf. Nov 14, 2007.

GPS SPS signal specification. Main document; 1995. Available at: http://www.navcen.uscg.gov/pubs/gps/sigspec/gpssps1.pdf. Nov 14, 2007.

Hein G, Avila-Rodriguez J-A, Wallner S. The Galileo code and others. InsideGNSS 2006;1(6):62–75.

Interface control document (ICD 200c). Available at: http://www.navcen.uscg.gov/pubs/gps/icd200/icd200cw1234.pdf. Nov 14, 2007.

Kaplan ED, Hegarty C. Understanding GPS: principles and applications. 2nd ed. Artech House; 2006. Norwood, MA, USA.

Lo S, Chen A, Enge P, Gao G, Akos D, Issler J-L, Ries L, Grellier T, Dantepal J. GNSS album — images and spectral signatures of the new GNSS signals. InsideGNSS 2006;1(4):46–56.

McCaskill TB. Effect of broadcast and precise ephemerides on estimates of frequency stability of GPS Navstar clocks. International technical meeting; Washington: Institute of Navigation; September 1993.

Parkinson BW, Spilker Jr. JJ. Global positioning system: theory and applications. American Institute of Aeronautics and Astronautics; 1996.

Ponsonby JEB. Spectrum management and the impact of GLONASS and GPS satellite systems on Radio Astronomy. Journal of Royal Institute of Navigation 1992;44(3).

Progri IF, Bromberg MC, Michalson WR, Wang J. A theoretical survey of the spreading modulation of the new GPS signals (L1C, L2C, and L5). Available at: http://www.gmat.unsw.edu.au/snap/publications/progri_etal2007b.pdf. ION NTM 2007 Proceedings, pp 561–569.

Sarwat DV, Pursley MB. Cross-correlation properties of pseudorandom and related sequences. Proceedings of the IEEE; New York Institute of Electrical and Electronic Engineers; May 1980.

Taylor J, Barnes E. GPS current signal-in-space performance. ION 2005 Annual Technical Meeting; San Diego (CA); 2005.

Ward P. GPS receiver search techniques. In: IEEE PLANS: Proceedings; Atlanta (GA); 1996.

Web Links

http://tycho.usno.navy.mil/gps.html. Nov 14, 2007.

http://www.Colorado.EDU/geography/gcraft/notes/gps/gps.html. Nov 14, 2007.

http://www.navcen.uscg.gov/gps. Nov 14, 2007.

CHAPTER 7

Acquisition and Tracking of GNSS Signals

An understanding of the way signals are acquired and tracked helps in the understanding of the real performances of a system. In the case of GNSS, this means code transmission, code correlations, but also the way positioning is carried out. In addition, specific measurement techniques allowing a better accuracy are described, together with some details on the current differential networks available for real-time accuracy down to the centimeter level.

This chapter is divided into three parts: "transmission," including how codes and signals are generated at the satellite end, "reception," including receiver architectures and the main implementations, and "measurement," describing the various positioning techniques usually implemented. In addition, links are drawn between the "hardware" part (codes, signals, and architectures) and the "software" part (calculations and methods). This should help the reader to understand the real needs for positioning and the principal ways this has been achieved within GNSS.

◼ 7.1 TRANSMISSION PART

This first section is intended to provide a simple view of the signals transmitted by the various GNSS constellations. It will only be an overview; further readings are required for the reader to go deeper into the details.

Introduction

The fundamental difference between the multiple access techniques of GLONASS on the one hand (FDMA) and GPS and Galileo on the other hand (CDMA) lies in the code requirements. For CDMA, codes should exhibit a very

Global Positioning: Technologies and Performance. By Nel Samama
Copyright © 2008 John Wiley & Sons, Inc.

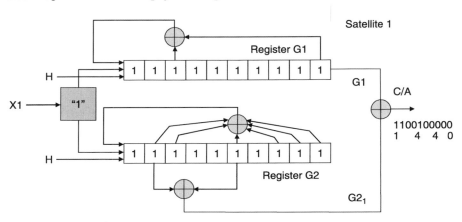

FIGURE 7.1   A C/A code implementation.

good characteristic for rejection of cross-correlation, whereas there is no such need for GLONASS. Therefore, the codes for GPS and Galileo are a little more sophisticated.

GPS

Two so-called PRN (Pseudo Random Noise) ranging codes are transmitted: the C/A code and the P code (which can be changed into the Y code for the case where the anti-spoofing (AS) mode is activated). From Block IIR-M, two additional codes are transmitted: the L2 civil-moderate (L2 CM) and the L2 civil-long (L2 CL). In subsequent satellites, an additional code will be available on L5.

For illustration purposes, let us deal with the C/A and P codes of GPS. Both are Gold codes, based on the combination of two (C/A) or four (P) maximal length linear feedback shift registers (LFSRs). A typical implementation for C/A code is given in Fig. 7.1. H is the clock, X1 a specific event within the whole transmission system that allows synchronization and that leads to the "all at 1" initialization of both G1 and G2 registers. Note that this feature is fundamental in many cases. It allows, for instance, the C/A and P codes to be perfectly synchronized. In addition, the navigation message is also synchronized in this way.

Thus, C/A code is a combination of two LFSRs of maximal length of ten bits each. At the 1.023 MHz clock rate, this means a 1 ms length for the complete code. For the P code, the philosophy is similar, but using four LFSRs of maximal length of 12 bits. This leads to a code with duration of 266 days (at a 10.23 MHz clock rate). Please refer to Chapter 6 for comments on this incredible length.

GLONASS

As mentioned, the GLONASS codes do not require the same cross-correlation features. The C/A code is indeed a maximal length LFSR of nine bits. The clock rate is 0.511 MHz (half that of GPS C/A code) and then lasts the same duration[1]: 1 ms

---

[1] Half the clock rate with one bit length less = the same duration.

(511 chips). This rather short code length allows fast acquisition, where only 511 code shifts are requested.

The P code is a military one and is not really described in any official documents, as opposed to its GPS equivalent, which is fully described. The clock rate is ten times that of the C/A code, namely 5.11 MHz (the same ratio between C/A and P as for GPS). The P code is a LFSR of maximal length of 25 bits, thus leading to a complete code length of 55,554,432 chips that lasts a total of 6.57 s. As for the GPS P code, the sequence is truncated to a shorter sequence. In the case of GLONASS, this latter sequence is 1 s in duration. The longer P code, with respect to the C/A one, allows better correlation characteristics, thus leading to improved tracking performance in noisy environments, but also means more complex acquisition as there are now, in 1 s, 5.11 million code shift possibilities. As for GPS, the algorithm consists of using first the C/A code for rapid acquisition and then the P code for fine tracking and correlation.

Galileo

The Galileo codes are also LFSR generated but are based on a structure with two successive layers consisting of a short-duration so-called "primary code," modulated (modulo two addition) by a long duration so-called "secondary code." The resulting code has an equivalent duration equal to the long code duration (secondary code) and then exhibits very good immunity to noise, still allowing fast raw acquisition using the primary sequences. This code structure is given in Fig. 6.8.

The primary codes are Gold codes of up to 25 bits, and the secondary codes are predefined sequences of lengths up to 100 chips. Table 7.1 summarizes the various lengths allocated to the different Galileo signals. Note that the so-called "pilot signals" (or "pilot tones") implement a 100 ms code length, except for the L1 signal without data, which has a duration of 25 ms, due to the reduced length of its primary code (only 4092 chips).

The associated characteristics of PRS are not given in the table because they are not available to the public. The various durations take into account the chip rates, which differ from one signal to another. For instance, the chip rate on L1 is

TABLE 7.1 Galileo code specifications.

Signal	Primary Code Length (chips)	Secondary Code Length (chips)	Total Duration (ms)
E5a data	10,230	20	20
E5a pilot	10,230	100	100
E5b data	10,230	4	4
E5b pilot	10,230	100	100
E6c data	5,115	NA	1
E6c pilot	5,115	100	100
L1 data	4,092	NA	4
L1 pilot	4,092	25	100

NA, Not Applicable.

1023 Mcps (Mega chips per second), leading to 4 ms for a full 4092 code length, as compared to the 20 ms of a complete sequence in the E5 band, where the chip rate is ten times higher, at 10,230 Mcps.

Structure and Generation of the Codes

The various signals available, or soon to be available in the near future, have been described in Chapters 3 and 6. The goal of this section is to illustrate the internal way the signal is constructed at the satellite end. As the aim is to describe the principles, GPS will be given as an example, for the three signals L1 C/A, L1 P, and L2 P.

The first step is the generation of the P code (Fig. 7.2), which will be used for synchronization purposes for all the code generation.

The second step (Fig. 7.3) consists of taking out a characteristic event of the P code, the X1 event, in order to achieve the global synchronization of all codes generated by the satellite. The code generation is also synchronized to the local atomic-clock-driven oscillator. Thus, the P code is generated (Fig. 7.4).

Both the X1 event and the oscillator are used in order to provide, through a "divide by ten" function, the Gold codes at the chip rate of 1.023 MHz that is required for the C/A code. It appears clearly that both P and C/A codes are thus perfectly synchronized. Furthermore, as will be described in the next paragraph, it appears that codes generated on the two frequencies are also perfectly synchronized at the transmitter end. This is of primary concern when dealing with dual-frequency ionosphere propagation, as a dual-frequency path is required that exhibits both the same transmitting location and the same transmitting time. See Fig. 7.5 for a diagrammatic representation.

Going a little further than just the codes, let us now shift to another important feature of GNSS: the navigation data. As previously described, the data rate is 50 Hz and there is a strong need for synchronization with the ranging codes for acquisition and tracking optimization. Figure 7.6 shows the way this new frequency is created from the C/A code and the so-called G event (the specific configuration of the C/A code that repeats every second).

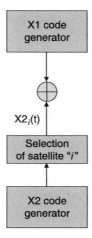

FIGURE 7.2 P code generation basics.

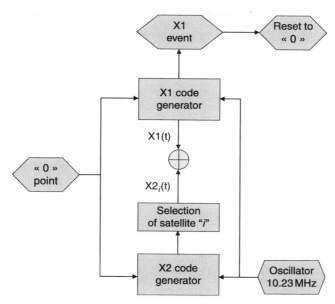

FIGURE 7.3 X1 event generation.

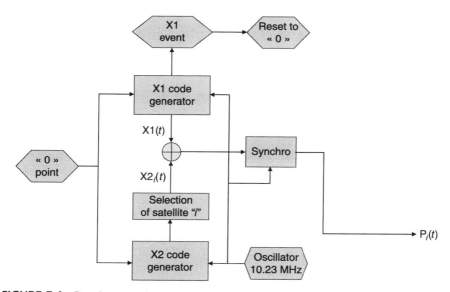

FIGURE 7.4 P code generation.

The global code plus data signals are then obtained by modulo two additions for the C/A and P codes for the L1 band, and for the P code on L2 (see Fig. 7.7). Note that at this stage, the P code is available either embedded with navigation data or not.

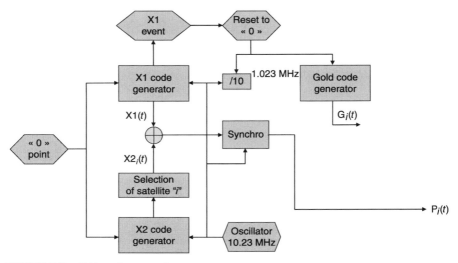

FIGURE 7.5 C/A code generation.

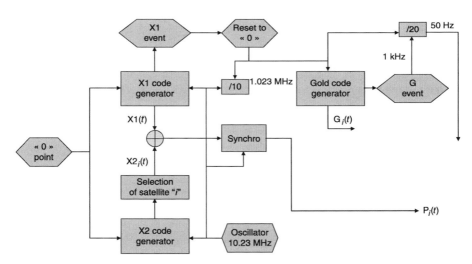

FIGURE 7.6 50 Hz navigation data bit rate generation.

Structure and Generation of the Signals

A simplified representation of the codes is given in Fig. 7.8. This is the starting point for the next step, which consists of using these various codes in order to modulate the carrier frequencies in the L1 and L2 bands.

Figure 7.9 shows how L1 and L2 GPS signals are generated. Many different features are of interest. First is the fact that all frequencies, carrier chip rates, code chip rates, and navigation data bit rates are obtained from the same oscillator. This clearly helps achieve synchronization between all the signals generated. The second

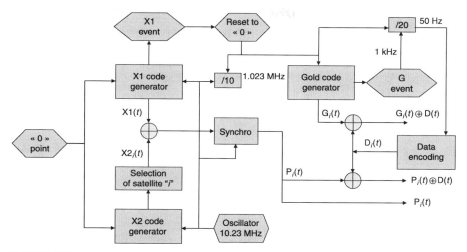

FIGURE 7.7 C/A, P, and navigation data generation.

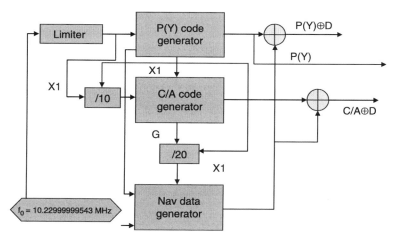

FIGURE 7.8 Code summary.

notable point is the oscillator frequency shown in Fig. 7.8. In fact, this frequency is not exactly 10.23 MHz because of relativistic effects. Indeed, at least two main effects have to be considered: the gravitational field and the speed of displacement of the satellite clocks with respect to the same clocks on the Earth's surface (the ground segment has the task of "following" the satellites' clock).[2] The combination of these two effects leads to a ratio of 4.45×10^{-10}, with the satellite's clock running faster than those on the ground. Rather than having to take this into account at the receiver end, it was decided to deal with relativistic effects at the satellite end, by having the

---

[2]The TIMATION program, in the early 1960s, dealt with this problem on a scientific basis.

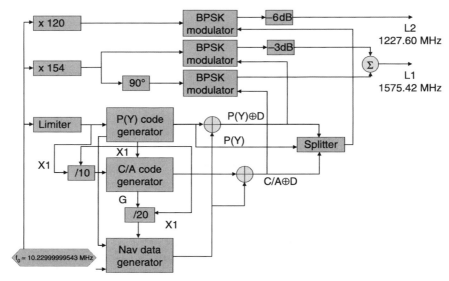

FIGURE 7.9   Signal generation diagram.

frequency of the central oscillator shifted in such a way that the behavior is equivalent to that which would have been observed with no relativity. Thus, 10.22999999543 MHz is used instead of the 10.23 MHz of theory. Of course, similar considerations have been taken into account for both GLONASS and Galileo (refer to Chapter 5 for details).

Interface Control Documents

The full description of the various signals, codes and modulations are available within the so-called Interface Control Documents (ICDs) relative to the particulars constellations. For GPS, the latest versions and history are available on the Navcen web site,[3] and the GLONASS ICD is available at the GLONASS web site.[4] For Galileo, the first version of the ICD concerning the Open Service was released on May 23, 2006, in a preliminary version (draft 0). It is available on the Galileo Joint Undertaking web site.[5]

Note that the preliminary version of the future signal description of GPS L1C is already also available on the Navcen web site (Interface specification IS-GPS-200, dated 7 March, 2006).

■ 7.2 RECEIVER ARCHITECTURES

The main difficulty in GNSS positioning is to make the best possible measurements, that is, the most accurate ones, in order to provide the positioning with the

---

[3]http://www.navcen.uscg.gov.
[4]http://www.Glonass-ianc.rsa.ru.
[5]http://www.Galileoju.com.

best resulting accuracy. Indeed, as long as the measurements are totally accurate, the positioning will be perfect. Unfortunately, as already described in Chapter 6, a lot of errors will lead to inaccuracies. Other possible errors are caused by the receiver's own electronic and processing unit, which could give rise to inadequate correlations because the signal is not as pure as one would like. Multipath, low power levels or cross channels (different satellites) can be the cause of this. In addition, the signal has to be found in a two-dimensional area that copes with Doppler shifts, due to both the satellite's motion and the receiver's displacement, on one hand, and time shift, due to the propagation delay from satellite to receiver, on the other. Of course, both searches (that is, Doppler and time) are fundamental to carrying out position calculation.

This section is devoted to the receiver architectures. In order to solve this difficult problem of acquisition and tracking of the GNSS signals, different approaches have been implemented throughout the years, leading to real improvements in the resulting quality of the positioning.

The Generic Problem of Signal Acquisition

The basic measurement is clearly a time measurement, which is required for distance estimations from the satellites to the receiver. This is achieved through the use of a code sequence that is also the identifier of the satellite (for GPS and Galileo, as for GLONASS, identification is achieved through different frequencies, but a code is still used). This code is duplicated at the receiver and shifted (in time) to correlate with the incoming code from the satellite. Once the correlation has been achieved, the corresponding time shift allows the pseudo range to be estimated (as will be explained in Chapter 8).

Unfortunately, the signal is distorted due to the Doppler effect, induced either by the satellite's motion, the receiver's motion, or both. The Doppler effect is a physical compression or dilatation of the signal due to the transmitter's and/or receiver's relative motion. It is a characteristic of the axial projection of the velocity vector on the axis from the transmitter to the receiver. As this distortion is a physical one, it means that it applies to the carrier and also to the code that modulates it, thus also leading to a distortion in the code, which must be taken into account. The effect is that the code can be either reduced in length or enlarged. If this effect is not taken into account, then the correlation might be non-optimal and the propagation time not accurately found. There is therefore a strong requirement to have a way to take into account the frequency of the incoming signal. Thus, the basic architecture of a receiver must implement both a frequency search (to cope with Doppler) and a time search (to cope with propagation time).

Let us estimate the respective sizes of these two search domains. The Doppler of a satellite, from the moment it appears on the horizon to the moment it disappears on the other side of the horizon, is given in Fig. 2.4. This typical curve describes the fact that when the satellite appears, it gets closer to the receiver (positive Doppler), then it passes over the receiver, thus exhibiting a zero Doppler (as the satellite displacement is tangent to the axis between the receiver and the satellite), and then the satellite goes away from the receiver, thus exhibiting a negative Doppler. Note that Fig. 2.4 shows

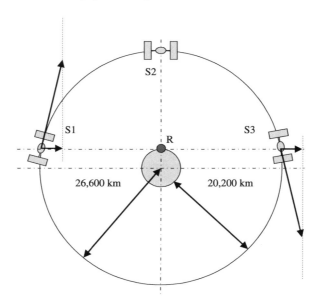

FIGURE 7.10   Calculation of the maximum Doppler values to be considered.

the value of the so-called Doppler frequency, which is the difference in frequencies compared to the initial transmitted value.

A typical interval of Doppler shift to be considered can be obtained by a simple calculation of the projection of the vector of maximum velocity (one should also add the Earth's rotation). Figure 7.10 describes the geometry to be considered. In such a case, the larger value of Doppler is found to be around ± 5 kHz.

In addition, the corresponding value of the receiver's displacement-induced Doppler must be added, as well as the effect of the drift of the receiver's local oscillator. Typical cumulative resulting values are ± 10 kHz.

The time aspect is a little different because it relies on the code structure, length, and rate. A first raw approach consists in considering that there are as many different time shifts as there are chips in the code.[6] In the case of GPS L1 C/A, this means 1023 possible time shifts. In the case of L1-B Galileo, there are 4096 possible time shifts. For GLONASS C/A, there are 511 possible time shifts. Another important point lies in the time duration corresponding to the code lengths. For GPS C/A, the code lasts 1 ms, the same as with GLONASS C/A, whereas the Galileo L1-B code lasts 4 ms, the chip rate being the same at 1.023 MHz. The direct impact is that the search domain is 1 ms for both GPS and GLONASS, for respectively 1023 and 511 time slots. The search domain for Galileo L1-B is 4 ms in length and 4092 possible time slots. This simple approach shows that the second axis of signal search, the time, will take different values for the various codes. In all cases, this time search is long.

---

[6]As a matter of fact, the correlator architectures allow a resolution of approximately 100 times better than one chip, as described in the following sections.

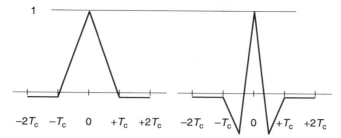

FIGURE 7.11 Correlation functions for GPS and GLONASS (left), and Galileo (right).

The real complete signal search is the combination of the Doppler and time searches. The difficulty of the signal search is that the two-dimensional search area is huge. When the signal has a sharp aspect it is rather a good configuration, but noise sources are bound to produce many different peaks that would lead to a much more difficult determination of the true peak.

Due to the structure of the global signal, being the combination of the ranging and identifying code plus the navigation message, there is another constraint in the acquisition process: the time available to find the signal. In the case of GPS, the navigation data rate is 50 samples per second, meaning that every 20 ms, the navigation message bit could change.[7] Such a change produces a complete change of the code sequence (a bit by bit inversion), because the ranging code and the navigation data are added (modulo two). This is clearly a problem for acquisition... unless the receiver can "know" the data bit changes. This is another approach that is further described in the next section. Thus, the receiver has only 20 ms to find the satellite. In the case of some Galileo signals, it is even more stringent as some navigation message data rates are as high as 250 or even 1000 samples per second.[8]

One can now easily understand that the correlation functions, the basis of the searches, are of prime importance. A good correlation function should allow

- Simple implementation in order to allow a fast response time;
- High efficiency in order to provide high conversion gain at the correlation;
- High rejection of non-correlated signals in order to reduce noise interference effects;
- High resolution in order to allow a high precision of the time shift.

The corresponding GPS, GLONASS, and Galileo correlation functions are shown in Fig. 7.11. Note that the case of GLONASS remains quite specific because the code is only required for ranging and not for identification purposes.

The next fundamental aspect is related to power levels. In Chapter 10, the reader will be given some views on the availability of GNSS signals indoors, and it will be

---

[7]Remember that as all the components of the signals are synchronized, this change is in phase with the carrier and the ranging code.

[8]Please read the discussion of the next sub-section, "Possible High Level Approaches," for solutions currently adopted.

shown that the main problem is the received power level, which is usually too low to allow good reception conditions. The power levels are a first concern when dealing with searching within the abovementioned two-dimensional search area. The correlation peak is of better quality when the incident power level is high. The guaranteed power levels on the Earth's surface range from -125 and -137 dBm for the various GNSS signals. Augmenting these power levels is not yet possible because the GPS signals are now present in space and it is not possible to consider getting rid of the current constellation and building a new one. A compromise has to be found in order to increase the power levels of new satellites without impacting too much on the already existing ones. At the first stage of the early implementation of GPS signals, the situation was quite different from the one GNSS is facing today. The constellation was almost alone in the sky (for satellite-based navigation) and the life-time of satellites was highly dependent on the power transmitted by the satellite. In order to preserve this latter fundamental parameter, the power level was set to the acceptable value of typically -130 dBm. Once again, the case of GLONASS is a little different, with the need to coexist with the radio astronomy community.[9] Thus, it is not possible to increase the power levels that much.

The other solution, in order to acquire the signal in degraded conditions (with reduced received power levels), would be to increase the code length in order to be able to lower the detection limit (remember that the correlation equivalent gain is a function of the length of the code), but at the expense of an icreased acquisition time. This is nevertheless the procedure adopted with some Galileo signals, together with an increased power level (of a few dB). However, increasing the length will certainly cause some difficulties regarding the navigation message bit duration, which is the next limit for acquisition. From all this arose the idea of duplicating the signals in order to provide so-called "pilot tones." The idea here is to provide the users with two similar signals, shifted by $90°$ in phase, both having the ranging code but with only one also being modulated by the navigation message, the second being without navigation data. This allows so-called long integration (no time limit for integration as there are no longer navigation data bits), far above the 20 ms described earlier for the case of GPS. These pilot tones are implemented in Galileo signals, but are also planned for GPS L2 and L5 signals.

Possible High Level Approaches

Let us now skip to the hardware part of the receiver and give some first possible approaches of real implementations. The first available receivers (GPS receivers for instance) were based on a sequential approach; the satellites were acquired one at a time in a row. This meant having a high quality oscillator, or a way of following with a high accuracy the drift of this oscillator, with respect to time.[10] Indeed, the first satellite was acquired and tracked, the corresponding pseudo range was stored,

---

[9]This co-habitation is the reason for the new definition of the GLONASS spectrum, in order to reduce the interference between navigation and astronomy communities (see Chapter 6).

[10]Please refer to Chapter 8 for details (clock bias calculation in the position, velocity, and time (PVT) solution).

and then the second satellite acquired and tracked, then the third one, and so on. Once enough satellites were acquired and data stored, the calculations of position, velocity, and time were carried out.

At the time the full constellation was in operation, the receiver's architecture made some progress, thanks to the electronics industry, and parallel channel receivers allowed multiple simultaneous acquisition and tracking capabilities. For some years, the "standard" was an eight-channel receiver. The reason for this is not well known, but it appeared that the industry had finally decided that eight was a "good number": it was a *de facto* accepted standard. With the advent of geostationary satellites and the improvement of receivers' sensitivity, things moved to, say, 12 channels for mass market and 14 to 16 for professional receivers. By channel, one has to understand a structure that deals with the time and Doppler search for one particular satellite. It is quite probable that this number will grow a little with Galileo, especially for mass market receivers using the L1-E1 band, in order to take advantage of the diversity of signals. This is bound to happen even if multi-constellation calculations are possible, allowing positioning with a total of four satellites from both constellations.

Pushing forward the logic, multiple channel receivers could be used in order to reduce the acquisition time by implementing parallelism. As a matter of fact, it is simply achieved by assigning a few correlators to a single satellite. Let us imagine that there are 1023 possible correlation times and that 1023 parallel correlators are available. Then, it is possible to determine the correlation time at once by "trying" all the time shifts on the various channels. This can be further implemented with several received satellites, on the condition that the receiver has enough channels. Such realizations have been implemented with, first, "only" 32,000 correlators, and then with as many as 200,000 parallel correlators. Companies such as Global Locate or u-blox have released such GPS chips. The main advantage is to achieve a quick acquisition phase. Indeed, once acquired, the tracking is much easier and does not require such a parallelism. Nevertheless, even in the tracking phase, this approach could also help in multipath mitigation. This is achieved at the expense of higher power consumption.

Another approach to deal with the complex two-dimensional search while in the acquisition phase of the GNSS signal is to implement heavy signal processing such as Fast Fourier Transform (FFT). The idea is to transfer the signal into the frequency domain in order to quickly find the Doppler shift of the signal. This can be achieved through the use of a Fourier transformation: unfortunately, this is also quite heavy in terms of power consumption.

Receiver Radio Architectures

A GNSS receiver is based on typical radio architectures, consisting of an antenna, the first component that allows the conversion of a radiated electric field (the incident wave) to an electric power signal. This electric field is the one generated by the satellite antenna, more than 20,000 km away. Following the antenna, the first electronic[11] component must be a low noise amplifier (LNA). In Chapter 6, it was seen that the

---

[11]The electronic components working in this high frequency range are called microwave components.

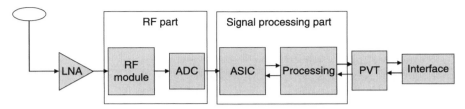

FIGURE 7.12 Diagram of a typical receiver.

signal-to-noise ratio at the antenna is somewhere between -15 and -20 dB, thus leading to the signal being significantly lower than the noise received by the antenna. In such a case, there is a great need not to degrade this incident signal too much in order to be able to acquire it in the best possible conditions. Out of the scope of this book, further readings concerning microwave systems would lead to the conclusion that a first electronic component that exhibits both high gain and low noise factor is the best choice.[12] Then the radio frequency part of the receiver, which is included to deal with the high frequency signals, usually downconverts them into a lower frequency band. Remember that the high frequency is just the carrier used because of the good propagation properties and relatively small-size antenna necessary for these bands. The "useful" signals are the ranging code and the navigation data.[13] provided at a much lower frequency (1 kHz for the code and 50 Hz for navigation data).

Once this downconversion has been achieved, the signals are digitalized and transferred to a digital signal processor (DSP) for pseudo range extraction, noise rejection, and so on. The next part is intended to provide the complete navigation solution, which is composed of position, velocity, and time (PVT solution).[14] The final data are passed to the interface for application specific purposes. Figure 7.12 shows this typical implementation.

Coming down to details concerning the radio frequency part, it is possible to show the downconverting components, that is, the mixer associated with a local oscillator and a frequency synthesizer, and the multichannel architecture. In Fig. 7.13, the multichannel receiving capabilities are located in the digital part; this could be available at the radio frequency part but this is no longer the current typical architecture.

Figure 7.14 shows a typical homodyne structure where only one downconversion is achieved. The so-called intermediate frequency, in this example, is 175.42 MHz. The analog-to-digital converter permits the transformation of the analog signal at 175.42 MHz to digital samples at a rate of 24.56 MHz. This latter sampling rate allows the rebuilding of a 12.28 MHz signal (refer to the Shannon criteria). Some

---

[12]The noise factor is a specific characteristic associated with microwave components that characterizes how the signal-to-noise ratio is degraded, from input to output of the component. Then, the application of the FRIIS formulae results in the mentioned component.

[13]Except for the phase receivers where the useful signal is actually the carrier, as explained later on in this chapter.

[14]Refer to Chapter 8 for details concerning the so-called PVT solution.

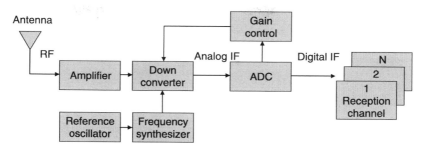

FIGURE 7.13   A diagrammatic representation of the receiver architecture.

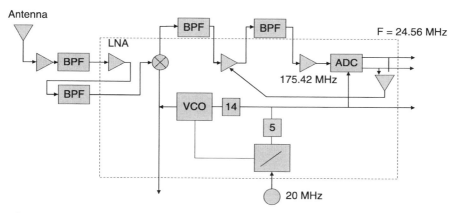

FIGURE 7.14   A homodyne conversion structure.

band pass filters are required at the input side and the local oscillator is at 1400 MHz, leading to the abovementioned intermediate frequency.

The local oscillator is a fundamental element in all radio receivers, but is even more important in GNSS receivers as it must exhibit a frequency agility in order to cope with the Doppler shift of the signal. Furthermore, it should forward this information, that is, the Doppler shift, to the signal processing unit in order to adjust the code replica shape. This is compulsory in order to be able to achieve a correct correlation. On the other hand, it has no need to be of high quality, as long as the architecture is multichannel. This means that the bias will be removed through calculations (refer to Chapter 8 for details). This assertion remains correct for both positioning and velocity determination.

Most receivers make use of a super-heterodyne architecture, where two successive downconversions are carried out, as shown in Fig. 7.15. The values given in this figure are typical examples. This receiver is a C/A code GPS receiver with an antenna, including a first LNA of 36 dB of gain with a bandwidth of 50 MHz. The final bandwidth of the last represented component is 2 MHz, the size of the main lobe of the C/A frequency spectrum.

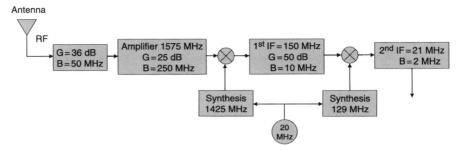

FIGURE 7.15   The super-heterodyne conversion structure.

The concept currently under development is the software radio approach. This is certainly the future of all radio receivers but still requires further research in order to be fully applied, although the first commercially available products are already impressive. The basic idea is to note that the computation capabilities of programmable microprocessors, even the smallest ones, are growing at a fast rate. This means that instead of implementing the digital signal processing into dedicated and hardwired structures, it is certainly possible to have it implemented under a software approach, that is, by programming the abovementioned microprocessors. The advantages of such an approach are numerous: portability, ease of modification, cost of change, versatility, and so on.

Figure 7.16 presents the allocation of the hardware and software in this concept. Current software receivers implement the analog radio frequency unit from the antenna received signal to the downconversion and the analog to digital sampling. Then, the software copes with all other processing.

When dealing with systems located in the same bandwidth, as is the case for GPS and Galileo in the L1 band, such an architecture has a tremendous advantage over classical hardware structures, as the addition of new signals or modification of existing signals is only a matter of updating the software. Thus, it appears that current receivers are already ready for Galileo open service signals, and more precisely those signals present in the L1 band (because the ICD, draft version, was released in May 2006).

The next steps towards a global software radio receiver are wider frequency band capability and the integration of many radio standards. The main idea is to deal with a single analog part for, say, the L1, E1, L2, E5, E6, and L5 bands, and why not also

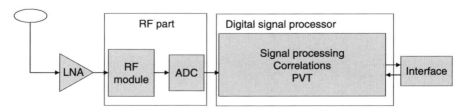

FIGURE 7.16   The software receiver concept.

include the GLONASS L1, L2, and L3 bands. The analog-to-digital converter (ADC) forwards samples to the programmable processor. It should then deal with the various modulation schemes, codes, and navigation data. Pushing the concept forward, the same receiver should also be able to cope with WiFi or 3G telecommunication signals. In fact, the same analog hardware, essentially composed of an antenna, a frequency mixer, and an ADC, is complemented by the software part, which is mainly software packages depending on the radio signal one wants to decode. Then, it is simply by adding the software components that a receiver can exhibit GNSS, either GPS or Galileo, or both, or even GLONASS, WiFi, Bluetooth, 3G, or TV capabilities.

Channel Details

The signal search within a receiver is achieved in two successive phases. The first tries to find rough code phase synchronization with code chips, and the second tries to achieve finer acquisition and tracking. During the raw first phase, the error of the delay between the incident code from the satellite and the internal receiver replica is usually reduced to around one chip. Then the tracking is intended to refine the resulting error to its minimum. The local duplication of the code and correlation with the incident signal is assigned to the so-called delay locked loop (DLL). It is in charge of achieving the best possible correlation, knowing that classical correlation functions are maximal when both codes are in phase, that is, perfectly superposed. A typical architecture of a DLL is shown in Fig. 7.17 where a so-called early-late structure is represented.

Such a correlation structure allows an improved performance is relation to the accuracy of the correlation; around one-hundredth of a chip length is achievable,[15] taking into account a feedback loop principle. The idea is to generate locally not only one code replica, but three code replicas. The first one, called the "prompt," is the one that will be directly compared to the incident signal from the satellite and

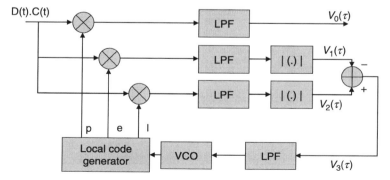

FIGURE 7.17 A typical DLL early-late architecture.

---

[15]Equivalent to about 3 m for the C/A GPS code for instance, as compared to 293 m for one chip (length equivalent to the chip duration).

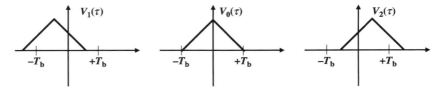

FIGURE 7.18 Early, prompt, and late shapes of the GPS correlation function.

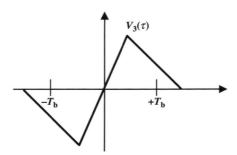

FIGURE 7.19 Early minus late shape of the correlation function.

with which the maximum correlation function is to be found. The second is switched with respect to the prompt by a given chip fraction, say T_c/N, where T_c is the chip time and N an integer. It is called the late replica. The third one, called the early replica, is switched by $-T_c/N$. When considering the search for the maximum value of the prompt replica with the incident code, the difficulty is to find a way to subjugate the correction to be applied to the replica time switch, either by advancing it or retarding it. As a matter of fact, it is not easy to know which direction it needs to be switched[16] if the correlation value is, say, 0.8 (as compared to the nominal 1 when perfectly in phase). The three replica codes are given a typical simple solution to this question. When having a look at the various shapes of the respective correlation functions, as shown in Fig. 7.18, it appears that the correlation peaks are switched away from each other by the same amount of time as the replica codes are switched from each other.

Having the three replicas allows the computation of the difference between the early and the late correlation functions in order to obtain the typical S-curve given in Fig. 7.19. What can be observed is that the maximum correlation peak is obtained when this latter curve crosses zero. Furthermore, when comparing the value of the prompt correlation function at the time the early minus late curve crosses the zero line can give the information about the value of the time switch needed to reach the real maximum of correlation. In addition, the fact that the early-late value is positive indicates the prompt should be retarded, whereas a negative indication indicates the need to switch the prompt forward.

---

[16]In other words, it is difficult to say whether one is on the left-hand side or on the right-hand side of the maximum peak on the correlation function curve (see Fig. 7.19)

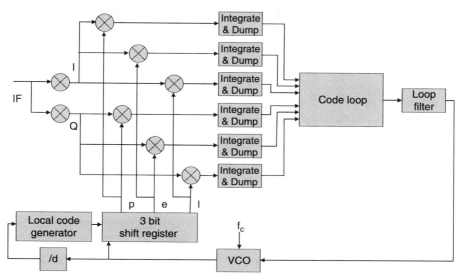

FIGURE 7.20 Complete DLL.

The final complete code loop is composed of two parts, respectively concerning the in-phase and quadrature-phase components of the incoming signal, as described in Fig. 7.20.

There is also the need, as already described, to adjust the frequency of the replica code in order to achieve the best possible correlation. This functionality is devoted to the so-called phase lock loop (PLL), which has the task of tracking the real frequency received at the receiver end. Note that it has to cope with two different effects: the first is the natural Doppler shift due to both satellite displacement and receiver motion, and the second is due to the natural frequency drift of the receiver's oscillator. Indeed, this latter effect is seen by the receiver in an identical way to a Doppler-induced frequency shift and must be taken into account. Unlike the Doppler due to satellite motions, the

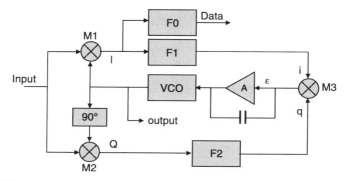

FIGURE 7.21 A typical Costa PLL.

receiver-induced Doppler and the internal oscillator drift cannot be predicted,[17] So there is the need for a PLL. A typical architecture is given in Fig. 7.21 showing a Costa loop.

Let us consider that the input signal is given by

$$S_{\text{input}} = \sqrt{2}\, A \cos(\omega_0 t + \theta_i), \tag{7.1}$$

and the output signal by

$$S_{\text{output}} = \sqrt{2}\, A \cos(\omega_0 t + \theta_o). \tag{7.2}$$

Then it is easy to state that

$$
\begin{aligned}
I &= A \cos(2\omega_0 t + \theta_i + \theta_o) + A \cos(\theta_i - \theta_o) \\
&= A \cos(2\omega_0 t + \theta_i + \theta_o) + A \cos(\Phi).
\end{aligned} \tag{7.3}
$$

By low pass filtering, it is possible to obtain the difference of phase Φ between the input and the output. In other words, it is possible to use such a loop to tune a controlled oscillator to the right frequency, that is, the input frequency. As a matter of fact, one can see that signal "i" (see Fig. 7.21) is in fact $A \sin(\phi)$, whereas signal "q" is indeed $A \cos(\phi)$. Then, the M3 multiplication leads to $A^2 \sin \phi \cos \phi$, or $A^2 \sin(2\phi)/2$. The feedback of the loop is then twice the value of the phase difference.

A complete PLL structure is obtained in a similar way as for the DLL and is shown in Fig. 7.22.

It is possible to help the PLL by an external aid, such as a predictive calculation of the contribution of the satellite's displacement to the Doppler or even oscillator drift considerations.[18] In such a case, the final PLL global structure is as in Fig. 7.23.

As there is the need for feedback from the PLL to the DLL, another structure of the DLL is presented in Fig. 7.24 that differs from Fig. 7.20. The Doppler shift

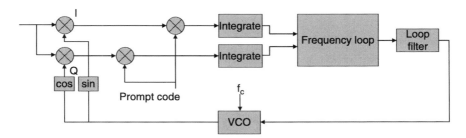

FIGURE 7.22 A typical PLL structure.

---

[17]For the receiver's Doppler, other sensors could help in finding it. Concerning the oscillator's natural drift, controlling it leads to more complicated components and to more expensive solutions.
[18]This can be achieved through modeling of the oscillator drift. Experiments have shown repetitive behavior that could be used as aiding values.

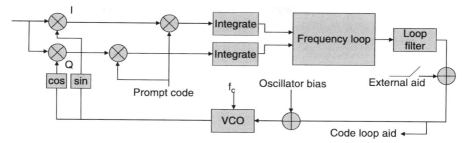

FIGURE 7.23   External aided PLL structure.

obtained from the PLL is applied to the DLL, taking into account the scale factor corresponding to the frequency ratio between the carrier frequency and the code frequency.

There are many different real implementations of frequency (or phase) loops and delay loops. Further readings are required to acquire the subtleties of all the discriminators and loop filter implementations. Nevertheless, they are typical feedback loops of various orders and the related characteristics are mainly the response time, the frequency band, and the accuracy. Depending on various parameters, such as the integration time, the order of the filters or the complete loop architecture, the associated performances will differ. If one wants to provide a very dynamic receiver, able to work even if it is subject to very high accelerations, then the equivalent band should be larger than in the case of the static mode. This is then achieved with an increased noise signal power at input, reducing the resulting accuracy of the measurements.

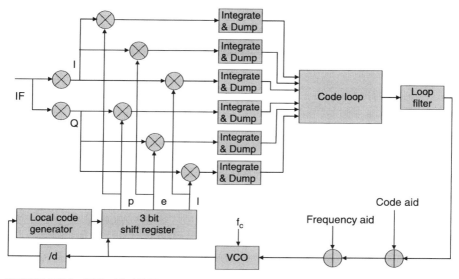

FIGURE 7.24   PLL-aided DLL structure.

Concerning the order of the filter, let us just state that a higher order[19] corresponds to better dynamic behavior, but is not always required.

The global architecture of a receiver is then composed of a DLL and a PLL that can be either a Costa loop or a frequency lock loop (FLL), or even a PLL assisted by an FLL, depending on the dynamic and final accuracy required. It depends on the specific application to which the receiver is devoted.

During the acquisition phase, where the receiver has no indication about the frequency received, the receiver should use an FLL. Once the raw acquisition has been achieved (typically within one chip), the architectures described above are implemented in order to carry out the tracking.

Another important parameter is the spacing between early, prompt, and late replicas of the code. The reduction of this spacing allows the impact of thermal noise to be reduced, but it should be complemented by a corresponding increase in the input bandwidth in order to avoid the induced flatness of the correlation function. This effect is mainly due to the truncated spectrum of the code, theoretically infinite. Thus, when using a reduced spacing, which could be envisaged to reduce the multipath impact, one should at the same time consider an increase of the incident spectrum.[20] Current receivers implement a variable spacing approach in order to cope with multipath (see Chapter 8 for details).

Now the basic operation mode of both acquisition and tracking is better understood, it is possible to come back to specific approaches implemented in order to provide a reduced acquisition time. There are indeed two different problems: the first is linked to the wide search domain, both in time and frequency, as described in this section, and the second is the need for navigation data in order to proceed to the positioning calculation (see Chapter 8). The approach proposed is to use a telecommunication transmission channel with a higher data rate in order to provide these data. Thus, the so-called Assisted-GNSS (A-GNSS) is a GNSS receiver that receives, through GSM, GPRS, or UMTS, assistance data in order to be helped. The principle, although detailed in Chapter 10, is based on a GNSS receiver located at a mobile network base station and which receives the GNSS data continuously. Furthermore, this receiver also computes the corresponding GNSS location of the base station. When the A-GNSS receiver is switched on, the telecommunication link to the base station allows the receiver to obtain the navigation data directly through the telecom channel, together with the location of the base station. This will help the receiver by means of an external aid; knowing the location of the base, considered as being the starting point of the location calculation for the receiver, allows it to determine approximate Doppler shifts of the satellites. In addition, the satellites to be searched for first can also be pointed out by the base station. Thus the reduction of acquisition time is important. It is possible to determine your position in a few

---

[19]Note that as for typical feedback systems, the order is related to the differential equation form that characterized the system. Resolution of this equation leads to the typical output form of the signal, that is, the response time, the dynamic transient behavior, the frequency response, and the resulting accuracy.

[20]The usual bandwidth for mass market GPS receivers, for instance, is typically 2 MHz. This is due to the shape of the main lobe of the C/A signal, which is around 2 MHz. Considering only this reduced band means a distortion in the shape of the peak of the correlation function. When the spacing is reduced, this effect is even amplified. One solution consists in widening the input bandwidth.

seconds when leaving the airport parking lot, for example, where about 40 s would have been required with classical GNSS.

■ 7.3 MEASUREMENT TECHNIQUES

In this chapter, measurement techniques refer to the way it is possible to carry out measurements related to range, that is, pseudo range, or to phase, that is, carrier phase. Ways of achieving location, speed, and time calculations are discussed in Chapter 8 (Positioning Techniques).

There are essentially two different ways of carrying out measurements. The first is the one discussed at the beginning, based on codes, and is called the "code phase," either C/A or P for the GPS system. The ranging codes that are going to be available in the next few years are numerous and include Galileo codes, the GLONASS codes, and also the GPS codes. Of course, new codes have been designed, such as the M-code, which will not be available, as is the case for PRS-associated ones. Nevertheless, various user communities, and more specifically the geodesy community, have found that it could be of great advantage to work with the carrier of the signals (leading to the so-called "carrier phase" technique) instead of codes. The first reason for this was the poor performances allowed by the C/A code. For example, let us consider that the chip is the length unit one is using when dealing with the C/A code. Then, remember that a C/A code chip corresponds to about 293 m, which is really a rough measuring device, even if it has been shown that correlation could be achieved with an accuracy 100 times better. Nevertheless, there is another measuring device that, if used, could lead to much better accuracy: the frequency of the carrier itself. As in the L-band this is around 1.575 GHz for L1-E1-E2 GPS and Galileo, this means a wavelength of 19 cm. This is thus the scale of the new measuring device, as compared to 293 m. From a few meters of accuracy, using phase measurements (consisting of playing with the carrier frequency) allows the accuracy to be calculated to a few centimeters in real time, and even lower for processing over a longtime.

This section is devoted to the explanation of these two principles of measurement.

Code Phase Measurements

Let us suppose the satellite has been acquired and navigation data have been obtained from the demodulation process, as described in the previous paragraphs. Let us also assume that the receiver is able to find out the exact transmission time of the signal it is currently receiving[21] from the satellite; thus it is able to convert

---

[21] In fact, the receiver recovers from the navigation message the parameters that will enable it to calculate the orbit of the satellite. In other words, this means that once decoded, the navigation message allows the receiver to carry out a calculation that gives the location of the satellite as a function of time. Thus, it is not the location of the satellite at any given time, nor the location of the satellite at time of transmission that is transmitted to the receiver, but rather the parameters that will allow it to carry out calculations.

the time difference, from the receiving instant to the transmitting one, into a distance by multiplying this time difference by the speed of light. The obtained value is the pseudo range. There are then two major difficulties: the timescales not being synchronized between the receiver and the satellites, and the fact that all the error sources described in Chapter 6 might modify the speed of light, thus complicating the transformation from the time of flight to the pseudo range.

Let us look at the time reference scales. Indeed, there are three different timescales, as follows:

- t, the GNSS time. This is the continuous timescale used for the GNSS constellation. Note that the GPS, GLONASS, and Galileo reference timescales are not identical. This is a potential problem when dealing with the so-called interoperability concept between GPS and Galileo for instance, where a bias of time reference scales is needed in order to achieve positioning. These biases are intended to be provided by the navigation messages.

- t^s, the satellite time reference scale. This is the time kept by the atomic clocks of the satellite.

- t_r, the receiver time reference scale, kept by the receiver's oscillator.

The satellite time is linked to the GNSS reference timescale by the relation

$$t^s = t + dt^s \tag{7.4}$$

as the receiver time is linked by the relation

$$t_r = t + dt_r. \tag{7.5}$$

Let t_e and t_e^s be the transmission time in the time scales of the GNSS and satellite, respectively, and t_r and t_r^r the receiving time in the GNSS reference scale and receiver timescale, respectively. The fundamental relation giving the distance between the satellite and the receiver is given by

$$c \cdot (t_r - t_e) = \rho + \Delta_{\text{Propag}}. \tag{7.6}$$

The real distance is given by ρ and corresponds to the distance considered from the transmission time to the receiving time. Note that the transmission time should be the one at the sending time of the signal received by the receiver at the receiving time, and not the one sent at the time of reception, of course. Δ_{Propag} includes all the propagation errors.

Considering Equations 1 and 2, one obtains

$$t_r = t_r^r - dt_r \tag{7.7}$$

and

$$t^e = t_e^s - dt^s \tag{7.8}$$

This leads to

$$c \cdot \left(t_r^r - dt_r - t_e^s + dt^s \right) = \rho + \Delta_{\text{Propag}} \tag{7.9}$$

or

$$c \cdot \left(t_r^r - t_e^s\right) + c \cdot (dt^s - dt_r) = \rho + \Delta_{\text{Propag}}. \tag{7.10}$$

From this expression is derived the definition of the pseudo range, as follows:

$$\text{PR} = c \cdot \left(t_r^r - t_e^s\right) = \rho - c \cdot (dt^s - dt_r) + \Delta_{\text{Propag}}. \tag{7.11}$$

As the various errors are biasing the pseudo range, the receiver should find a way to output the real distance of interest, that is, ρ:

$$\rho = c \cdot \left(t_r^r - t_e^s\right) + c \cdot (dt^s - dt_r) - \Delta_{\text{Propag}} \tag{7.12}$$

where dt^s is obtained from the navigation message and dt_r will be the fourth coordinate of the navigation vector in the solution of the trilateration process (detailed in Chapter 8).

The following "ultimate" expression gives a summary of all the corrections that must also be taken into account in order to improve the distance evaluation:

$$\text{PR}_{r,m}^s = \rho_{r,m}^s - cdt_m^s + cdt_{r,m} + \rho_{r,\text{ion},m}^s + \rho_{r,\text{trop},m}^s + \rho_{r,\text{rel},m}^s. \tag{7.13}$$

where

- $\rho_{r,m}^s$ is the real distance from the satellite "s" and the receiver "r";
- dt_m^s is the clock bias of the satellite with respect to the GNSS time reference frame, at time t_m;[22]
- $dt_{r,m}$ is the clock bias of the receiver with respect to the GNSS time reference frame, at time t_m;
- $\rho_{r,\text{ion},m}^s$ is the distance equivalent of ionosphere propagation correction;
- $\rho_{r,\text{trop},m}^s$ is the distance equivalent of troposphere propagation correction;
- $\rho_{r,\text{rel},m}^s$ is the relativistic-effect-related correction.[23]

The accuracy of the code phase measurement is typically a few meters, leading to the current stand-alone mass market receiver, which has a range of $2-5$ m.

Carrier Phase Measurements

The major drawback of the code is the length of a chip, which limits the accuracy of the distance measurement. It is possible to use another representation, as follows. Let us first consider that the general form of a sinusoid function of amplitude A and frequency f can be written as follows:

$$x(t) = A \cos(2\pi f t + \phi_0). \tag{7.14}$$

---

[22]This remark is fundamental as the clock biases are not constant. One must absolutely take the time-related variation of the clock biases into account; this is known as the clock drift.

[23]Refer to Chapter 5 for a few details on these corrections. Remember that it is compulsory to take them into account. If not considered, the accuracy of positioning would shift by a few kilometers each day!

The corresponding phase of this signal, $2\pi f t + \Phi_0$ can be given in "cycles" as:

$$\phi(t)_{\text{cycle}} = f \times t + \frac{\Phi_0}{2\pi}. \tag{7.15}$$

The principle of carrier phase measurement is to carry out the distance measurement from the satellite to the receiver through the difference of phase between the received signal (transmitted by the satellite) and a local replica, generated at the receiver end. Unfortunately, this difference can only be evaluated within a cycle, that is, within 19 cm for L1 and within 24.4 cm for L2. This is indeed only the fractional part of the signal that can be compared.[24]

The phase difference can be given by

$$\Delta\phi = \phi(t_r) - \phi(t_e), \tag{7.16}$$

and the phase difference measured can be written as

$$\Delta\phi_{\text{mes}}(t_r) = \Delta\phi - N_r^s(t_r), \tag{7.17}$$

where $N_r^s(t_r)$ is the number of complete cycles in the distance from the satellite to the receiver. It is called the ambiguity. When dealing with the L1 band and considering a satellite to receiver distance of 21,000 km, the corresponding ambiguity is about 70 when dealing with code phase, as compared to 110,289,525 for carrier phase. The ambiguity, for each satellite considered, is an additional unknown that must be calculated in the computation of the positioning. Of course, the ambiguity of code phase is much easier to determine than the one relative to the carrier phase.

Furthermore, in the case of the carrier phase, the receiver is able to "follow" the number of cycle slips from one measurement time to the next. As shown in Fig. 7.25, this allows the receiver to evaluate the variation of the satellite-to-receiver distance as the initial ambiguity remains constant while following the carrier phase.

This remains true as long as the carrier phase is not lost (because of heavy foliage or other obstacles in the satellite-to-receiver path). In such a case, the calculation is no longer possible.[25] In the computations one has to consider the ambiguity as an additional unknown, but one has to remember that there is one such unknown per frequency and per satellite. Thus the calculation is much more complex than with the code (but also much more accurate. . .).

The observation measurement of the phase, taking into account the various error sources, is given as follows. (Note that in Eq. 7.18 there are minus symbols whereas in the code phase equation there were plus symbols. This is due to the "carrier advance" through the ionosphere.)

$$\lambda\phi_{r,m}^s = \rho_{r,m}^s - cdt_m^s + cdt_{r,m} - \lambda N_r^s - \rho_{r,\text{ion},m}^s - \rho_{r,\text{trop},m}^s - \rho_{r,\text{rel},m}^s \tag{7.18}$$

---

[24]Note that the same applies for the code but for a total equivalent distance of 1 ms, corresponding to 300 km (the total length of the C/A code for instance for the GPS constellation). See Chapter 8 for details.
[25]Nevertheless, advanced mechanisms for carrier phase jump have been implemented to cope with this problem.

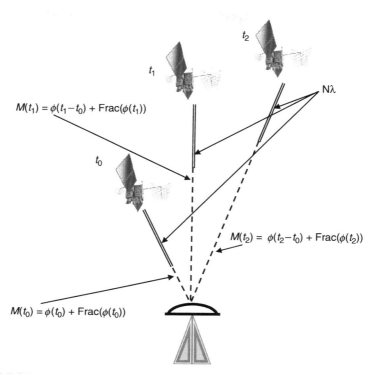

FIGURE 7.25 Carrier phase measurements.

where

- $\phi_{r,m}^s$ is the number of cycles observed at instant t_m;
- N_r^s is the ambiguity in carrier phase (unknown) at the initial time;
- λ is the wavelength of the carrier frequency.

The other parameters have already been described (in the previous section on "Code Phase Measurements").

Note that there is a fundamental difference between code phase and carrier phase measurements in the sense that carrier phase measurements are purely differential measurements. This means that it is not a direct distance measurement that is carried out, but a difference of distance. Thus, the quality and accuracy of the initial point must be of high grade in order to maintain the positioning accuracy to a few centimeters. It is also noticeable that propagation errors must also be eliminated as thoroughly as possible (to maintain accuracy). Indeed, differential techniques are used in order to reduce these errors in the navigation solution, as described in the next section.

Relative Techniques

At this point in the reader's overall understanding of positioning systems, some issues have to be addressed concerning positioning accuracy. When dealing with code correlation functions that exhibit a few meters of accuracy, it is "simple" to accept that bias errors of the same order of magnitude, namely propagation- or clock-related ones, have to be carefully modeled. Things are totally different when dealing with carrier phase measurements, where centimeter accuracy is reachable. With this level of accuracy, some questions have to be answered.

Concerning the representation, on a map for instance, of such an accurate position, one has to question the real accuracy of the map being used. Could it be possible this map has this level of accuracy or should this positioning only be considered as a sort of relative positioning with respect to the initial point located in a local reference frame? Some parts of the answer are given in Chapter 2 when dealing with a geographical representation of the world.

The other questions are relative to the errors to be considered in carrier phase measurement. First, let us recall briefly the major bias sources: propagation and clock biases. Of course, one now has to deal also with integer ambiguities. Furthermore, for clock biases, one should take into account both the satellite and the receiver clock biases, as the corrections broadcast by the navigation message are not precise enough to allow centimeter accuracy.[26] Thus the approach of carrier phase measurements, unlike that implemented for code phase measurements, is based on physical methods that allow all possible biases to be eliminated. The first steps of such an approach are illustrated in Figs 7.26 to 7.28, showing the so-called simple,

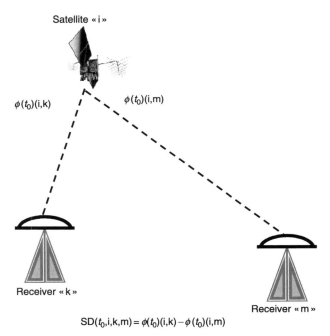

Satellite « i »

$\phi(t_0)(i,k)$ $\phi(t_0)(i,m)$

Receiver « k »

Receiver « m »

$$SD(t_0,i,k,m) = \phi(t_0)(i,k) - \phi(t_0)(i,m)$$

FIGURE 7.26 Illustration of a simple difference measurement technique.

---

[26]Note that 1 cm is equivalent to 33 ps. This is not accuracy for the clock bias provided by the satellite and associated ground control segment.

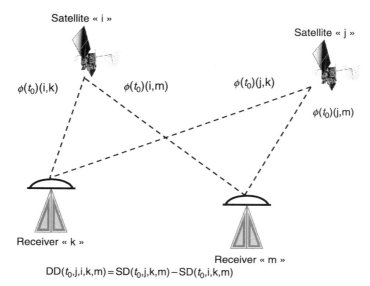

FIGURE 7.27 Illustration of a double difference measurement technique.

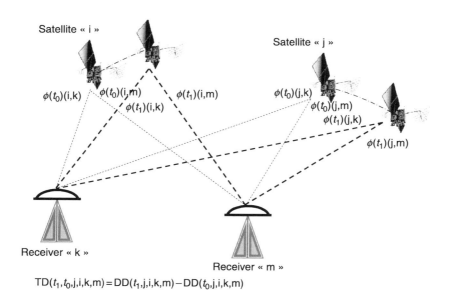

FIGURE 7.28 Illustration of a triple difference measurement technique.

double, and triple differences. The main idea is to implement differential approaches in order to physically remove certain biases by carrying out difference of quantities.

The simple difference, obtained by using two receivers and a single satellite, allows the clock bias of the satellite, to be completely removed by achieving the difference, at any given instant t, of the pseudo distances resulting from the signal processing of the two receivers. This is possible because the two receivers are not "too far" from each other and given the fact that the same satellite is considered for both receivers (see Fig. 7.26 for details).

The double difference is the difference of simple differences, for two satellites. It allows the clock bias of the receivers to be removed because the differences are carried out at the same instant for the two satellites (this aspect is fundamental; see Fig. 7.27 for details).

The triple difference (see Fig. 7.28) allows the removal of the integer ambiguities common to the preceding measurements (refer to the previous section for details), where only the changing cycles are accounted for. This is achievable by considering the difference of two double differences, at two successive measurement instants (this is the basis of the carrier phase measurement technique).

Furthermore, as one knows that the propagation errors are at least two orders of magnitude higher than the accuracy being searched for (around 1 m where one is seeking centimeter accuracy), there is the need for a method to eliminate these effects thoroughly. This can only be achieved by a local differential method, such as the one described above for troposphere effects,[27] and by the use of a dual-frequency method for removing the ionosphere effects.

In summary, it is possible to state that code phase based measurements allow an accuracy in the "few meters" range and are a compromise of basic techniques, both physical measurements and acceptable modeling. When dealing with more accurate positioning, there is the need for advanced methods, both in terms of measurement techniques and also physical methods, in order to remove all error factors.[28]

Another interesting approach of local and regional augmentation systems concerns the high accuracy carrier phase measurement networks intended to provide professionals with a high density of reference base stations, allowing centimeter accuracy positioning measurements in real time, over large areas. Such networks are based on specific so-called kinematic techniques and are available in many countries. They are usually operated by private companies that offer differential local accurate services.[29] The four main techniques are as follows:

- **The Static Geodesic Mode.** This consists in having fixed geodesic receivers carrying out carrier phase measurement and resolving ambiguity. The positioning is relative and requires enough time in order to solve the ambiguity. This time depends on meteorological conditions and the length of the base line between the two receivers (typically 1 h for a base line of 15 km).

---

[27]This is subject to the distance between the two receivers being close in order to avoid there being different meteorological conditions at both locations.

[28]Note that this could lead to different approaches depending on the environment considered. For instance, when dealing with indoor positioning using satellite-like signals, there is no longer any requirement concerning the dual-frequency need. Indeed, indoor propagation is very complex, but not affected by a dispersive medium like the ionosphere. Thus, there is no need for a dual-frequency receiver (or transmitter) when implementing an indoor phase measurement centimeter accuracy positioning system (see Chapter 10 for details).

[29]This has to be compared with global differential services provided by international providers such as EutelSat or Qualcomm. Refer to Chapter 3 for details on these satellite-based services.

- *The Rapid Static Mode.* This consists in using specific algorithms that allow a rapid ambiguity resolution once at least four satellites are tracked (a few minutes are enough). This is usually achieved in dual-frequency mode, although other modes are possible. Note also that a pivot is usually employed carrying out continuous measurements at the same location while other receivers are moving around (a few minutes on each new location to be found).

- *The Kinematic Mode.* This is the one described above in this chapter, which consists in having a fixed receiver and achieving positioning with the carrier phase differential method. The carrier phase of the rover receiver is followed by the electronics and once the integer ambiguities have been solved, only differential phase measurements are carried out. This gives the evolution of the relative displacement with very high accuracy. This works only when the carrier phase of the signal is not lost from one measurement to the next.

- *The Real-Time Kinematic Mode.* This is a special implementation of the previous mode, where moving resolution techniques are employed. They are quite efficient as long as at least five or six satellites are available. This last technique is the one implemented by the real time networks with base lines of up to 70–100 km. These quite large base lines are made possible by the use of interpolated data obtained from fixed base stations. The modeling thus realized is achieved on a grid that is updated with real measurements when available.

Precise Point Positioning

In addition to these two basic principles of measurement (code phase and carrier phase) as well as the relevant techniques, described above, some other techniques are also available. Among them is the so-called Precise Point Positioning (PPP).[30] This technique is based on the use of currently available precise GPS orbit and clock data. The International GNSS Service (IGS), for example, provides such data with various accuracies, depending on the delay of availability from real time. So-called "Ultra-rapid" products, available in real time, give around 10 cm of accuracy for the orbits and 5 ns for the clocks. Better accuracies are possible, with increased latency. One advantage of PPP compared to relative techniques is that only a single receiver is required, although there is the need for additional correction methods in order to mitigate some systematic effects that cannot be eliminated through differentiation. Moreover, the positioning refers to the global reference frame of the GNSS rather than to a local relative frame.

✅ 7.4 EXERCISES

Exercise 7.1 Give the basic principles of synchronization of the various components (codes, navigation messages, and carrier frequencies) of the GNSS signals. Consider the GPS signals for illustration purposes. Add comments on how synchronization is maintained over a long period.

---

[30]Do not confuse with "Private Public Partnership"!

Exercise 7.2 Give qualitative explanations of the huge search domain of satellite navigation signals. In particular, specify the importance of searching in both time and frequency directions.

Exercise 7.3 What is the difference between the code correlation function and the code discriminator? What are the possible code discriminators and their respective performance? (You need to do further reading in order to give a precise answer.)

Exercise 7.4 What solutions have been found to facilitate the acquisition of GNSS signals, both in terms of signals and receiver architectures?

Exercise 7.5 Describe in detail the code phase and the carrier phase measurement techniques. What are the major differences? Why are carrier phase measurements only carried out in a differential way? Is it possible, with carrier phase measurements, to use the estimation of the satellite clock bias obtained from the navigation message? Why?

Exercise 7.6 In what sense can the frequency locked loop help the delay locked loop? Why is it important to achieve such help? In which cases will the performance of a receiver be limited without this help? What are then the effects on positioning?

Exercise 7.7 Is there an interest in using a feedback loop for both frequency and delay? In order to answer one needs to analyze the real needs (time and frequency measurements) together with the characteristics of the signals (Doppler shift and constant variation of pseudo ranges). Give quantitative details for specific cases. For example, consider a low elevation satellite and a high elevation one, with static, dynamic, and high dynamic receiver modes.

Exercise 7.8 In order to increase sensitivity, one solution is to implement a long integration technique. Briefly describe the concept and interest of long integration. Can you also explain the different ways this could be achieved, either with data or dataless signals. What is the foreseeable difficulty of this technique when dealing with dynamic conditions? Give quantitative values of the corresponding limitations.

Exercise 7.9 What are the advantages of terrestrial differential networks? Would it be possible for you to imagine such networks for mass market applications? (maybe in an approach as not yet defined).

Exercise 7.10 Give a summary of your understanding of the carrier phase differential measurement method. In particular, clarify the issues underlying the use of simple, double and triple differences. Some people imagine using pseudolites (terrestrial GNSS-like transmitters) to reconstruct local positioning systems. How does the carrier phase technique of measurements apply to pseudolites? Give details of the three abovementioned differences in this specific case.

■ BIBLIOGRAPHY

Braasch MS, Van Dierendonck AJ. GPS receiver architectures and measurements. Proc IEEE 1999;87(1).

Gao Y. What is precise point positioning (PPP), and what are its requirements, advantages and challenges? InsideGNSS 2006;1(8):16–18.

Gill WJ. A comparison of binary delay-lock loop implementations. IEEE Trans Aerosp Electron Syst 1966;2:415–424.

Hartmann HP. Analysis of a dithering loop for PN code tracking. IEEE Trans Aerosp Electron Syst 1974;10(1):2–9.

Hein G, Avila-Rodriguez J-A, Wallner S. The Galileo code and others. InsideGNSS 2006;1(6): 62–75.

Jwo D-J. Optimization and sensitivity analysis of GPS receiver tracking loops in dynamic environments. IEE Proc.-Radar Sonar Navig 2001;148(4).

Kaplan ED, Hegarty C. Understanding GPS: principles and applications. 2nd ed. Artech House; 2006. Norwood, MA, USA.

Kouba J, Héroux P. GPS precise point positioning using IGS orbit products. GPS Solutions 2000;5(2):12–28.

Leick A. GPS satellite surveying. John Wiley and Sons; 2004, Hoboken, NJ, USA.

Meyr H, Ascheid G. Synchronization in digital communications. Vol. 1. John Wiley and Sons; 1990. New York, USA.

Oehler V. The Galileo integrity concept. In: ION GNSS 2004: Proceedings Long Beach (CA) September 2004.

Parkinson BW, Spilker Jr. JJ. Global positioning system: theory and applications. American Institute of Aeronautics and Astronautics; 1996.

Polydoros A, Weber CL. Analysis and optimization of correlative code-tracking loops in spread spectrum systems. Los Angeles (CA): Communication Sciences Institute; University of Southern California; August 1982. Rep. CSI-82-11-05.

Polydoros A, Weber CL. A unified approach to serial search spread-spectrum code acquisition — Parts I and II. IEEE Trans Commun 1984;32(5):542–560.

Polydoros A, Weber CL. Analysis and optimization of correlative code tracking loops in spread spectrum systems. IEEE Trans Commun 1985;33(1):30–43.

Simon MK. Noncoherent pseudonoise code tracking performance of spread spectrum receivers. IEEE Trans Commun 1977;25(3):327–345.

Spilker JJ. Delay-lock tracking of binary signals. IEEE Trans Space Electron Telem 1963;9:1–8.

Strazyk JA, Zhu Z. Average correlation for C/A code acquisition and tracking in frequency domain. Athens; School of Electrical Engineering and Computer Science; Ohio University. 2000.

Van Nee DJR, Coenen AJRM. New fast GPS code-acquisition technique using FFT. Electron Lett 1991;27(2):158–160.

Won J-H, Pany T, Hein GW. GNSS software defined radio — real receiver or just a tool for experts? InsideGNSS 2006;1(5):48–56.

Yost RA, Boyd RW. A modified PN code tracking loop: its performance analysis and comparative evaluation. IEEE Trans Commun 1982;30(5):1027–1036.

Web Links

http://www.Colorado.EDU/geography/gcraft/notes/gps/gps.html. Nov 14, 2007.

CHAPTER 8

Techniques for Calculating Positions

So far, the signal-related characteristics have been described and the main parameters to be dealt with are understood. It is quite clear that minimizing the measurement errors is of prime importance. But what exactly is the impact of a pseudo range error on the calculated position? To answer this fundamental question, one needs to know how position calculations are carried out. The complete solution is now explained, showing that GNSS are not only positioning systems, but can also provide a velocity determination of the receiver in a three-dimensional approach and for a precise time. Differences between the three constellations are highlighted.

This chapter starts with the "calculation" part that allows the full position, velocity, and time (PVT) result to be provided. The need for satellites' locations is obvious, and a short description of the calculations required is given. This will lead to an understanding of the real advantage of Satellite-Based Augmentation Systems (SBAS), dealt with in Section 8.6. Another very important point, linked to the uncertainties of the measurements, are the so-called Dilution of Precision (DOP) parameters. They indeed are indicators giving the impact of the spatial distribution of the satellites on the final accuracy of the position (described in Section 8.5). Finally, the effect of multipath is analyzed through the observation of the deformation of the correlation function.

■ 8.1 CALCULATING THE PVT SOLUTION

The basic method of calculating the PVT solution consists in exploiting the correlation output data in order to deal with corrected pseudo ranges. By "corrected," one has to understand that the distances used in the PVT module are considered as the

Global Positioning: Technologies and Performance. By Nel Samama
Copyright © 2008 John Wiley & Sons, Inc.

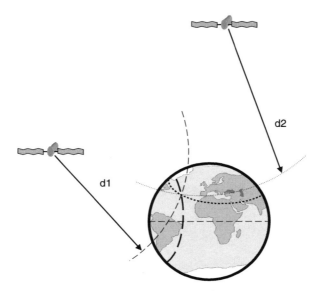

FIGURE 8.1 A typical pseudo range measurement configuration.

real Euclidian distances from the satellites to the receiver. Note also that velocity computations are obtained from Doppler measurements (and are not only the first derivative of position).

Basic Principles of Trilateration

The basic principle of trilateration has already been described in Chapter 3, but let us come back to an important point when dealing with mathematical methods of resolution, illustrated in Fig. 8.1. Two satellites are used in this figure with the associated d1 and d2 pseudo ranges. The first circles represented are centered on the satellites, respectively satellite 1 and 2, with radii of d1 and d2 respectively (note that the circle associated with satellite 1 is the dashed line and the one associated with satellite 2 is the dotted line). Let us now consider, for each individual satellite, the track of the d1 and d2 circles on the Earth's surface. The two tracks are represented as bold lines in Fig. 8.1 (with the dashed line and dotted lines corresponding to satellites 1 and 2). These intersect at two points (the second point is behind the represented face of the Earth in the present case). In this case, as we are not assuming a known altitude, another measurement is required.

Thus, there is a need for a third measurement. Note that this third pseudo range allows a non-ambiguous geometrical determination of the Earth position because there is an additional piece of information[1] (the fact that the position being searched for lies on the Earth's surface). Nevertheless, this is not used directly in the

---

[1] Remember that from a mathematical point of view, the intersection of three spheres (centered on the satellites) gives indeed two physical points, one being above the plane including the three satellites, thus not being on the Earth's surface.

FIGURE 8.2   Illustration of the expanding mechanism of position finding.

computations as it does not provide real simplification. In addition, there is still the need for a fourth satellite to allow time bias determination between the GNSS time and the receiver's time.

The other specific point to consider is related to measurement noise. The pseudo range is "noisy" and so it is possible to be in a situation where the actual mathematical computation of the three-sphere intersection does not exist. Thus, the method implemented for position finding must be able to compensate for this problem, roughly illustrated in Fig. 8.2. In the left-hand diagram of Fig. 8.2, the measurement errors are zero, leading to a perfect match of the mathematical solution. Of course, this configuration is far from being the actual situation — the two other graphs are indeed the reality, is that either the spheres do not intersect or the intersection is a sort of a spherical triangle. The latter case is the one that is sought by the resolution method in order to provide a positioning zone where the receiver is bound to be. In the case of the intermediate situation, the method has the role of enlarging the spheres by an equal quantity for the three satellites in order to reach the previous situation where the positioning is defined by a geographical zone.

Finally, one of the most subtle approaches for position finding is certainly associated with time bias processing. As already described in Chapter 3 and also detailed in the following section, the idea is to consider the receiver clock bias as a common unknown variable for all the signals,[2] as the measurements are carried out within the same millisecond, making them simultaneous with regard to the receiver clock drift (therefore negligible).

Coordinate System

For positioning calculation purposes, it is useful to use the Earth Centered Earth Fixed (ECEF) coordinate reference system (described in Chapter 2) that has the characteristic of turning with the Earth's rotation. Note that as the Earth is an ellipsoid, it means that the vertical of a terrestrial position does not cross the center of the Earth, unless it is on the equatorial plane or at the poles.

Furthermore, the (x, y, z) representation is not very useful (except for calculations) in the real world. Thus, the longitude, latitude, and altitude referential is more convenient. Nevertheless, it requires a way to model the Earth, and this is

---

[2]Note that considering the receiver clock bias as a path "common to all satellites" is an unusual approach, but is physically very meaningful.

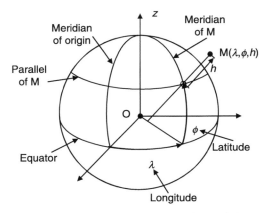

FIGURE 8.3 Longitude, latitude, and altitude representation and the associated ellipsoid.

commonly achieved by using ellipsoids (see Fig. 8.3) whose main characteristics, relative to the three main constellations, are described in Chapter 2.

An ellipsoid is completely characterized by its two semi-axes. In the case of GNSS-related ellipsoids, the larger semi-axis is always the radius of the Earth in the equatorial plane and the smaller semi-axis is the one corresponding to the distance from the pole to the center of the Earth. Calling a the greater and b the smaller semi-axis, it is possible to calculate the eccentricity e and the flatness parameter f of the ellipse as follows:

$$e = \sqrt{1 - \frac{b^2}{a^2}} \tag{8.1}$$

$$f = 1 - \frac{b}{a}. \tag{8.2}$$

It is then simple to obtain the longitude, latitude, and altitude coordinates of any receiver's coordinates (x_r, y_r, z_r) given in the ECEF referential. The longitude λ, being the angle measured in the equatorial plane, between the X-axis and the reported user position through a meridian onto the equatorial plane, is then given by

$$\lambda = \begin{cases} \arctan\left(\dfrac{y_r}{x_r}\right) \leftrightarrow x_r \geq 0 \\[2mm] 180° + \arctan\left(\dfrac{y_r}{x_r}\right) \leftrightarrow x_r \leq 0, \quad y_r \geq 0 \\[2mm] -180° + \arctan\left(\dfrac{y_r}{x_r}\right) \leftrightarrow x_r < 0, \quad y_r < 0 \end{cases} \tag{8.3}$$

(note that the longitude is counted positive when east of the reference meridian and negative when west of this meridian).

The latitude ϕ is the angle between the normal vector to the ellipsoid at the user's position and the equatorial plane. It is usually counted positive when north of the equatorial plane and negative when south.

The altitude is finally the shortest distance between the user's position and the ellipsoid (see Chapter 2 for the care required in the use of altitude data).

Sphere Intersection Approach

The basic method for calculating a position is given below. It is based on a Taylor's series development[3] of the first order of the pseudo ranges obtained from at least four satellites in order to cope with the three spatial coordinates, plus the clock time bias of the receiver. This resolution method can be seen as a sphere intersection related method, as compared to the approach described in the next section, which describes a hyperboloid intersection approach.

The system to be resolved is composed of the four pseudo range expressions (ρ_i) given below:

$$\rho_i = \sqrt{(x_i - x_r)^2 + (y_i - y_r)^2 + (z_i - z_r)^2} + ct_r, \tag{8.4}$$

where (x_r, y_r, z_r) is the position of the receiver (searched for) and (x_i, y_i, z_i) is the position of satellite "i" (which has to be calculated from the navigation data), and t_r the receiver's clock bias with respect to the constellation reference time. The locations of the satellites are those at the moment the signals are sent from the satellite to the receiver. Therefore, an iterative process has to be implemented. In fact, at the first iteration of the process, only an estimated value of the propagation time can be used (based on an estimation of the receiver's position), and must be checked at the end of this first iteration. In the case where a mismatch is observed, a new iteration is then required, and so on.

Thus, the principle of the resolution is to state an initial estimated position for the receiver.[4] This initial position is denoted $(\hat{x}_r, \hat{y}_r, \hat{z}_r)$. Furthermore, the solution vector of GNSS positioning is not only a spatial vector, but includes the time bias of the internal clock of the receiver leading to a real first estimate of the form $(\hat{x}_r, \hat{y}_r, \hat{z}_r, c\hat{t}_r)$. The last time coordinate is multiplied by the speed of light in order to obtain a homogeneous four-coordinate vector, that is, where all coordinates are given in meters.

This first position is then used to estimate the relative displacement from one iteration to the next. This quantity will be used as a convergence criterion for the algorithm. The difference between the real position and the estimated one is then represented by the vector $(\Delta\hat{x}_r, \Delta\hat{y}_r, \Delta\hat{z}_r, c\Delta\hat{t}_r)$.

One can define the function f as follows:

$$\rho_i = \sqrt{(x_i - x_r)^2 + (y_i - y_r)^2 + (z_i - z_r)^2} + ct_r = f(x_r, y_r, z_r, t_r). \tag{8.5}$$

---

[3]This is achieved for linearization purposes.
[4]Note that this is one of the interesting improvements offered by the Assisted GNSS (A-GNSS) approach. As the mobile telecommunication network is able to provide an approximate position (see Chapter 4 for details), it can be used as an estimated position for the receiver. Note that the accuracy of mobile network positioning is very good with regard to the current first position accuracy requirement.

It is also possible to define the function f for the estimated position as

$$\rho_i = \sqrt{(x_i - \hat{x}_r)^2 + (y_i - \hat{y}_r)^2 + (z_i - \hat{z}_r)^2} + c\hat{t}_r = f(\hat{x}_r, \hat{y}_r, \hat{z}_r, \hat{t}_r). \qquad (8.6)$$

Considering that the actual position of the receiver is given by

$$\begin{cases} x_r = \hat{x}_r + \Delta x_r \\ y_r = \hat{y}_r + \Delta y_r \\ z_r = \hat{z}_r + \Delta z_r \\ t_r = \hat{t}_r + \Delta t_r, \end{cases} \qquad (8.7)$$

then it follows that

$$f(x_r, y_r, z_r, ct_r) = f(\hat{x}_r + \Delta x_r, \hat{y}_r + \Delta y_r, \hat{z}_r + \Delta z_r, c\hat{t}_r + c\Delta t_r). \qquad (8.8)$$

Considering that near the convergence $(\Delta\hat{x}_r, \Delta\hat{y}_r, \Delta\hat{z}_r, c\Delta\hat{t}_r)$ will be small in comparison with $(\hat{x}_r, \hat{y}_r, \hat{z}_r, c\hat{t}_r)$, a first-order Taylor series development is possible and leads to

$$f(\hat{x}_r + \Delta x_r, \hat{y}_r + \Delta y_r, \hat{z}_r + \Delta z_r, \hat{t}_r + \Delta t_r)$$

$$= f(\hat{x}_r, \hat{y}_r, \hat{z}_r, \hat{t}_r) + \frac{\partial f(\hat{x}_r, \hat{y}_r, \hat{z}_r, c\hat{t}_r)}{\partial \hat{x}_r} \Delta x_r + \frac{\partial f(\hat{x}_r, \hat{y}_r, \hat{z}_r, c\hat{t}_r)}{\partial \hat{y}_r} \Delta y_r \qquad (8.9)$$

$$+ \frac{\partial f(\hat{x}_r, \hat{y}_r, \hat{z}_r, c\hat{t}_r)}{\partial \hat{z}_r} \Delta z_r + \frac{\partial f(\hat{x}_r, \hat{y}_r, \hat{z}_r, c\hat{t}_r)}{\partial \hat{t}_r} \Delta t_r.$$

When considering the intermediate variable,

$$\hat{r}_i = \sqrt{(x_i - \hat{x}_r)^2 + (y_i - \hat{y}_r)^2 + (z_i - \hat{z}_r)^2}. \qquad (8.10)$$

The equality gives

$$f(\hat{x}_r + \Delta x_r, \hat{y}_r + \Delta y_r, \hat{z}_r + \Delta z_r, \hat{t}_r + \Delta t_r)$$

$$= \hat{r}_i + c\hat{t}_r - \frac{x_i - \hat{x}_r}{\hat{r}_i} \Delta x_r - \frac{y_i - \hat{y}_r}{\hat{r}_i} \Delta y_r - \frac{z_i - \hat{z}_r}{\hat{r}_i} \Delta z_r + c\Delta t_r \qquad (8.11)$$

The partial derivatives are given by

$$\begin{cases} \dfrac{\partial f(\hat{x}_r, \hat{y}_r, \hat{z}_r, c\hat{t}_r)}{\partial \hat{x}_r} \Delta x_r = -\dfrac{x_i - \hat{x}_r}{\hat{r}_i} \Delta x_r \\[3mm] \dfrac{\partial f(\hat{x}_r, \hat{y}_r, \hat{z}_r, c\hat{t}_r)}{\partial \hat{y}_r} \Delta y_r = -\dfrac{y_i - \hat{y}_r}{\hat{r}_i} \Delta y_r \\[3mm] \dfrac{\partial f(\hat{x}_r, \hat{y}_r, \hat{z}_r, c\hat{t}_r)}{\partial \hat{z}_r} \Delta z_r = -\dfrac{z_i - \hat{z}_r}{\hat{r}_i} \Delta z_r \\[3mm] \dfrac{\partial f(\hat{x}_r, \hat{y}_r, \hat{z}_r, c\hat{t}_r)}{\partial \hat{t}_r} \Delta t_r = c. \end{cases} \qquad (8.12)$$

The final relation between estimated and actual pseudo ranges is then given by

$$\rho_i = \hat{\rho}_i - \frac{x_i - \hat{x}_r}{\hat{r}_i}\Delta x_r - \frac{y_i - \hat{y}_r}{\hat{r}_i}\Delta y_r - \frac{z_i - \hat{z}_r}{\hat{r}_i}\Delta z_r + c\Delta t_r, \tag{8.13}$$

or also

$$\hat{\rho}_i - \rho_i = \frac{x_i - \hat{x}_r}{\hat{r}_i}\Delta x_r + \frac{y_i - \hat{y}_r}{\hat{r}_i}\Delta y_r + \frac{z_i - \hat{z}_r}{\hat{r}_i}\Delta z_r - c\Delta t_r. \tag{8.14}$$

By defining a new set of intermediate variables,

$$\begin{cases} \Delta\rho = \hat{\rho}_i - \rho_i \\ a_{xi} = \dfrac{x_i - \hat{x}_r}{\hat{r}_i} \\ a_{yi} = \dfrac{y_i - \hat{y}_r}{\hat{r}_i} \\ a_{zi} = \dfrac{z_i - \hat{z}_r}{\hat{r}_i}, \end{cases} \tag{8.15}$$

the equation, for any given satellite, that has to be solved is now

$$\Delta\rho = a_{xi}\Delta x_r + a_{yi}\Delta y_r + a_{zi}\Delta z_r - c\Delta t_r. \tag{8.16}$$

When considering the four satellites required for a three-dimensional positioning, one has to deal with a system composed of four equations and four unknowns [being now the vector $(\Delta\hat{x}_r, \Delta\hat{y}_r, \Delta\hat{z}_r, c\Delta\hat{t}_r)$]. The system can be fully described by

$$\begin{cases} \Delta\rho_1 = a_{x1}\Delta x_r + a_{y1}\Delta y_r + a_{z1}\Delta z_r - c\Delta t_r \\ \Delta\rho_2 = a_{x2}\Delta x_r + a_{y2}\Delta y_r + a_{z2}\Delta z_r - c\Delta t_r \\ \Delta\rho_3 = a_{x3}\Delta x_r + a_{y3}\Delta y_r + a_{z3}\Delta z_r - c\Delta t_r \\ \Delta\rho_4 = a_{x4}\Delta x_r + a_{y4}\Delta y_r + a_{z4}\Delta z_r - c\Delta t_r. \end{cases} \tag{8.17}$$

Such equations take advantage of the matrix representation. So introducing the following

$$\Delta\rho = \begin{bmatrix} \Delta\rho_1 \\ \Delta\rho_2 \\ \Delta\rho_3 \\ \Delta\rho_4 \end{bmatrix} \qquad H = \begin{bmatrix} a_{x1} & a_{y1} & a_{z1} & 1 \\ a_{x2} & a_{y2} & a_{z2} & 1 \\ a_{x3} & a_{y3} & a_{z3} & 1 \\ a_{x4} & a_{y4} & a_{z4} & 1 \end{bmatrix} \qquad \Delta x = \begin{bmatrix} \Delta x_r \\ \Delta y_r \\ \Delta z_r \\ -c\Delta t_r \end{bmatrix}. \tag{8.18}$$

The system is finally expressed as

$$\Delta\rho = H\Delta x, \tag{8.19}$$

the solution of which is given by

$$\Delta x = H^{-1}\Delta\rho. \tag{8.20}$$

The reader should understand that one is interested in making the Δx vector equal to zero in order to find the position of the receiver, which, in this case, is for the last iteration the position considered as the initial estimate. It is thus an iterative method that has to be given a convergence criterion (as it is usually not possible to reach zero because of measurement uncertainties).

Given an initial point not too far away from the actual one and with pseudo ranges of acceptable quality (not too noisy), the convergence is typically achieved in fewer than ten iterations.[5]

All these calculations are carried out in the code phase related measurement approach. The carrier phase technique is described later in this chapter. Nevertheless, the same problem of ambiguity occurs for code phase measurements, as the distance between the satellites and the receiver is far more than just a code length. Thus, the pseudo range is not just the fractional part of the code (which is the actual measurement of the receiver signal processing part), but this fractional part plus a whole number of complete code equivalent distances.[6] So, the real pseudo range has the following complete form (note that n_i and "Code Phase Measurement" are given in ms):

$$PR_{\text{satellite i to receiver}} = n_i \times 299{,}792.458 + \text{Code Phase Measurement}$$
$$\times \, 299{,}792.458. \tag{8.21}$$

This leads to the equations set to be solved as being[7]

$$\begin{bmatrix} PR_{\text{satellite i to receiver}} \\ \quad = n_i \times 299{,}792.458 + \text{Code Phase Measurement (i)} \times 299{,}792.458 \\ PR_{\text{satellite j to receiver}} \\ \quad = n_j \times 299{,}792.458 + \text{Code Phase Measurement (j)} \times 299{,}792.458 \\ PR_{\text{satellite k to receiver}} \\ \quad = n_k \times 299{,}792.458 + \text{Code Phase Measurement (k)} \times 299{,}792.458 \\ PR_{\text{satellite m to receiver}} \\ \quad = n_m \times 299{,}792.458 + \text{Code Phase Measurement (m)} \times 299{,}792.458 \end{bmatrix}$$
$$\tag{8.22}$$

This is indeed not that different from the one solved in the first part of this section, giving some values of the ambiguities. These ambiguities can be determined using the Hand Over Word (HOW, refer to Fig. 6.11), which contains the time of the first data bit transition at the beginning of the next TLM (first bits of the next sub-frame, refer to Fig. 6.11). In fact, this data bit is synchronized with the C/A code, allowing an estimate of the transit time from the satellite to the receiver. Nevertheless, there are still 20 C/A codes in every TLM data bit (of 20 ms). Further computations are required to check (and eventually adjust) the perfect alignment between the 1 ms C/A code and the 20 ms TLM data bit.

These ambiguities can also be evaluated from the estimated initial position considered for the computation of the position, that is, from $(\Delta\hat{x}_r, \Delta\hat{y}_r, \Delta\hat{z}_r, c\Delta\hat{t}_r)$. In fact, the computation is then a double iteration process, where the first loop consists of evaluating a possible set of integer ambiguities and then calculating,

---

[5]In the case where the initial position is too far from the real point, the algorithm would not converge. Then, the receiver often "asks" the user for help, such as the country he/she is in. For not too large countries it is quite enough to consider the central position of this country for resolution of integer ambiguities.

[6]Remember that the code is 1 ms in time, equivalent to a little bit less than 300 km. The actual distance from satellite to receiver is typically between 19,100 km and 28,920 km (for all constellations).

[7]Note that the real speed of light that must be considered is 299,792,458 m/s, and nothing else!

through the second iterative loop, the resulting position. Then, a verification of the initial hypothesis concerning the integer ambiguities is required, possibly leading to a new updated set of integer ambiguities, thus a new second loop process.

Analytical Model of Hyperboloids

The method presented above is very powerful and always converges;[8] this has the advantage of having an output of the clock bias. However, it requires iterative computing. Another method allows an analytical solution of the receiver's position. The general case will be presented and then a discussion of the technique is given.

The system to be solved can be stated as follows (note that as we are going to carry out differences, there is no impact in not considering the clock bias effect at the start):

$$\begin{cases} d_1^2 = (x_1 - x_r)^2 + (y_1 - y_r)^2 + (z_1 - z_r)^2 \\ d_2^2 = (x_2 - x_r)^2 + (y_2 - y_r)^2 + (z_2 - z_r)^2 \\ d_3^2 = (x_3 - x_r)^2 + (y_3 - y_r)^2 + (z_3 - z_r)^2 \\ d_4^2 = (x_4 - x_r)^2 + (y_4 - y_r)^2 + (z_4 - z_r)^2, \end{cases} \tag{8.23}$$

where (x_i, y_i, z_i) is the position of satellite "i". Any given difference between any two equations is then easily obtained as

$$d_2^2 - d_1^2 = (x_2 - x_r)^2 - (x_1 - x_r)^2 + (y_2 - y_r)^2 - (y_1 - y_r)^2$$
$$+ (z_2 - z_r)^2 - (z_1 - z_r)^2. \tag{8.24}$$

This can be written in another form:

$$d_2^2 - d_1^2 = 2x_r(x_1 - x_2) + (x_1 + x_2)(x_2 - x_1)$$
$$+ 2y_r(y_1 - y_2) + (y_1 + y_2)(y_2 - y_1) \tag{8.25}$$
$$+ 2z_r(z_1 - z_2) + (z_1 + z_2)(z_2 - z_1).$$

By introducing intermediate variables that can be obtained from the satellite locations, the above expression is simplified as follows:

$$d_2^2 - d_1^2 = 2x_r\Delta X_{12} + \Sigma X_{12}\Delta X_{21} + 2y_r\Delta Y_{12} + \Sigma Y_{12}\Delta Y_{21} + 2z_r\Delta Z_{12}$$
$$+ \Sigma Z_{12}\Delta Z_{21} \tag{8.26}$$

where

$$\begin{array}{lll} \Delta X_{12} = x_1 - x_2 & \Delta Y_{12} = y_1 - y_2 & \Delta Z_{12} = z_1 - z_2 \\ \Delta X_{21} = x_2 - x_1 & \Delta Y_{21} = y_2 - y_1 & \Delta Z_{21} = z_2 - z_1 \\ \Sigma X_{12} = x_1 + x_2 & \Sigma Y_{12} = y_1 + y_2 & \Sigma Z_{12} = z_1 + z_2. \end{array} \tag{8.27}$$

This expression can then be obtained by making differences for all the equalities of the starting system, with respect to the first equation, that is,

---

[8]The only cases where it does not are mentioned in Chapter 10 concerning indoor positioning using repeaters.

considering the equation of satellite 1 as reference. The newly obtained system is thus given by

$$
\begin{cases}
d_2^2 - d_1^2 = 2x_r\Delta X_{12} + \Sigma X_{12}\Delta X_{21} + 2y_r\Delta Y_{12} + \Sigma Y_{12}\Delta Y_{21} + 2z_r\Delta Z_{12} \\
\qquad + \Sigma Z_{12}\Delta Z_{21} \\
d_3^2 - d_1^2 = 2x_r\Delta X_{13} + \Sigma X_{13}\Delta X_{31} + 2y_r\Delta Y_{13} + \Sigma Y_{13}\Delta Y_{31} + 2z_r\Delta Z_{13} \\
\qquad + \Sigma Z_{13}\Delta Z_{31} \\
d_4^2 - d_1^2 = 2x_r\Delta X_{14} + \Sigma X_{14}\Delta X_{41} + 2y_r\Delta Y_{14} + \Sigma Y_{14}\Delta Y_{41} + 2z_r\Delta Z_{14} \\
\qquad + \Sigma Z_{14}\Delta Z_{41},
\end{cases}
\tag{8.28}
$$

which can also be written as

$$
\begin{cases}
2x_r\Delta X_{12} + 2y_r\Delta Y_{12} + 2z_r\Delta Z_{12} = \left(d_2^2 - d_1^2\right) \\
\qquad - \left(\Sigma X_{12}\Delta X_{21} + \Sigma Y_{12}\Delta Y_{21} + \Sigma Z_{12}\Delta Z_{21}\right) & \text{(8.29a)} \\
2x_r\Delta X_{13} + 2y_r\Delta Y_{13} + 2z_r\Delta Z_{13} = \left(d_3^2 - d_1^2\right) \\
\qquad - \left(\Sigma X_{13}\Delta X_{31} + \Sigma Y_{13}\Delta Y_{31} + \Sigma Z_{13}\Delta Z_{31}\right) & \text{(8.29b)} \\
2x_r\Delta X_{14} + 2y_r\Delta Y_{14} + 2z_r\Delta Z_{14} = \left(d_4^2 - d_1^2\right) \\
\qquad - \left(\Sigma X_{14}\Delta X_{41} + \Sigma Y_{14}\Delta Y_{41} + \Sigma Z_{14}\Delta Z_{41}\right). & \text{(8.29c)}
\end{cases}
$$

By introducing the following new intermediate variables,

$$
\begin{cases}
D_{12} = \left(d_2^2 - d_1^2\right) - \left(\Sigma X_{12}\Delta X_{21} + \Sigma Y_{12}\Delta Y_{21} + \Sigma Z_{12}\Delta Z_{21}\right) \\
D_{13} = \left(d_3^2 - d_1^2\right) - \left(\Sigma X_{13}\Delta X_{31} + \Sigma Y_{13}\Delta Y_{31} + \Sigma Z_{13}\Delta Z_{31}\right) \\
D_{14} = \left(d_4^2 - d_1^2\right) - \left(\Sigma X_{14}\Delta X_{41} + \Sigma Y_{14}\Delta Y_{41} + \Sigma Z_{14}\Delta Z_{41}\right),
\end{cases}
\tag{8.30}
$$

it is possible to directly calculate these variables. Let us carry on with the calculation. From Equations 8.29a–c and through intermediate simple linear combinations, it is possible to get rid of z_r. The new system is as follows (where (a) to (c) denote the parts of Eq. 8.29):

$$
\begin{aligned}
\Delta Z_{13} \times \text{(a)} - \Delta Z_{12} \times \text{(b)} \Rightarrow & 2x_r\Delta X_{12}\Delta Z_{13} + 2y_r\Delta Y_{12}\Delta Z_{13} + 2z_r\Delta Z_{12}\Delta Z_{13} \\
& - 2x_r\Delta X_{13}\Delta Z_{12} - 2y_r\Delta Y_{13}\Delta Z_{12} - 2z_r\Delta Z_{13}\Delta Z_{12} \\
& = \Delta Z_{13}D_{21} - \Delta Z_{12}D_{31}
\end{aligned}
$$

$$\tag{8.31}$$

and

$$
\begin{aligned}
\Delta Z_{14} \times \text{(a)} - \Delta Z_{12} \times \text{(c)} \Rightarrow & 2x_r\Delta X_{12}\Delta Z_{14} + 2y_r\Delta Y_{12}\Delta Z_{14} + 2z_r\Delta Z_{12}\Delta Z_{14} \\
& - 2x_r\Delta X_{14}\Delta Z_{12} - 2y_r\Delta Y_{14}\Delta Z_{12} - 2z_r\Delta Z_{14}\Delta Z_{12} \\
& = \Delta Z_{14}D_{21} - \Delta Z_{12}D_{41}.
\end{aligned}
$$

$$\tag{8.32}$$

Introducing the following new intermediate variables,

$$\Delta XZ_{12}^{13} = \Delta X_{12}\Delta Z_{13} - \Delta X_{13}\Delta Z_{12}$$

$$\Delta YZ_{12}^{13} = \Delta Y_{12}\Delta Z_{13} - \Delta Y_{13}\Delta Z_{12}$$

$$\Delta XZ_{12}^{14} = \Delta X_{12}\Delta Z_{14} - \Delta X_{14}\Delta Z_{12}$$

$$\Delta YZ_{12}^{14} = \Delta Y_{12}\Delta Z_{14} - \Delta Y_{14}\Delta Z_{12},$$

(8.33)

leads to

$$\begin{cases} 2x_r\Delta XZ_{12}^{13} + 2y_r\Delta YZ_{12}^{13} = D_{21}\Delta Z_{13} - D_{31}\Delta Z_{12} \\ 2x_r\Delta XZ_{12}^{14} + 2y_r\Delta YZ_{12}^{14} = D_{21}\Delta Z_{14} - D_{41}\Delta Z_{12}. \end{cases}$$

(8.34)

This allows the calculation of the first two spatial variables, namely x_r and y_r. The solution is then

$$\begin{cases} x_r = \dfrac{1}{2}\dfrac{\left[\Delta YZ_{12}^{14}(D_{21}\Delta Z_{13} - D_{31}\Delta Z_{12}) - \Delta YZ_{12}^{13}(D_{21}\Delta Z_{14} - D_{41}\Delta Z_{12})\right]}{\left[\Delta XZ_{12}^{13}\Delta YZ_{12}^{14} - \Delta XZ_{12}^{14}\Delta YZ_{12}^{13}\right]} \\ y_r = \dfrac{1}{2}\dfrac{\left[\Delta XZ_{12}^{14}(D_{21}\Delta Z_{13} - D_{31}\Delta Z_{12}) - \Delta XZ_{12}^{13}(D_{21}\Delta Z_{14} - D_{41}\Delta Z_{12})\right]}{\left[\Delta YZ_{12}^{13}\Delta XZ_{12}^{14} - \Delta YZ_{12}^{14}\Delta XZ_{12}^{13}\right]}. \end{cases}$$

(8.35)

In order to obtain z_r, one can use Equation 8.29c. It follows that

$$z_r = \frac{1}{2}\frac{[D_{41} - 2x_r\Delta X_{14} - 2y_r\Delta Y_{14}]}{[\Delta Z_{14}]}.$$

(8.36)

Of course, if one wants to use a matrix-based approach, it is possible to directly solve the Equations 8.29 in the form (using Eq. 8.30)

$$\begin{bmatrix} \Delta X_{12} & \Delta Y_{12} & \Delta Z_{12} \\ \Delta X_{13} & \Delta Y_{13} & \Delta Z_{13} \\ \Delta X_{14} & \Delta Y_{14} & \Delta Z_{14} \end{bmatrix} \cdot \begin{bmatrix} x_r \\ y_r \\ z_r \end{bmatrix} = \frac{1}{2}\begin{bmatrix} D_{21} \\ D_{31} \\ D_{41} \end{bmatrix},$$

(8.37)

which leads to the solution

$$\begin{bmatrix} x_r \\ y_r \\ z_r \end{bmatrix} = \frac{1}{2}\begin{bmatrix} \Delta X_{12} & \Delta Y_{12} & \Delta Z_{12} \\ \Delta X_{13} & \Delta Y_{13} & \Delta Z_{13} \\ \Delta X_{14} & \Delta Y_{14} & \Delta Z_{14} \end{bmatrix}^{-1} \cdot \begin{bmatrix} D_{21} \\ D_{31} \\ D_{41} \end{bmatrix}.$$

(8.38)

The potential problem of this latter method is that direct distance measurements are not always available, as in the case of indoor positioning or mobile telecommunication network positioning, where only differences are known accurately. In such cases, this approach is still applicable but requires an assumption to be made on the initial value of d1 for example. Then, an iterative process should take place in order to find the ultimate solution.

Usually, more than four satellites are available for positioning. Note that in the Paris region there are always between 8 and 12 GPS satellites in the sky. In addition, with the current GLONASS constellation of around 15 satellites, there are also 4 or more satellites visible for about 80% of the time. One can easily imagine that typically 10 more satellites will be available when Galileo will be in operation, leading to a total of 25 to 30 signals for positioning. This will certainly allow improved "pick up" algorithms in order to use only the best ones, but there will be a lot more than four. A least-square technique is usually implemented in order to cope with more than four satellites; this approach is described in the following section "least-square method," for the linearization method. It is also applicable for the analytical one but will not be described.

Angle of Arrival-Related Mathematics

The basic idea in using angle measurements is to allow a reduction in the number of satellites required (or transmitting elements). In fact we can imagine applications or environments where the availability of four satellites to achieve three-dimensional positioning could be difficult. This number can be reduced to only two when using DOA determination. Because of the relative complexity and size of the antennas required, this approach is not intended to replace the classical one.

This section is devoted to theoretical aspects, including the basic equations of the problem, together with comments concerning the raw data required. Let us first define the positioning technique.

Let us consider Cartesian referential (Oxyz) in which all the positions are given. The position being sought is that of the mobile terminal and is defined as (x_r, y_r, z_r). So, it looks like the unknowns of the problem are only these three coordinates. Unfortunately, when dealing with DOA, one has to keep in mind that the measured angles need a reference frame, the coordinate system in which the angles are going to be given. This is very easy to carry out when the receiving antenna (the one where the angle measurements are carried out) presents a fixed attitude, in both horizontality and orientation. This is the case for GSM base antennas for example, but this is obviously not the case for a mobile terminal. By definition, the types of terminal we are dealing with are PDAs or cellular phones that do not exhibit any fixed attitude. In a car system, horizontality could be established more easily (which has a non-negligible impact on the system). In our case, dealing with the ultimate electronic terminal of the future, we have to add the three unknown rotation angles θ_x, θ_y, and θ_z, (respectively around the x-, y-, and z-axes) to the three unknown coordinates. Thus, we have to cope with a system with six unknowns; and we need to find at least six independent equations to be able to find the position.

Global Theory

The first assumption is that the locations of the two satellites S1 and S2 are available, say respectively (x_{s1}, y_{s1}, z_{s1}) and (x_{s2}, y_{s2}, z_{s2}). The calculations have to end with (x_r, y_r, z_r) and $(\theta_x, \theta_y, \theta_z)$. The basic measurements carried out are DOA ones, that is, (θ_1, Φ_1) from S1 and (θ_2, Φ_2) from S2. This means that for the six unknowns, DOA only provides four equations; and we shall need to find two others somewhere. This can be either from a third satellite or additional data from other sensors.

The measurements (θ_1, Φ_1) and (θ_2, Φ_2) are obtained in the local referential defined by $O'(x_m, y_m, z_m)$ and the local axes x', y', and z'. Within this referential, the directing vectors defined by the DOAs are

$$\begin{pmatrix} \sin\theta_1 \cos\Phi_1 \\ \sin\theta_1 \sin\Phi_1 \\ \cos\theta_1 \end{pmatrix} \quad \text{and} \quad \begin{pmatrix} \sin\theta_2 \cos\Phi_2 \\ \sin\theta_2 \sin\Phi_2 \\ \cos\theta_2 \end{pmatrix}. \tag{8.39}$$

The rotation matrices corresponding to θ_x, θ_y, and θ_z are given by

$$\mathbf{M}_{\theta_x} = \begin{pmatrix} 1 & 0 & 0 \\ 0 & \cos\theta_x & -\sin\theta_x \\ 0 & \sin\theta_x & \cos\theta_x \end{pmatrix}$$

$$\mathbf{M}_{\theta_y} = \begin{pmatrix} \cos\theta_y & 0 & -\sin\theta_y \\ 0 & 1 & 0 \\ \sin\theta_y & 0 & \cos\theta_y \end{pmatrix} \tag{8.40}$$

$$\mathbf{M}_{\theta_z} = \begin{pmatrix} \cos\theta_z & -\sin\theta_z & 0 \\ \sin\theta_z & \cos\theta_z & 0 \\ 0 & 0 & 1 \end{pmatrix}.$$

It is then possible to return to the direction vectors in the $Oxyz$ referential by simply applying the following transformations to the initial directing vectors [note that the new DOAs are now defined by (θ_1', Φ_1') and (θ_2', Φ_2')]:

$$\begin{pmatrix} \sin\theta_1' \cos\Phi_1' \\ \sin\theta_1' \sin\Phi_1' \\ \cos\theta_1' \end{pmatrix} = \mathbf{M}_{\theta_x}^{-1}\mathbf{M}_{\theta_y}^{-1}\mathbf{M}_{\theta_z}^{-1} \begin{pmatrix} \sin\theta_1 \cos\Phi_1 \\ \sin\theta_1 \sin\Phi_1 \\ \cos\theta_1 \end{pmatrix}$$

and

$$\begin{pmatrix} \sin\theta_2' \cos\Phi_2' \\ \sin\theta_2' \sin\Phi_2^i \\ \cos\theta_2' \end{pmatrix} = \mathbf{M}_{\theta_x}^{-1}\mathbf{M}_{\theta_y}^{-1}\mathbf{M}_{\theta_z}^{-1} \begin{pmatrix} \sin\theta_2 \cos\Phi_2 \\ \sin\theta_2 \sin\Phi_2 \\ \cos\theta_2 \end{pmatrix}. \tag{8.41}$$

These two equations allow us to calculate the different angles in the $Oxyz$ referential (note that θ_x, θ_y, and θ_z are embedded in the previous equations). From this point, it is possible to define the equations of the straight lines from satellites 1 and 2 to the mobile position, respectively:

$$a_1 x_r + b_1 y_r + c_1 z_r + d_1 = 0$$
$$A_1 x_r + B_1 y_r + C_1 z_r + D_1 = 0$$

and

$$a_2 x_r + b_2 y_r + c_2 z_r + d_2 = 0$$
$$A_2 x_r + B_2 y_r + C_2 z_r + D_2 = 0, \tag{8.42}$$

where

$$\begin{pmatrix} a_1 \\ b_1 \\ c_1 \end{pmatrix} = \begin{pmatrix} -\cos \theta_1' \cos \Phi_1' \\ -\cos \theta_1' \sin \Phi_1' \\ \sin \theta_1' \end{pmatrix} \quad \text{and} \quad \begin{pmatrix} A_1 \\ B_1 \\ C_1 \end{pmatrix} = \begin{pmatrix} \sin \Phi_1' \\ -\cos \Phi_1' \\ 0 \end{pmatrix}, \quad (8.43)$$

and

$$d_1 = -[a_1 x_{s_1} + b_1 y_{s_1} + c_1 z_{s_1}]$$
$$D_1 = -[A_1 x_{s_1} + B_1 y_{s_1} + C_1 z_{s_1}]$$

and

$$d_2 = -[a_2 x_{s_2} + b_2 y_{s_2} + c_2 z_{s_2}]$$
$$D_2 = -[A_2 x_{s_2} + B_2 y_{s_2} + C_2 z_{s_2}]$$

$$(8.44)$$

The system is now made up of equations (8.42), including our six unknown variables (x_r, y_r, z_r) and $(\theta_x, \theta_y, \theta_z)$. At this point different strategies could be followed, as two equations are missing:

- The use of an inertial aiding system, in order to obtain two pieces of complementary data. For example, this could be achieved with a two-dimensional accelerometer or gyroscope (or both) to define the horizontality of the mobile terminal.

- The combined use of a one-dimensional accelerometer and a satellite measurement that allows us to obtain the distance separating a satellite from the mobile terminal.

- The implementation of some orientation devices to obtain the absolute orientation of the terminal (namely magnetometers). We know that electronic compasses can be disturbed in certain environments, depending on the amount of metallic mass near the compass.

- The fact that the terminal is built in a known environment (a car for instance), where the horizontality is defined (the two DOAs allow us to calculate the position of the car).

- Certainly some other possibilities, depending on specific application requirements.

Least-Square Method

When more than four satellites are available, the linearization method can be specified by

$$\Delta\rho = \begin{bmatrix} \Delta\rho_1 \\ \Delta\rho_2 \\ \cdots \\ \Delta\rho_n \end{bmatrix} \quad H = \begin{bmatrix} a_{x1} & a_{y1} & a_{z1} & 1 \\ a_{x2} & a_{y2} & a_{z2} & 1 \\ \cdots & \cdots & \cdots & \cdots \\ a_{xn} & a_{yn} & a_{zn} & 1 \end{bmatrix} \quad \Delta x = \begin{bmatrix} \Delta x_r \\ \Delta y_r \\ \Delta z_r \\ -c\Delta t_r \end{bmatrix}, \quad (8.45)$$

where $\Delta\rho$ is an $N \times 1$ vector, H an $N \times 4$ matrix, and Δx an 4×1 vector. The relation $\Delta\rho = H\Delta x$ is still valid. Given the fact that measurements are noisy, it is possible to introduce a residual vector calculated as follows:

$$r = H\Delta x - \Delta\rho. \tag{8.46}$$

The basic idea of the least-square method is to minimize the sum of the square of the residuals. This sum has the form

$$r_1^2 + \cdots + r_n^2 = (H\Delta x - \Delta\rho)^2 = (H\Delta x - \Delta\rho)^T(H\Delta x - \Delta\rho). \tag{8.47}$$

Minimizing this quantity is achieved when its gradient is zero, leading to the expression

$$\nabla\left(r_1^2 + \cdots + r_n^2\right) = 2\Delta x^T \cdot H^T \cdot H - 2\Delta\rho^T \cdot H = 0, \tag{8.48}$$

which is, finally,

$$\Delta x = (H^T \cdot H)^{-1} \cdot H^T \cdot \Delta\rho. \tag{8.49}$$

Calculation of Velocity

Although the receiver's velocity can be obtained from a derivation of the position at different times, this is not the actual way velocity is currently calculated. In fact, the receiver must carry out two separate measurements in order to find the correlation: time and Doppler. The second is closely linked to the respective velocity of the receiver and the satellite.[9] The induced Doppler shift is given by

$$\Delta f = f_{ui} - f_{ri} = \frac{v_{rr}}{c} f_{ti} = \frac{(v_{si} - v_r) \cdot a_i}{c} f_{ti}, \tag{8.50}$$

where f_{ti} is the transmitted frequency of satellite i, f_{ri} the received frequency from satellite i, v_{rr} the radial projection of the relative velocity between the satellite and the receiver (the projection onto the line of sight between the satellite and the receiver), v_{si} the velocity of the satellite, v_r the velocity of the receiver, and a_i the pointing vector along the line of sight from receiver to satellite i. Finally, c is the speed of light, assumed to be the speed of the signal.

Although the real transmitted frequency is not exactly the oscillator's nominal value, due to bias at the satellite, this value is corrected by the ground segment and transmitted to the receiver through the navigation data, and we are not going to take it into account. The receiver biases are more important in quantity and cannot be neglected. As for the position, the calculation method must take into account the internal receiver clock drift as a variable. This is achieved by considering that the received frequency f_{ri} differs from the purely Doppler-induced received frequency f_i as follows:

$$f_{ri} = f_i(1 + \dot{t}_r), \tag{8.51}$$

where $\dot{t}_r$ is the receiver's clock drift, that is, the derivative of the receiver's clock bias considered in the positioning method. One can then quite easily obtain the following

---

[9]Refer to Chapter 6 for details concerning the Doppler effect and related considerations with respect to GNSS acquisition of signals.

expression, valid for each satellite:

$$v_{si} \cdot a_i - c\frac{f_{ti} - f_i}{f_{ti}} = v_r \cdot a_i - c\dot{t}_r\frac{f_i}{f_{ti}}, \tag{8.52}$$

of which another possible form is

$$v_{xi}a_{xi} + v_{yi}a_{yi} + v_{zi}a_{zi} - c\frac{f_{ti} - f_i}{f_{ti}} = \dot{x}_r a_{xi} + \dot{y}_r a_{xi} + \dot{z}_r a_{xi} - c\dot{t}_r\frac{f_i}{f_{ti}}. \tag{8.53}$$

By introducing the following intermediate variable and considering that $v_{si} = (v_{xi}, v_{yi}, v_{zi})$, $v_r = (\dot{x}_r, \dot{y}_r, \dot{z}_r)$, and $a_i = (a_{xi}, a_{yi}, a_{zi})$, this can also be written as

$$d_i = v_{xi}a_{xi} + v_{yi}a_{yi} + v_{zi}a_{zi} - c\frac{f_{ti} - f_i}{f_{ti}}. \tag{8.54}$$

Note that d_i is fully determined once the receiver's loops have achieved the correlation process, because the real received frequency has been defined quite accurately. The fundamental equation to deal with is now

$$d_i = \dot{x}_r a_{xi} + \dot{y}_r a_{xi} + \dot{z}_r a_{xi} - c\dot{t}_r\frac{f_i}{f_{ti}}. \tag{8.55}$$

Taking into account both the Doppler shift due to the satellite displacement (which lies within ± 5 kHz) and the additional shift induced by the receiver clock drift, even for large drift, the total shift of the received frequency is usually within ± 10 kHz. This leads to the simplified form:

$$d_i = \dot{x}_r a_{xi} + \dot{y}_r a_{xi} + \dot{z}_r a_{xi} - c\dot{t}_r, \tag{8.56}$$

where, as for the positioning approach, there are four unknowns $(\dot{x}_r, \dot{y}_r, \dot{z}_r, c\dot{t}_r)$. Thus, four measurements are enough to find a solution. The following matrices can be introduced:

$$d = \begin{bmatrix} d_1 \\ d_2 \\ d_3 \\ d_4 \end{bmatrix} \qquad H = \begin{bmatrix} a_{x1} & a_{y1} & a_{z1} & 1 \\ a_{x2} & a_{y2} & a_{z2} & 1 \\ a_{x3} & a_{y3} & a_{z3} & 1 \\ a_{x4} & a_{y4} & a_{z4} & 1 \end{bmatrix} \qquad g = \begin{bmatrix} \dot{x}_r \\ \dot{y}_r \\ \dot{z}_r \\ -c\dot{t}_r \end{bmatrix}. \tag{8.57}$$

The system to be solved is then simply (note that the H matrix is identical to that used for positioning)

$$d = Hg, \tag{8.58}$$

or for the velocity vector g, the general form

$$g = H^{-1}d. \tag{8.59}$$

This system can be solved through a similar iterative method as that developed for positioning. Note that this allows the determination of the clock drift, which can help in specific applications where there is a need to follow the clock bias accurately (a typical example is indoor positioning using repeaters).

An important point to understand is that in order to compute the velocity vector one needs to know the position of the receiver (through the H matrix). As a matter of fact, this is required because one has to deal with the line of sight projection of the relative velocity. This can only be achieved when making an assumption about the receiver's position. Usually, the velocity calculation is carried out once the positioning has been determined.

Another important point is to understand that this velocity is not a mean value of position evolution, but rather instantaneous velocity estimation. Indeed, the Doppler measurements are based on instantaneous shift of the oscillator frequency (provided for by the phase lock loop, PLL), thus leading to the determination of the velocity. The accuracy of this velocity vector is linked to the accuracy of the oscillator, which is typically a fraction of Hertz. The corresponding accuracy of the radial velocity determined is then a fraction of 0.19 m/s.[10] The error induced by the positioning inaccuracy can also be estimated but remains negligible.

Calculation of Time

The navigation message provides corrections to apply to the GNSS time in order to synchronize with UTC, either USNO for GPS, SU for GLONASS, or BIPM for Galileo. We have already dealt with the corrections to be applied to each individual satellite clock in order to match the satellite's time with the GNSS time. In addition, the receiver also exhibits the capability to provide users with very accurate time information. The main advantage of this is that, wherever located, as soon as it can receive GNSS signals, any receiver can provide a reference of time. Moreover, this may be used in many scientific applications as well as for Internet synchronization for example. The accuracy lies typically within 10 ns.

There are two ways to obtain this time information, using Equation 8.4. The data one wants to obtain is, of course, t_r. If the receiver is located at a known position, then only one satellite is required because the navigation message will provide all the data required in order to process t_r, which is the receiver clock bias with respect to the GNSS time. Then, once again using the navigation data, it will be possible to return to the real UTC time, if needed. Where the position is not known, then the complete computation of the position must be carried out, thus requiring at least four satellites.

■ 8.2 SATELLITE POSITION COMPUTATIONS

The locations of the satellites are of prime importance and their accuracy is also a major concern.[11] The complete algorithm used in order to compute such locations is not within the scope of this book and can be found in the respective Interface Control Documents (ICDs) for the GPS, GLONASS, and Galileo constellations. Nevertheless, let us describe the principles.

---

[10]Obtained by applying the equation $\Delta f = f \cdot c v_r / c$.

[11]It is even more important in the case of carrier phase based positioning when accuracy to the centimeter is needed.

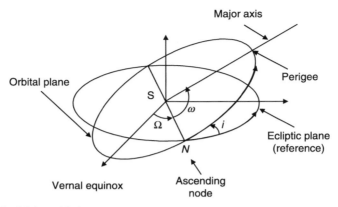

FIGURE 8.4 Major orbital parameters.

Within the navigation message some parameters are included that allow the receiver to compute the location of the satellite at any given time. Furthermore, the time to consider is the sending time and not the receiving time. The propagation time, which is precisely the measurement achieved, is not known. Thus, this approach allows multiple calculations and corrections when required. The part of the navigation message devoted to the orbital parameters is called the ephemeris. It is composed of 16 parameters as follows (illustration of parameters is given in Fig. 8.4).

The seven elliptic orbital parameters are

- e, eccentricity of the ellipse;
- $\sqrt{a}$, square root of the larger semi-axis of the ellipse;
- i_0, inclination angle of the orbit at instant t_{oe} (time of ephemeris);
- t_{oe}, the time all the abovementioned parameters were last updated (where this time is too far from current time, the corresponding satellite is excluded from the set of acceptable satellites for positioning purposes);
- Ω_0, longitude of the ascending node (corresponding to the angle between the x-axis in ECEF and the ascending node in the equatorial plane);
- ω, argument of the perigee at time t_{oe} (corresponding to the angle between the ascending node and the direction of the perigee in the orbital plane of the satellite);
- M_0, the mean anomaly at time t_{oe}.

There are also nine correction parameters:

- di/dt, the variation of the inclination angle with time;
- $\dot{\Omega}$, the variation of the longitude of the ascending node with time;
- Δn, the correction of $n = dM/dt$;
- C_{uc}, the amplitude of the correction of the latitude cosine;
- C_{us}, the amplitude of the correction of the latitude sine;

- C_{rc}, the amplitude of the correction of the orbital radius cosine;
- C_{rs}, the amplitude of the correction of the orbital radius sine;
- C_{ic}, the amplitude of the correction of the inclination angle cosine;
- C_{is}, the amplitude of the correction of the inclination angle sine.

Note that i_0, Ω_0, ω, M_0, di/dt, Ω, and Δn are given in "semi-circles" in the navigation message. To convert them into radians, one has to multiply these values by π.[12]

The calculations are then carried out in ECEF at any time t (details of calculations are not provided here). Note that the first six parameters are those used for the so-called almanac data, which are designed in order to provide the receiver with a rough idea of the constellation for satellite search choice purposes (in first acquisition mode). Ephemeris is specific to a given satellite and must be regularly updated by the ground segment for increased positioning accuracy.

■ 8.3 QUANTIFIED ESTIMATION OF ERRORS

From all the formulas considered above, it is possible to consider the remaining errors for both the pseudo ranges and the positions, once the positioning has been calculated, as being given by

$$\begin{aligned} \varepsilon_\rho &= \Delta\rho - (\rho_{\text{real}} - \rho_x) \\ \varepsilon_x &= \Delta x - (x_{\text{real}} - x_x), \end{aligned} \tag{8.60}$$

where ρ_{real} is the pseudo range and ρ_x is the one finally used, as x_{real} and x_x are the corresponding values of the position. Then it can be shown that

$$\varepsilon_x = \left[\left(H^T H \right)^{-1} H^T \right] \varepsilon_\rho = K \varepsilon_\rho, \tag{8.61}$$

where the K matrix only depends on the respective geometries of the satellites and the receiver. It is then possible to define the covariance matrix (cov) of the position error as being.[13]

$$\text{cov}(\varepsilon_x) = E\left[\varepsilon_\rho (\varepsilon_\rho)^T \right], \tag{8.62}$$

where E represents the expectation operator. Considering the geometry as fixed, one obtains

$$\text{cov}(\varepsilon_x) = \left(H^T H \right)^{-1} \text{cov}(\varepsilon_\rho). \tag{8.63}$$

With the additional assumption that the components of $d\rho$ are identical and equally distributed for all satellites, with a variance equal to the square of the satellite

---

[12]The value of π that must be used is 3.1415926535898 exactly!
[13]This is possible as it is usually considered that $d\rho$ has components that are Gaussian distributed with a zero mean.

TABLE 8.1 Relative importance of error sources on pseudo range measurement (given in meters, for GPS).

Error Sources	PPS Service 1σ (m)	SPS Service 1σ (m)
Spatial Segment		
Clock stability	3.0	3.0
Acceleration uncertainties	1.0	1.0
Other	0.5	0.5
Ground Segment		
Ephemeris	4.2	4.2
Other	0.9	0.9
Propagation		
Ionosphere	2.3	4.9–9.8
Troposphere	2.0	2.0
Multipath	1.2	2.5
User Segment		
Receiver	0.2	1.5
Other	0.5	0.5
UERE (user equivalent range error)	6.3	8.1–11.7

PPS, Precise Positioning Service; SPS, Standard Positioning Service.

user equivalent range error (UERE), this gives

$$\mathrm{cov}(\varepsilon_x) = (\mathrm{H}^{\mathrm{T}}\mathrm{H})^{-1}\sigma_{\mathrm{UERE}}^2 = M\sigma_{\mathrm{UERE}}^2. \qquad (8.64)$$

This expression indicates that the components of the M matrix allow translation from the pseudo range errors into the covariance of the position errors. This is also going to be used to define the so-called Dilution Of Precision coefficients (see Section 8.5). This mathematical expression should nevertheless be linked to real values in order to allow an understanding of the major trends.

Estimation of the impact of error sources on the pseudo range measurements is available for the GPS constellation, because it has been in operation for many years. Error sources were described in Chapter 6, but induced errors on the pseudo ranges are given in Table 8.1, classified by segment contributions. The two proposed services, the Precise Positioning Service[14] (PPS) and the Standard Positioning Service[15] (SPS) are presented and errors are given at 1σ. It appears quite clearly from this table that a better accuracy can be reached by finding methods that allow reduction of the most important budget lines, namely the ephemeris and the iono-sphere propagation. The reader can now fully understand the reasons why most research is leading in these two directions, notably with the SBAS systems that address both problems. The other remarkable point is the fact that even with PPS, centimeter accuracy is not achievable without specific measurement procedures in order to physically eliminate some error sources (see Chapter 7 for details).

---

[14]Based on a dual-frequency mode.
[15]Based on a single-frequency mode.

TABLE 8.2 Position error versus pseudo range error for pseudo range errors greater than 10'm.

$\varepsilon\rho$	0	10	20	40	60	80	100
0	0	19.6	39.2	78.4	117.6	156.7	195.9
10	9.1	15.0	33.7	72.5	111.6	150.7	189.9
20	18.1	15.2	29.9	67.3	106.0	145.0	184.0
30	27.2	20.0	28.7	63.1	101.0	139.6	178.5
40	36.3	27.1	30.3	59.9	96.6	134.7	173.3
50	45.4	35.2	34.3	58.0	92.8	130.2	168.4
60	54.5	43.6	40.0	57.5	89.9	126.2	163. 8
70	63.5	52.3	46.8	58.4	87.7	122.7	159.6
80	72.6	61.1	54.2	60.6	86.5	119.8	155.9
90	81.7	69.9	62.1	64.1	86.2	117.6	152. 6
100	90.8	78.8	70.3	68.6	86.9	116.0	149.8

■ 8.4 IMPACT OF PSEUDO RANGE ERRORS ON THE COMPUTED POSITIONING

In such a situation, it could be interesting to also carry out some computations to "feel" the way in which a position is modified with respect to a given error in the pseudo ranges. A typical case is considered in Tables 8.2 and 8.3 (for example purposes).

Tables 8.2 and 8.3 provide, for a specific and identical satellite configuration, the induced position error when considering pseudo range errors on two satellites (out of four used for the position computation, assuming the two remaining satellites to be free of error). The results of Table 8.2 could be due to multipath and the results from Table 8.3 could be typical errors due to ionosphere propagation or clock bias. What can be observed is that there is effectively a linear behavior described by the

TABLE 8.3 Position error versus pseudo range error for pseudo range errors between 1 and 10 m.

$\varepsilon\rho$	0	1	2	4	6	8	10
0	0	2.0	3.9	7.8	11.8	15.7	19.6
1	0.9	1.5	3.4	7.2	11.2	15.1	19.0
2	1.8	1.5	3.0	6.7	10.6	14.5	18.4
3	2.7	2.0	2.9	6.3	10.1	14.0	17.9
4	3.6	2.7	3.0	6.0	9.7	13.5	17.3
5	4.5	3.5	3.4	5.8	9.3	13.0	16.8
6	5.5	4.4	4.0	5.7	9.0	12.6	16.4
7	6.4	5.2	4.7	5.8	8.8	12.3	16.0
8	7.3	6.1	5.4	6.1	8.6	12.0	15.6
9	8.2	7.0	6.2	6.4	8.6	11.8	15.3
10	9.1	7.9	7.0	6.9	8.7	11.6	15.0

equations of Section 8.3, but multiple errors can compensate for each other. When errors from various satellites are not identical, the resulting error can be lower than expected. The ultimate case is obtained when the errors of pseudo ranges are identical for the four pseudo ranges. In such a case, the position is not affected as the common bias slips directly to the clock bias component of the navigation solution.

■ 8.5 IMPACT OF GEOMETRICAL DISTRIBUTION OF SATELLITES AND RECEIVER (NOTION OF DOP)

Remember that the induced position error is directly linked to the pseudo range errors by Equation 8.64, where the M matrix only depends on the geometrical configuration of the constellation and the receiver. Let us introduce the following notations:

$$
\mathbf{M} = \begin{bmatrix} M_{11} & M_{12} & M_{13} & M_{14} \\ M_{21} & M_{22} & M_{23} & M_{24} \\ M_{31} & M_{32} & M_{33} & M_{34} \\ M_{41} & M_{42} & M_{43} & M_{44} \end{bmatrix} \quad \mathrm{cov}(\varepsilon_x) = \begin{bmatrix} \sigma^2_{x_r} & \sigma^2_{x_r y_r} & \sigma^2_{x_r z_r} & \sigma^2_{x_r ct_r} \\ \sigma^2_{x_r y_r} & \sigma^2_{y_r} & \sigma^2_{y_r z_r} & \sigma^2_{y_r ct_r} \\ \sigma^2_{x_r z_r} & \sigma^2_{y_r z_r} & \sigma^2_{z_r} & \sigma^2_{z_r ct_r} \\ \sigma^2_{x_r ct_r} & \sigma^2_{y_r ct_r} & \sigma^2_{z_r ct_r} & \sigma^2_{ct_r} \end{bmatrix}.
$$

$$(8.65)$$

It is then possible to define some DOP (dilution of precision) terms as being the square root of some components of the trace of the M matrix. This represents the impact of the geometric scattering of the satellites, with respect to the receiver's position, on the position error, and hence on the accuracy of the positioning. In other words, it means that there is a linear relation between the DOP values and the resulting position accuracy for a given pseudo range error value. Thus, the DOP coefficients are of prime importance and the corresponding geometrical distribution of the satellites that are used for the measurements is of fundamental importance. Five DOP parameters are usually considered, depending on the type of data one wants to deal with:

$$ \mathrm{GDOP} = \sqrt{M_{11}^2 + M_{22}^2 + M_{33}^2 + M_{44}^2}, \tag{8.66a} $$

$$ \mathrm{PDOP} = \sqrt{M_{11}^2 + M_{22}^2 + M_{33}^2}, \tag{8.66b} $$

$$ \mathrm{HDOP} = \sqrt{M_{11}^2 + M_{22}^2}, \tag{8.66c} $$

$$ \mathrm{VDOP} = \sqrt{M_{33}^2}, \tag{8.66d} $$

$$ \mathrm{TDOP} = \sqrt{M_{44}^2}/c, \tag{8.66e} $$

where GDOP is the Geometrical DOP, PDOP the Position DOP, HDOP the Horizontal DOP, VDOP the vertical DOP, and TDOP the Time DOP. These can

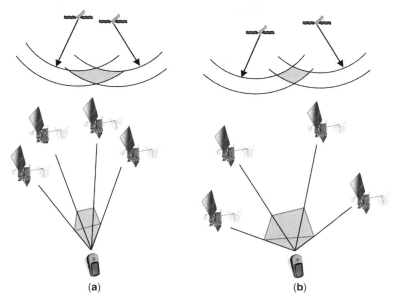

(a)   (b)

FIGURE 8.5   Poor (**a**) and good (**b**) constellation geometries.

also be written using the coefficient of the covariance matrix as follows:

$$\text{GDOP} \cdot \sigma_{\text{UERE}} = \sqrt{\sigma_{x_r}^2 + \sigma_{y_r}^2 + \sigma_{z_r}^2 + \sigma_{ct_r}^2}, \tag{8.67a}$$

$$\text{PDOP} \cdot \sigma_{\text{UERE}} = \sqrt{\sigma_{x_r}^2 + \sigma_{y_r}^2 + \sigma_{z_r}^2}, \tag{8.67b}$$

$$\text{HDOP} \cdot \sigma_{\text{UERE}} = \sqrt{\sigma_{x_r}^2 + \sigma_{y_r}^2}, \tag{8.67c}$$

$$\text{VDOP} \cdot \sigma_{\text{UERE}} = \sigma_{z_r}, \tag{8.67d}$$

$$\text{TDOP} \cdot \sigma_{\text{UERE}} = \sigma_{ct_r}. \tag{8.67e}$$

It is also possible to graphically show the impact of the geometrical scattering of the satellite (the DOP) on the resulting accuracy by the way of two different geometries. Figure 8.5 illustrates this matter. In summary, one can consider that if two satellites are too close to each other then the corresponding pseudo ranges (with their associated errors) will not give sufficient discrimination to decrease the uncertainties of position (the ultimate absurd situation is obtained if the two satellites were at the same location).

So, one can say that if the satellites are evenly distributed around the receiver, then the DOP will be quite good, that is, a small value.[16] So, there is a clear difference

---

[16]Note that nominal performances are given for a typical DOP of less than 6. When the receiver is located in a very good environment, namely on a roof with no surrounding buildings for example, the DOP can be one or even less. On the other hand, in urban canyons, the DOP can reach a value of more than 50. This is one of the main reasons (with multipath) that urban canyon positioning, although sometimes possible, is often of poor accuracy.

between the HDOP (horizontal, that is, two-dimensional or "surface DOP") and the VDOP (vertical, that is, relative to the altitude). Indeed, with respect to HDOP, satellites can be situated all around the receiver — in front, behind, to the left, and to the right — thus leading to very good HDOP. For the VDOP, things are totally different, as it is not possible to provide the receiver with a uniform distribution of satellites. It is not possible to receive signals from behind the receiver because of the Earth.[17] Thus, the distribution is truncated to one half space (the upper one), reducing the achievable HDOP. This is one of the principle reasons for limited vertical accuracy in comparison with the achievable horizontal accuracy.[18]

■ 8.6 BENEFITS OF AUGMENTATION SYSTEMS

Now that the principle of acquisition methods and techniques of position calculations are familiar, let us come back to augmentation systems. These systems are designed to compensate for major limitations of the GNSS. The three domains where one would like to improve the GNSS are:

- Pseudo range measurement error;
- Coverage;
- Integrity (the fact of knowing the quality of a signal), dealt with in Section 8.7.

For the first domain, augmentations could help in providing improved models and differential data. This has been the approach of the Satellite Based Augmentation Systems (SBAS) like WAAS, EGNOS, and MSAS[19] (refer also to Chapter 5). The ground-based facilities gather a lot of data that are used in order to provide better parameters for an embedded model of ionosphere and troposphere propagation. A representation of the EGNOS message is given in Table 8.4 for illustration purposes.

For the second domain, coverage, approach followed by augmentation systems is rather based on local or regional base stations that give support to the GNSS constellations. It can include other systems, like LORAN or GSM stations, used in order to provide positioning data in difficult environments for GNSS signals (indoors, urban canyons), but also satellite signal-based augmentations in order to provide a continuity in providing the satellite signal. This is notably the case of pseudolites, which can be used either indoors or outdoors. In addition to providing better accuracy under certain conditions, this is mainly used for providing better coverage (see Chapters 10 and 11 for further discussions).

---

[17]This is an important difference for air navigation where better HDOP can be obtained by using satellites with lower elevations.

[18]This problem can be overcome when using local augmentation systems like pseudolites or repeaters, as it is then possible to improve the "constellation" thus achieved. The idea is simply to locate such augmentations below the receiver's position. This has an immediate effect on vertical accuracy (see Chapter 10 for details).

[19]Note also that future systems like QZSS or GAGAN can provide such complementary data.

TABLE 8.4 **Structure of the EGNOS message. Note the various correction data available.**

Message 0	Don't use this SBAS signal for anything (for SBAS testing)
Message 1	PRN Mask assignments, set up to 51 of 210 bits
Messages 2–5	Fast corrections
Message 6	Integrity information
Message 7	Fast correction degradation factor
Message 8	Reserved for future messages
Message 9	GEO navigation message (X, Y, Z, time, and so on)
Message 10	Degradation parameters
Message 11	Reserved for future messages
Message 12	SBAS network time/UTC offset parameters
Messages 13–16	Reserved for future messages
Message 17	GEO satellite almanacs
Message 18	Ionospheric grid point masks
Messages 19–23	Reserved for future messages
Message 24	Mixed fact corrections/long-term satellite error corrections
Message 25	Long-term satellite error corrections
Message 26	Ionospheric delay corrections
Message 27	SBAS outside service volume degradation
Messages 28–61	Reserved for future messages
Message 62	Internal test message
Message 63	Null message

■ 8.7 DISCUSSION ON INTEROPERABILITY AND INTEGRITY

Discussions Concerning Interoperability

The position calculation methods described in this chapter are applicable to all the GNSS constellations. It is therefore possible to imagine using satellites from different constellations in order to carry out positioning. For example, one can think of two GPS satellites with two Galileo satellites, or two GPS satellites with one GLONASS and one Galileo, or any combination.[20] This is called "interoperability" and is much stronger that just a superposition of constellations. Of course, it is possible to carry out multiple positioning from each constellation and then make some comparisons, but interoperability is indeed an embedded method that will allow, for example, a three-dimensional positioning in an urban canyon where no positioning is possible with any single constellation.

The fundamental equation to be used for a given satellite remains the pseudo range one given by Equation 8.4. The only point common to the different constellations is the fact one is interested in the position of the receiver, given in an ECEF reference frame (x_r, y_r, z_r). All the other variables are different; the locations of the satellites

---

[20]However, the most probable first implementations will certainly deal with GPS and Galileo, especially in the E1-L1-E2 band for receiver electronics reasons (see Chapter 7 for details).

are given in different reference coordinate frames, namely WGS84 for GPS, PZ90 for GLONASS, and GTRF for Galileo. This is not a real difficulty as it is possible to define transformation matrices to achieve the conversions (see Chapter 2). This latter assertion is true when considering that all the satellites, from all the constellations, are of an identical "quality." This also leads to the problem of the choice of satellites to use in the case where there are more than four. The least-square method is usually based on identical hypotheses for each satellite, which is not obvious within a constellation, and not even realistic for different constellations. Nevertheless, the method can certainly be extended by individual measurement weights in order to account for different qualities.

In addition, there is also the problem of receiver clock bias. Of course, as this is a physical bias, it is identical for the three constellations, but unfortunately, the time reference frames are not. This can be summarized by the following set of equations for three satellites of the various constellations:

$$\rho_{\text{GPS}i} = \sqrt{(x_{\text{GPS}i} - x_r)^2 + (y_{\text{GPS}i} - y_r)^2 + (z_{\text{GPS}i} - z_r)^2} + ct_{r\text{GPS}} \tag{8.68}$$

$$\rho_{\text{GLONASS}i} = \sqrt{(x_{\text{GLONASS}i} - x_r)^2 + (y_{\text{GLONASS}i} - y_r)^2 + (z_{\text{GLONASS}i} - z_r)^2} \\ + ct_{r\text{GLONASS}} \tag{8.69}$$

$$\rho_{\text{Galileo}i} = \sqrt{(x_{\text{Galileo}i} - x_r)^2 + (y_{\text{Galileo}i} - y_r)^2 + (z_{\text{Galileo}i} - z_r)^2} + ct_{r\text{Galileo}}. \tag{8.70}$$

It is quite clear that the clock biases are not the same. Two approaches are then possible: either to take into account the clock corrections provided by the navigation message of GLONASS and Galileo, which should be able to characterize the bias between the considered constellation time with respect to the GPS time, or to use additional satellites in order to remove these biases (by implementing some kind of differential approach). The first method is much simpler but needs to be qualified in terms of induced errors.

Of course, when considering the pseudo ranges, usual corrections should be applied in order to take into account the various corrections, different for each constellation, but one can consider that once implemented in a receiver, this is not a real difficulty. Navigation messages have to be dealt with in different ways depending on the constellation (this is just a processing complexity, not an issue).

The next comment on interoperability concerns the DOP. In some cases, like urban canyons, positioning will be achievable while using interoperability because, say, two satellites of each constellation will be available. Thus, no positioning is possible with any single constellation, but possible with a combination of constellations. One still has to remember that the positioning, while possible, is bound to be of reduced accuracy because of the poor DOP available.

Discussions Concerning Integrity

Integrity can be defined as a reliability indicator of the quality of positioning. Given the various error sources, it is obvious that the estimated user range error, even taking into account the DOP, is a "passive" indicator. Let us imagine that the signals transmitted from one satellite are totally wrong, then the user range error will still provide the user with the same value.[21] In order to make such a quality factor available, there is the need for other mechanisms. The three we are going to briefly discuss in this section are

- Receiver Autonomous Integrity Monitoring (RAIM);
- The SBAS integrity concept;
- The Galileo and future GPS III integrity concepts.

As already described, there are very often more than four satellites available in the sky for positioning purposes. The main idea of RAIM is to conduct independent position computations with various sets of chosen satellites. The need for at least one supplementary satellite is obvious, and the actual availability of more than five allows the erroneous signal, if any, to be defined in addition to the detection of the problem. The basic mechanism consists of computing six positions (in the case of five satellites), the first one with all five satellites (using the least-square method for instance), and the five others considering the possible five sets of four satellites. Then, comparisons are carried out between the various positions found and an analysis of the dispersion leads to the detection of problems relative to one satellite, if any. The problem could be linked to a navigation message error or even specific propagation conditions. Having an additional satellite allows any faulty satellite to be determined.

This technique was first implemented for applications requiring integrity, usually linked with safety, as in air or rail transportation systems. RAIM receivers are thus mainly available in these particular contexts.

The SBAS have been developed with two main goals: accuracy improvements and requirements for integrity. The second is clearly identified as the most important and has been the main guide for system definition. The concept has also been supported by the International Civil Aviation Organization (ICAO) and is thus clearly applicable to civil aviation. The definition of integrity is the ability of a system (infrastructure and user) to provide positioning with an associated level of confidence. Thus, it is not an intrinsic characteristic of the system, but related to a specific application and context. This is achieved through the definition of some parameters (supposing the parameters of integrity are available):

- An alarm limit XAL (X = V for vertical, as opposed to H for horizontal);
- A maximal alarm time TTA (Time To Alarm);
- An integrity-associated risk.

For civil aviation, for instance, the requirement is that the risk that a positioning error (XPE) exceeds the alarm limit (XAL) without the user being advised within 6 s (TTA) is less than, for example, 2×10^{-7} for any operation lasting 150 s (the case for certain "Approach Procedures with Vertical Guidance").

---

[21] The User Range Error (URE) is based on the measured quality of the incoming signal. This quality can be good even with "wrong" signals.

At the user level, integrity is characterized by the comparison of three values:

- The positioning error XPE (only available to users being at reference positions);
- The protection limit XPL, which keeps the XPE within a certain confidence level (depending on the application requirements) and is locally calculated at the receiver end;
- The alarm limit XAL, which depends on the application.

The protection limit is calculated by the user from an estimation of errors on all the variables that are used in position computation.

The principle of integrity is then stated as follows:

- When XPL > XAL, the navigation is not available with integrity (integrity violation);
- When XPL < XAL, three cases are possible:
  - XPL < XAL < XPE indicates that the navigation is no longer safe (with respect to the integrity criteria) for the user. The probability that this situation would occur without an alarm being sent should be less than $2 \times 10^{-7}/150\,\text{s}$ (for the abovementioned application).
  - XPL < XPE < XAL indicates that the system is no longer safe (with respect to the integrity criteria) but can still be used.
  - XPE < XPL < XAL indicates that the positioning is safe.

It is interesting to note that implementation of the integrity concept constitutes a real difference between Galileo and the two other constellations, GPS and GLONASS. Integrity is currently not available within these two latter systems, although GPS is planning to implement it within the third generation of the system, the GPS III.

Finally, note that the most important consideration about integrity is to remember that it is closely associated with an application and hence a context. The integrity concept implemented in EGNOS, WAAS, and MSAS is associated with the civil aviation context and should be very carefully extended to other applications, contexts, or environments. A typical example of a situation where case should be taken is when stating that urban applications will benefit from the integrity provided by the SBAS. However, the errors associated with multipath, which are certainly one of the main problems in urban canyons, are not at all taken into account in the SBAS integrity concept currently implemented.

■ 8.8 EFFECT OF MULTIPATH ON THE NAVIGATION SOLUTION

Let us now return to the problem of multipath. First, one should keep in mind that under certain environmental conditions, multipath effects have a great impact on the resulting positioning. Such environments include, for example, urban canyons or indoors. This section intends to show, using the results of a limited-scale simulation, the real effect of a multipath on the correlation function, and hence on the pseudo range measurement.

The graph in Fig. 8.6 shows a typical correlation curve obtained for a GPS C/A code shift equal to 400 chips (corresponding to 0.391 ms). Note that this curve is obtained on a simulation basis without any noise being considered (thus a very simplified case indeed, but enough for the present purpose).

The correlation peak is clearly visible for a time shift equal to 400 chips, as expected. This has been obtained with a single "prompt" correlator. As described in Chapter 7, the real correlation function used (the early-late architecture) is indeed slightly modified and involves three correlations. Such a technique allows one to determine the moment the zero line is crossed instead of the peak (which is more difficult to find and not really useful for feedback purposes). Thus, the early-late correlation is based on two local replicas in addition to the prompt:

- The early one, which is in advance with respect to the prompt replica by a fraction of a chip (one chip for first simulations);
- The late one, which is the same fraction of a chip, but late with respect to the prompt replica.

By carrying out the difference between the early and the late replicas, it is possible to improve the accuracy of the determination of the precise correlation time. In addition, this time corresponds to the zero line crossing described earlier, which is very interesting for the feedback code loop. Around the correlation time roughly defined by the prompt, if the early-late correlation value is positive, then it means that the receiver should delay the correlation time (the opposite in the case where

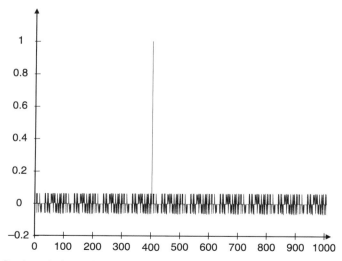

FIGURE 8.6 A typical correlation result (correlation function in vertical axis and chip offset in horizontal axis).

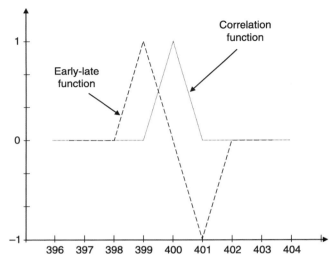

FIGURE 8.7 A typical early-late correlation result (correlation function in vertical axis and chip offset in horizontal axis).

the difference is negative). The curve in Fig. 8.7 shows the simulation result of prompt and early-late correlations. The delay is still equal to 400 chips (the scale has been changed).

The next question is related to the behavior of such correlations when multipath occurs. Here the basic idea is to consider that the satellite signal is in fact duplicated

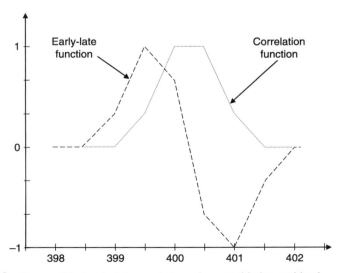

FIGURE 8.8 The modified early-late correlation when considering multipath.

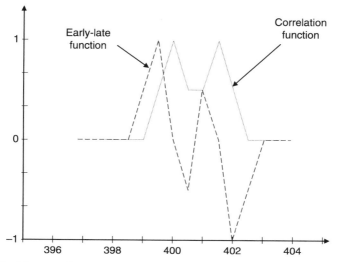

FIGURE 8.9 The modified early-late correlation when considering a 1.5 chip delayed multipath.

by a reflection onto a surface surrounding the receiver. This reflected surface could, for instance, be a building. The signal is thus composed of two superimposed codes that are delayed by a quantity equivalent to the corresponding additional distance traveled by the reflected signal. Figure 6.15 illustrates this configuration.

"How is the correlation peak influenced by the simultaneous presence of both paths" is the question we are trying to answer (when considering a reflected path of the same amplitude as the direct path with a delay of half a chip corresponding to roughly 150 m). The reflected path is late compared to the direct path, which is, of course, the shortest one. Figure 8.8 shows the resulting correlation function for an early-late architecture with one chip spacing (that is, early is half a chip earlier and later is half a chip later with respect to prompt).

The following remarks can be made.

- The prompt correlation function is no longer a triangle and exhibits a plate shape between chip 400 and 400.5.
- The early-late correlation value is no longer zero at 400 chip delay.
- The zero crossing now occurs at 400.25 chips.

The induced error of the reflected path is then equal to 0.25 chip (about 73 m) in comparison with the direct path only case. Let us now see what happens when the delay is 1.5 chips (around 440 m). Figure 8.9 shows the resulting form of the correlation functions. It can be observed that the first early-late zero crossing is good at 400 chips, even if multiple zero crossings occur. As a matter of fact, the impact of multipath depends on the value of the delay of the reflected path.

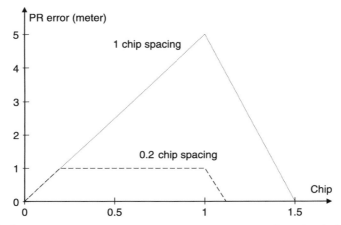

FIGURE 8.10 Induced multipath error on pseudo range versus reflected path delay (1 chip spacing and 0.2 chip spacing).

So, it is possible to plot the curve giving the correlation-induced error (in meters) versus the delay (in chips) between the direct and the reflected path. This is given in Fig. 8.10 for the following parameters:

- Early-late correlator spacing of one chip and 0.2 chip;
- Same amplitude of direct and reflected path (worst case);
- Error measured with respect to the initial delay of 400 chips.

The next step would have been to consider real multipath exhibiting modified amplitudes with respect to the direct path. This is not carried out here because our main goal, understanding the problem, has already been reached.

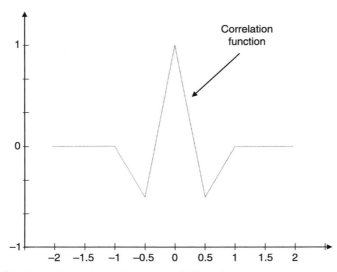

FIGURE 8.11 Typical correlation function of Galileo signals.

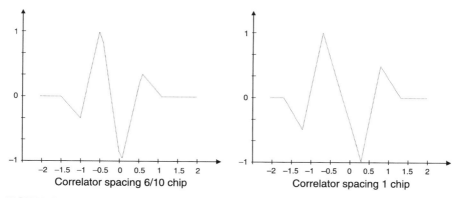

FIGURE 8.12 Typical early-late forms.

It appears clearly that the impact of multipath is tremendous. The problem is that all the techniques described above concerning integrity and error reduction are not applicable to multipath. Furthermore, this problem is not predictable without having a model of the environment and using it in real time; this approach is far from being a current practice because it requires high computation capabilities and a complete description of the real world. Some simulation programs allow such approaches but are mainly used for industrial or research purposes.

Mitigation techniques are thus being developed by all GNSS manufacturers with specific correlators intended to reduce multipath effects. Thus "strobe correlators," "sliding correlators," or "edge correlators" are designed to reduce the multipath problem.

Note that the correlation function of the Galileo signals is quite different from that of GPS. Figure 8.11 shows such a function, while Fig. 8.12 gives the early-late forms

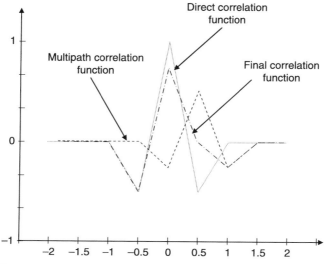

FIGURE 8.13 Typical correlation function of Galileo signals.

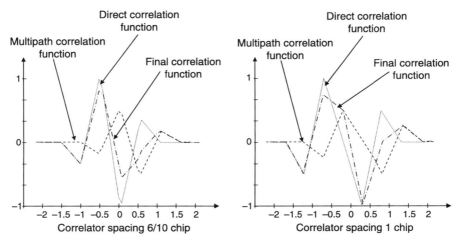

FIGURE 8.14 Typical early-late forms.

relative to the early-late spacing. The above analysis could be carried out with Galileo: the results would greatly depend on the situation considered, as for GPS.

For the case of a reflected path of amplitude that is half that of the direct path and switched by half a chip, the new resulting signal is given in Fig. 8.13. Figure 8.14 gives the results of the early-late computations for two different early-late spacings. Note that the results are acceptable, because the sum of the two considered paths remains close to the direct path (see Fig. 8.13).

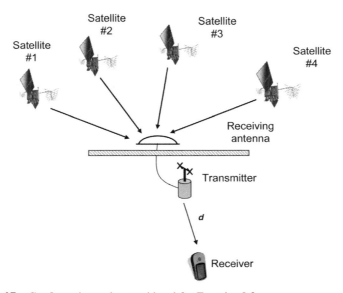

FIGURE 8.15 Configuration to be considered for Exercise 8.3.

✅ 8.9 EXERCISES

Exercise 8.1   What is the positioning technique that is used with GNSS? Answer the same question for a landmark triangulation-based technique.

Exercise 8.2   Among the positioning systems described in Chapter 4, make the distinction between those using a sphere intersection approach and those using a hyperboloid intersection approach. In addition specify those systems that require additional calculations. Also specify the calculations.

Exercise 8.3   Consider the configuration of Fig. 8.15. Calculate the position given by the receiver with respect to the one obtained at the receiving antenna (note that the transmitter just transmits the signals received by the receiving antenna). Justify your answer with calculations.

Exercise 8.4   Extract from Exercise 8.3 the real physical significance of t_r, the fourth component of the navigation solution vector. Refer to Chapter 10 (repeater part) to see how to take advantage of this feature.

Exercise 8.5   When is the least-square method applied? What is its principle? Can it be used to select the right satellites? (Give comments on the reliability of the determination of these "right" satellites.)

Exercise 8.6   Make a comparison between the systems of equations used for positioning and for velocity calculations (see Sections "Sphere Intersection Approach" and "Calculation of Velocity"). Give an explanation of both the so-called clock bias (t_r) and clock drift (dt_r/dt). Considering Exercise 8.4, give another physical meaning of the clock drift.

Exercise 8.7   Explain the difference between distance and velocity measurements in a radar system and in GNSS. Describe the difference in the way each system deals with synchronization (consider only the case of a mono-static radar).

Exercise 8.8   In the classical sphere intersection (linearization) method for GPS, classify those variables that are

- Calculated;
- Obtained from the navigation message;
- Physical measurements.

Include in your analysis all the corrections that must be taken into account in the pseudo range evaluation (namely satellite-related errors and propagation-related errors).

Exercise 8.9   Give a synthesized definition of the Dilution of Precision (DOP). Why are there so many different DOPs in GNSS? Describe situations where this diversity of indicators is important. Considering the positioning technique described in Chapter 4, carry out a review of those concerned with DOP and give a qualitative analysis for each.

Exercise 8.10   Describe the basic concept of integrity. Specify why its current implementation (in EGNOS and WAAS) is not likely to provide real integrity in urban canyons, for example.

Exercise 8.11 Explain the reasons for

- Reduced pseudo range resulting error when multipath occurs with reduced early-late spacing;
- Reduction of pseudo range resulting error when multipath delay is greater than one chip.

Exercise 8.12 Imagine and describe theoretical methods to eliminate

- The effect of a reflected path;
- The detection of NLOS path.

The idea of this question is to evaluate the real positioning problem of multipath and to analyze some solutions (detection of the first arriving signal, use of digital model of the environment, Doppler measurement coherence analysis between receiver and satellite, and so on) in terms of complexity and efficiency from a qualitative point of view.

■ BIBLIOGRAPHY

Bancroft S. An algebraic solution of the GPS equations. IEEE Trans Aerosp Electron Syst 1985;21(7):56–59.

Brown A. Navigation satellites. Encyclopedia of Physical Science and Technology, Volume 8. Academic Press; 1987.

El-Mahy MK. Efficient satellite orbit determination algorithm. Proceedings of the Eighteenth National Radio Science Conference; 2001; p 225–232. Mansoura, Egypt.

Hahn J, Powers E. GPS and Galileo timing interoperability. In: GNSS 2004: Proceedings; May 2004; Rotterdam. The Netherlands.

Hegarty CJ. Multipath performance of the new GNSS signals. In: ION NTM: Proceedings; January 2004; San Diego, CA, USA.

Hoshen J. The GPS equations and the problem of Apollonius. IEEE Trans Aerosp Electron Syst 1996;32(3):1116–1124.

Kaplan ED, Hegarty C. Understanding GPS: principles and applications. 2nd ed. Artech House; 2006. Norwood, MA, USA.

Kerneves D, Huyart B, Begaud X, Bergeault E, Jallet L. Direct measurement of direction of arrival of multiple signals. European Microwave Week; Wireless 2000; Paris, October 2000.

Langley R. The mathematics of GPS. GPS World 1991;2:45–50.

Leick A. GPS satellite surveying. John Wiley & Sons, 2004. Hoboken, NJ, USA.

Leva J. An alternative closed form solution to the GPS pseudorange equations. In: ION NTM: Proceedings; January 1995; Anaheim (CA).

Moudrak A. GPS Galileo time offset: how it affects positioning accuracy and how to cope with it. In: ION GNSS 2004: Proceedings; Long Beach (CA), 2004.

Parker T, Mataskis D. Time and frequency dissemination advances in GPS transfer techniques. GPS World 2004:32–38.

Parkinson BW, Spilker Jr JJ. Global positioning system: theory and applications. American Institute of Aeronautics and Astronautics; 1996.

Sturza MA. Navigation system integrity monitoring using redundant measurements. Navigation: Journal of the Institute of Navigation 1988–9;35(4).

Van Diggelen F. Receiver autonomous integrity monitoring using the NMEA 0183 message: $GPGRS. In: ION GPS-93; 6th International Technical Meeting: Proceedings; September 1993; Salt Lake City (UT).

Van Dyke K. GPS integrity failure modes and effects analysis (IFMEA). In: ION NTM: Proceedings; January 2003; Anaheim (CA).

Web Link

http://www.Colorado.EDU/geography/gcraft/notes/gps/gps.html. Nov 14, 2007.

CHAPTER 9

Indoor Positioning Problem and Main Techniques (Non-GNSS)

In previous chapters, we have described many of the different techniques and systems for positioning. Satellite-based positioning systems are the best known, but there have been numerous other approaches that allow large-area positioning, mainly dealt with in Chapter 4 (telecommunication network based, radio transmitter beacons, and so on). All these systems, together with satellite-based ones, have one point in common: a lack of accuracy in complex "radio-electric" environments. Propagation is so difficult that even better location-finding methods do not give acceptable results. The goal of this chapter is to outline the state of the art of indoor positioning techniques and to explain the main current efforts in this field. The choice of the indoor field is based on two factors: it is a very important topic when considering the continuity of a location-based service, and it is one of the most demanding environments.

Indoor positioning is a very important topic, mainly in terms of continuity of services. This has led to many theoretical and experimental works in this field using a large range of techniques, from purely GNSS approaches to networks of physical sensors or Wireless Local Area Telecommunication Networks (WLAN). Among all these techniques, GNSS-based ones present the advantage of making better use of the satellite receiver, which is considered to be the "best" solution for outdoor applications (even with the current limitations in urban canyon environments). Thus, techniques such as High Sensitivity GPS (HS-GPS) or Assisted GPS (A-GPS) have been widely investigated within the satellite community: the results are interesting but do not seem to give a definitive answer to indoor positioning. Pseudolites and repeaters are solutions that could now help in the development of a final system with good

Global Positioning: Technologies and Performance. By Nel Samama
Copyright © 2008 John Wiley & Sons, Inc.

accuracy and wide coverage. Studies are in progress and show encouraging results for both approaches. Because of the large deployment of WLANs for communication purposes, a great deal of work is being carried out on location finding with WLANs in order to find a way to "complement" the outdoor GNSS-based systems with indoor WLAN-based positioning. Among other techniques that will be described later in this chapter, one has to highlight the Ultra Wide Band (UWB) technology, based on radar concepts and now implemented for proximity high data rate communication. As it uses a time-based approach, it is possible to see it as being a good accurate solution for indoors.

Two very important aspects that one has to keep in mind are the availability and accuracy required. GNSS systems allow a global coverage and almost permanent availability. This is very good, but for indoors, where such systems do not yet provide equivalent performance, the questions are Does one need the same (or higher) level of accuracy? and What kind of availability is required? Also, is the permanent location-finding capability of GNSS compulsory? The next stages will show that specifications are very important in indoor positioning systems, especially as almost all technical requirements can be achieved, by one or another technology. The difficulty arises when one wants to combine technical requirements: accuracy and simplicity, terminal cost, infrastructure cost, autonomous mode, and so on. Users' requirements are also of uppermost importance, although one can imagine that future applications will certainly be put forward by imaginative people, not yet expert in positioning techniques. Thus, most current indoor applications in the telecommunications domain, like Location Based Services (LBS), do not need a permanent positioning capability, but require such a positioning on demand with a reduced delay. Accuracy depends clearly on the application; for example if need not be very high for service finding, but needs to be quite precise (to within a few meters) for navigation purposes. This requirement should certainly be even more stringent indoors, due to the usually reduced size of the places concerned.

■ 9.1 GENERAL INTRODUCTION TO INDOOR POSITIONING

With the advent of greater mobility, a significant need for localization has emerged. This is true not only for automotive applications, but also for personal purposes, thus leading to the necessity of having a technical solution for indoor positioning. This latter point appears to be of prime importance for telecommunication-related applications, as revealed by the United States' efforts concerning the emergency call E911.[1] The European Union provides such an emergency call, the E112, but has decided not to put any legal constraints on call localization; operators are simply asked to make their best effort to obtain a good location.[2] These regulations require

---

[1] The FCC regulation states that succeeding communications to E911 should be located with an accuracy of 50 m 67% of the time and 150 m 95% of the time.
[2] Directive 2002/22 on E112.

development, specifically in the areas not covered by GNSS. As a confirmation of this, the Galileo program includes a specific domain called "local elements" that specifically includes the indoor domain. It is quite clear that indoor positioning is a challenging future technical aspect of global navigation. If GNSS systems are the right candidates for global positioning in the places where it works well, that is, where the sky is free enough for the receiver to acquire enough satellites, many possibilities exist concerning both urban canyons and indoor environments. There are typically two directions that can be taken. The first relies on the use of satellite navigation constellation signals in order to reduce the number of different electronic systems required to achieve the positioning function (see Chapter 10 for details). The second tries to implement a different technique indoors and the final system will be made up of the integration of GNSS for outdoors and this newly developed one indoors.

The Basic Problem: Example of the Navigation Application

Let us take as an example of the limitation of current systems the navigation function. This allows a guidance application to be provided, and has to be available in different environments, namely outdoors and indoors. Let us also compare two of the major positioning techniques in use: the telecommunication network cell identification (the so-called Cell-Id described in Chapter 4), and a GNSS triangulation method. Table 9.1 summarizes the simple situation regarding the continuity of the associated service.

The proposed navigation service is achievable with neither of these two techniques because of the lack of coverage in the case of GNSS, and of the lack of positioning accuracy in the case of the Cell-Id approach. If one wants to propose such a service for a pedestrian, this is a real problem (although this is a very simple representation of reality, which is far more complex). In comparison, the automotive domain is easier in terms of real constraints on the positioning engine: no power restrictions, location that can only be on predefined tracks (that is, roads), constant attitude of the platform, and so on, leading to a globally satisfactory GNSS-based system. The situation is very different when dealing with a pedestrian, who is the typical target of telecommunication-based services and applications, and in particular the location-based services. Note that some in-vehicle GNSS systems add inertial techniques and advanced map-matching to overcome the drawbacks of the limited coverage of GNSS. The direct transposition from car to pedestrian navigation is not

TABLE 9.1 The "navigation" function and the continuity of service.

Localization Technique	Cell-Id	GNSS
Indoors	Yes	No
Outdoors	Yes	Yes
Navigation function	No	Yes
Continuity of service	No	No

straightforward, although some trials have been carried out in this area regarding inertial concepts, specific map matching to define indoor tracks, and so on.

In fact, the localization, which is a fundamental of all navigation-related applications, of all location-based services, and of all applications requiring location data, should exhibit the following characteristics:

- Be available in various types of environments (countryside, urban areas, indoors, and so on);
- Give an accuracy that clearly depends on the application; it is clear that an accuracy of 1 m is not needed for numerous applications, whereas it is insufficient in other applications;
- Allow continuity of service as a basic concept.

The "Perceived" Needs

The specifications of localization are thus very different for the various possible applications. Furthermore, there are no current techniques that cover a large range of specifications. For just three requirements — accuracy, indoor, and outdoor needs — Table 9.2 shows the diversity of specifications, considering classification by several main domains. Of course, other requirements are also of prime importance, such as the cost of infrastructure and terminals, and so on.

This table is not very "accurate" — the accuracy figures are very loose and the environmental requirements are also not clear. Unfortunately, this is the reality; the technical needs are tremendous and the situations cover a very large range of possibilities. Furthermore, this is only a very small part of the real complexity. For instance, let us take any line in Table 9.2. It is possible to divide each one into many new lines with more precise specifications. As an example of this we could consider the "Tourism" line. In this domain, there has not been a great demand for positioning, but some applications are already working in a location-based way. One can, for instance, book mobile electronic assistants that include both a positioning engine and specific points of interest. Navigation is then possible from one point of interest to the next. It is quite easy to imagine that if the localization engine is also able to

TABLE 9.2 Specification by main domains.

Domains	Accuracy	Indoors	Outdoors
Assistance	~100 m	Not compulsory	Essential
Comfort	<100 m	Not compulsory	Useful
Displacements	1 – 100 m	Useful	Essential
Games	1 – 100 m	Not compulsory	Useful
Health	1 – 100 m	Important	Important
Services	1 – 100 m	Useful	Essential
Tourism	1 – 100 m	Useful	Useful
Transport	1 – 10 m	Important	Essential
Emergency	1 m	Essential	Essential

work indoors, then an extension to museum visits or even to mall navigation would be quite immediate. Of course, this new feature would be considered as an improvement (which would be true). For museum visits, the technical requirements could be an accuracy of around 1 m, but also the output of the absolute orientation of the terminal, so that it could be possible to determine whether the user is looking at a given sculpture or has his back turned to it.

However, also concerning tourism, a really useful application could be to help travelers in finding their connections in a city center or in an airport. In such a case, accuracy is no longer so stringent and the absolute orientation, although useful, is not compulsory.

The real problem is that it is possible to divide all the lines up in such a way, and that there is no current technique comparable to GNSS for outdoors. For outdoor applications, this exhibits such good global performances that it can cover a very large range of needs. Indoor techniques are unfortunately not so versatile, nor are the indoor applications' specifications. This complexity is certainly the reason for the limited extension of current location-based services.

Other classifications are possible, such as, for instance, one that relies on places where the services could be proposed. Table 9.3 shows such a summary. Once again, the main conclusions remain the same. Many studies have been carried out in order to define the correct classification, but have not led to any simplification of the initial problem and there are no real specifications that can be applied to a group of applications.

The Wide Range of Possible Techniques

An analysis of possible techniques is carried out in this section, leading to the highlighting of major trade-offs that have to be made. The techniques to be considered are taken from the following categories:

- Networks of sensors (infrared, ultrasound, pressure sensors, and so on);
- Mobile telecommunication networks (GSM, UMTS, and so on);

TABLE 9.3 Specification by main places.

Places	Accuracy	Indoors	Outdoors
Airport/station	~ 10 m	Essential	Essential
Country/mountain	< 100 m	Not compulsory	Useful
Mall	$<$ a few m	Essential	Not compulsory
Conference center	$<$ a few m	Essential	Not compulsory
Warehouse	~ 1 m	Essential	Useful
Sea/port	$1 - 100$ m	—	Essential
Museum	$<$ a few m	Useful	—
Attraction park	~ 10 m	Useful	Useful
Road	~ 10 m	—	Essential
Lane	~ 10 m	—	Essential
Storage zone	< 10 m	Not compulsory	Essential

- Additional sensors (odometer, accelerometer, gyroscope, magnetometers, and so on);
- Wireless local area networks (WLAN: Bluetooth, Wi-Fi, UWB, and so on);
- GNSS-based (pseudolites, repeaters, and so on).

The "networks of sensors" category includes infrared, ultrasound sensors, but also pressure sensors distributed throughout a building. The main disadvantage of these techniques is the need for a wide deployment of a heavy infrastructure, which is balanced by the greater accuracy sometimes achievable (down to a few centimeters). This is no longer seen as a real viable option, although actual implementations have been set up.

Concerning the "mobile networks of telecommunication" category, much has been said about the possibility of implementing techniques (see Chapter 4) such as TDOA or E-TDOA, and even AOA (angle of arrival). It is well known that none of these techniques has ever been commercially available, except Cell-Id.[3] The reason for this is that for telecommunication purposes, the Cell-Id is a built-in facility (thus nothing more has to be done for positioning). However, for all the other techniques, there is a specific strong requirement for positioning approaches; a minimum of three base stations have to be seen from the mobile terminal. This is of course not the way a mobile network is set up.

"Additional sensors" are all that one can imagine in order to allow positioning at the terminal end using autonomous means. One of the most significant examples of this is the pedestrian navigation module (PNM) designed by EPFL. This PNM uses a combination of GPS and inertial sensors, together with pedestrian behavior models for walking, that finally allows pedestrian navigation in many environments. The principle is to use GPS when available, and to switch to inertial navigation when GPS is not available. The main difficulty is that it requires very accurate modeling of the pedestrian's movements.

For the WLAN method, one can consider that the infrastructure required is free, as long as it has been deployed anyway for other purposes (mobile Internet access or wireless telecommunication). Indeed, this is not absolutely true when thinking in terms of usual methods. If the technique is to make time measurements, then you need to seriously upgrade the time reference compared to current WLAN time capabilities. If you use received power levels as the main data, then once again you will need to increase the number of "access points" to a level higher than that needed for telecommunication purposes. The received signal strength (RSS) method consists in establishing databases of received power from various base stations for a given place. In fact, it is necessary to set up a grid over the whole place with a 1-m step (for example) in both X and Y directions. Then, measurements of the RSS from base stations are carried out. In this approach, it is interesting to be aware of the accuracy achievable, versus the number of base stations. First results show that one needs at least three base stations to achieve a 3–4 m accuracy. In fact, you will certainly need more stations than that, and at least five to achieve such a goal in a real environment. The principle of positioning is to search the databases (one for each base station) for

---

[3]The case of the Matrix approach (see Chapter 4) is the only exception.

the corresponding values, considering a 1–2 dB uncertainty, of the RSS measurements. This gives possible areas for each base station. Then, by merging these various areas one finally obtains the position. This leads to some questions. What happens if the real environment changes — new walls or the moving of desks and cupboards. The other difficulty is the orientation and inclination of the terminal. A mobile terminal is, by definition, handheld in a position that cannot be predicted. This position has a significant impact on the received power level. So, the database value research can lead to the wrong area. This aspect is not so important because of the number of measurements that are made (typically five), but is still there. So, the WLAN approach still has to be upgraded to be a really valuable solution.

For the GNSS-based techniques, the most famous is the Assisted-GNSS. As already stated, almost everybody agrees that it will not work satisfactorily in "deep" indoor conditions. The other well-known GNSS-based technique is the one that makes use of pseudo satellites (pseudolites), where the basic idea is to create a local terrestrial constellation of a few satellites (generators for instance). This is a good idea and the only difficulty arises in the synchronization required between pseudolites. Chapter 10 will address the current achieved performances of this technique in indoor environments.

Another GNSS-based technique uses the so-called "GNSS repeaters," which can be seen as a cheap local element that could also have network functionality (such as Galileo differential stations). A repeater is a simple component that includes an outdoor antenna to collect GNSS signals, a microwave amplifier, and an indoor transmitting antenna to transmit the signals. The implementation of a system that uses three such repeaters has been carried out and the results in various indoor configurations show an average accuracy in the 1–2 m range. Current results have been obtained with a single frequency L1 standard GPS receiver. Galileo and modernized GPS exhibit very interesting features such as multiple civil frequencies, pilot tones, and more sophisticated codes and modulations than existing systems. The indoor positioning could take advantage of this situation and further improve accuracy, availability, and integrity. Work is planned in these areas in order to evaluate the possible improvements of performances when using the second generation of repeaters that will be available in the future.

It is now possible to draw up a table of techniques, comparable to those given in Tables 9.2 and 9.3. Table 9.4 shows all the achievable performances. It can be seen that almost all accuracies are possible, but it is once again the other specifications that will direct the choice of the resulting final technical solution.

Comments on the Best Solution

The scientific and industrial communities consider it important to provide an answer to the problem of positioning continuity from outdoors to indoors. This should be achieved for the large diversity of possible environments that are bound to be faced by users. Many techniques, as described above, have been developed in order to achieve such a goal, but it would be interesting to analyze the real application requirements in positioning, specifically indoors.

TABLE 9.4 Specification by techniques.

Techniques	Indoors	Outdoors
Network of sensors	1 cm to 5 m	Not suitable
RF ID	<1 m	<1 m
WLAN	Few m	Not suitable
UWB	~10 cm	Not suitable
Cell-Id	500 m to 10 km	100 m to 10 km
E-OTD (2G)/TDOA (3G)	>>200 m	<100 m
GNSS	Not available	Few m
A-GNSS	10 m to not available	Few m
Pseudolites	~10 cm	Few m
Repeaters	~1 to 2 m	Few m
Inertial	<1 m (time dependent)	<1 m (time dependent)

If we think in terms of navigation within an office building, it is certainly enough to propose a simplified WLAN positioning system. If we want to implement a navigation system in an exhibition hall, a 2D-repeater approach could be an accurate candidate. Last but not least, a guidance system in railway stations or airport terminals would probably take advantage of a combination of these two techniques, depending on the specific requirements of the users (staff or customers, and so on).

The various systems available for indoor positioning can be classified in many ways, depending on criteria such as the localization type (symbolic, relative, or absolute), the coverage (indoors, outdoors, or ideally both), the fact that the infrastructure calculates the position of the mobile terminal or that the calculation is carried out at the receiver.

Systems as varied as bar codes, magnetic detectors, imaging systems, or infrared, ultrasound or radio systems have been considered and implemented. Those systems reliant on radio wave propagation are based on techniques already described in previous chapters; such as direction of arrival, flight time measurements, and received power levels. This is the case for systems such as the GSM, the "Active Badge," the GNSS, and many others.

Another raw classification is also possible (Table 9.5), in order to understand the current major trends towards the achievement of continuity of service. Note that a precise definition of this continuity would be required, as it depends highly on the application and can certainly change for different communities of users (even for the same generic application).

Each technique in Table 9.5 is split into two lines. The upper line characterizes the indoor positioning, and the lower characterizes the outdoor positioning. The "+" and "−" symbols qualify the relative interest or difficulty, as appropriate. Dots indicate where the technique is not relevant today.

The various columns give the following information:

- Pos: type of positioning (R for relative, A for absolute);
- Acc: positioning accuracy;

TABLE 9.5 **Brief classification of techniques.**

	Pos	Acc	Avail	Infra	Term	Σ
Sensor networks	R	+++	++	− − −	+	+3
	•	•	•	•	•	•
WLAN	A	+	++	− −	+	+2
	•	•	•	•	•	•
Telecommunication networks	A	− −	+	++	+	+2
	A	−	+	++	+	+ 3
GNSS	•	•	•	•	•	•
	A	++	+	++	++	+7
Inertial systems	R	+	++	+++	−	+5
	R	+	+++	+++	−	+6

See text for definitions and abbreviations.

- Avail: availability of both signals and positioning;
- Infra: complexity of the required infrastructure, if any;
- Term: complexity of the mobile terminal.

Table 9.5 allows an appreciation of the performances of the different classes of techniques. This table helps us to understand the two main directions of current work on indoor positioning:

- The first approach tries to integrate a few technologies (and techniques) in order to achieve both indoor and outdoor positioning in the same handset. For instance, GPS and WLAN or GPS and inertial systems, such as for the PNS (the Pedestrian Navigation System of the Ecole Polytechnique Fédérale de Lausanne).
- The second approach tries to develop a single technology so that it is more versatile: GSM/UMTS or GNSS with high sensitivity receivers.

The main advantage of using a GNSS receiver in order to achieve positioning in areas where the GNSS signals are not available, lies in the fact that the localization system thus obtained is really "universal." Of course, this universal receiver will then give positioning in all the various environments. Thus, these approaches really are a good alternative solution to all the proposed techniques, described above. In the case of a need for an infrastructure, the goal should clearly be to have it as light as possible.

Local or Global Coverage

The idea of the coverage is comparable to that of the infrastructure discussed in the following pages. A global coverage means that there is no need for local components, thus involving no additional cost to the primary system. For GNSS constellations, this means that no local elements are required in order to provide positioning in all conditions. This is clearly not the case, because urban areas and indoors are not well covered. Thus, there is a need for local elements. Nevertheless, the coverage of the GNSS is really global. This should also be the case, even with some limitations in the coverage, for GSM or UMTS systems.[4]

When considering indoor positioning, the large range of possible techniques makes this differentiation quite interesting. The main reason for this is the fact that one would appreciate getting this indoor feature for free, or almost free. Thus, all the techniques that will require a local infrastructure are bound to be "less interesting" than others that will not have this requirement, even at the cost of reduced performance. GNSS-based techniques, such as HS-GNSS or A-GNSS are, from this point of view, very attractive approaches.[5] The same would apply to inertial-based positioning, as the system is then operational in all environments without any calibration or deployment. On the other hand, techniques like those based on UWB, infrared, or ultrasound are purely local systems, and also require an important level of local deployment. Of course, this is a drawback with these approaches, but it can sometimes easily be balanced by the low cost of the basic elements required and the simplicity of deployment. This is particularly applicable to the case of RFID for instance, where the price tag is only a few cents, but the coverage is very limited. Between global and proximity coverage, there are those techniques that use global signals or receivers but that still require the addition of "augmentation" devices, such as pseudolites or repeaters. The coverage is clearly local for these additional components. Finally, there are some systems that, even if they require local deployment (and thus coverage), are subject to a large spread in the coming years for purposes other than positioning. The case of WLAN is one of them. The coverage of the indoor system is quite clearly local, but the deployments are so widely spread that the actual coverage is, if not global, much larger than local.

With or Without Local Infrastructure

From this first analysis, several directions of investigation have been the subject of development in the scientific community, based on a local terrestrial infrastructure intended to reproduce indoor conditions equivalent to outdoors. These conditions are typically spatial diversity and power at a similar level. In fact, it is possible to adopt one of two approaches: with or without infrastructure. The latter case is illustrated by the so-called high sensitivity GNSS (HS-GNSS), where the idea is to decrease the detection level of the satellite signal down to -150 or -160 dBm (as compared to the typical -130 dBm of outdoor received power level). Although this approach has yielded real improvements, it does not appear to be the final answer to indoor positioning (see Chapter 10 for details). GSM techniques, on the other hand, can also be considered as "no added infrastructure"; unfortunately, they do not provide accuracy to the level required for many applications. WLAN could also be considered as a "no added infrastructure" solution but once again we have seen that the telecommunication deployment is not enough to achieve positioning to a sufficient accuracy.

---

[4]It is not our present purpose to go over the respective advantages and disadvantages of terrestrial and satellite-based systems.

[5]Even if in the case of A-GNSS there is the need for specific equipment at the base station end.

Thus, one should take into consideration the need for a local infrastructure. The present chapter and Chapter 10 will describe many techniques. The main issue is then to find the "lightest" possible infrastructure: either the least expensive or the easiest to deploy. The coverage and complexity of the system, as described in Table 9.5, are then of great importance. Pseudolites and repeaters, GNSS-based techniques fully described in Chapter 10, are potentially good candidates when considering these aspects.

The GNSS Constellations and the Indoor Positioning Problem

Following the previous discussion concerning the solutions with or without infrastructure, the case of GNSS constellations is very important and quite specific. These constellations, GPS among others, have had fantastic success, far beyond the pioneers' hopes. The reader has to remember that original GPS specifications were derived from the TRANSIT system to expand military maritime requirements to a larger range of military applications (that is, land and aviation typically). No one, at that time, imagined that there would be a need for indoor positioning, as military conflicts were all based on potential conventional wars between countries. Terrorism or urban guerrilla battles are only modern wars. Of course, if the GPS was designed nowadays for military purposes, it is quite clear the specifications would be different.[6]

A compromise has to be found for a lot of contradictory features of GPS. The first one concerns the transmitted power level, and hence the life-time of the satellite, and the penetrating power of the signals. A typical figure of 30 dB can be considered as a rule of thumb when dealing with attenuation between outdoors and indoors. As such an increase in the required transmitted power level of the satellites is clearly not possible, therefore in theory the positioning is not achievable with GPS.[7] The way this problem, which can be seen as a problem of sensitivity at the receiver, has been dealt with for GPS, is to find receiver architectures that allow lower power level detection, as will be described later in this chapter.

Another way to achieve positioning is to use satellites whose signals are detected through windows. Unfortunately, in order to achieve a good positioning, there is the need for a good geometrical distribution of satellites (see the Dilution of Precision concept in Chapter 8). This is very difficult to achieve indoors with current constellations. One could imagine it to be easier when the three main constellations, GPS, GLONASS, and Galileo, are fully deployed, with typically 85 satellites[8] in the sky. To be able to access three or four satellites through the windows is then certainly possible. Obviously, the resulting accuracy would certainly be better in the case of a large open window in a top-floor New York apartment than on the first floor.

---

[6]But a satellite constellation may not be the definitive answer to these requirements anyway.
[7]Note that telecommunication networks that exhibit an acceptable reception indoors use power levels of about 30–40 dB, at least, greater than GPS signals.
[8]Considering 28 GPS, 24 GLONASS, 27 Galileo, and 6 Geo Satellites.

■ 9.2 A BRIEF REVIEW OF POSSIBLE TECHNIQUES

Introduction to Measurements Used

As has been seen in Chapters 1 and 2, the twentieth century did not really provide us with improved calculation techniques, but rather with improved technical means (satellites, signal processing, and so on). This is also the case with indoor positioning solutions that rely very often on old techniques that have been modified.

The four usual methods used for localization purposes, whatever the techniques used, are the time of arrival (TOA) or time difference of arrival (TDOA), the angle of arrival (AOA), the radio signal strength (RSS) measurements, and the non-radio-based (NRB) measurements. Let us briefly describe these four approaches.

Time and Distance

For TOA/TDOA the measurements made are essentially the propagation time of the signal, from the transmitter to the receiver. The transmitter (respectively receiver) could be the mobile terminal (respectively base station) or the base station (respectively mobile terminal). This approach requires being able to deal with a very precise time synchronization of the global system. The required synchronization is usually not achieved in telecommunication systems, and the implementation of such a TOA solution will need to add time capabilities to the WLAN embedded capabilities. Nevertheless, some companies, such as Hitachi, propose this approach with an announced accuracy of between 1 and 3 m for open-space indoor environments.

Angles

For AOA the measurements are now angles of arrival at the antenna. Once again, this can be achieved at the mobile terminal end — but is very complex to implement mainly because of the dynamics of the mobile terminal attitude — or at the base station end (used in GSM-based localization techniques, for instance). The major drawback of this method is the complexity of the antenna, leading to quite expensive solutions.

Power Level

For RSS the measurements are now power levels received by the module. The basic idea is to establish the received power level map (the geographical map of the received power levels throughout the considered area) for all the transmitters considered. From these results, two methods are possible. The first tries to establish a propagation model within the building, and the second develops algorithms to allow the merging of the various maps for a given set of measured RSSI (that is specific to a physical location within the area).

Non-Radio (Pressure, Presence, Acceleration, and so on)

The last technique consists in using some physical properties of the environment. For example, if the place is equipped with pressure sensors installed in the ground, one can consider carrying out pressure measurements in order to define whether someone is there. Another possibility would be to use accelerometers in order to measure the acceleration of a mobile terminal. Then, by integrating this acceleration twice, it is possible to calculate the displacement. Note that this approach is typical of that used by inertial systems (in a considerably more sophisticated way). Of course, a large range of physical measurements can be used for localization: capacity, light intensity, reflection, presence, and so on.

Comments on the Applicability of these Techniques to Indoor Environments

Indoor environments present some very specific characteristics that make it difficult for a wide range of potential techniques. Among the most important is the reduced size of the locations and the variety of situations. For example, if one just considers the way houses are constructed throughout the world, there are almost as many architectures and materials used as there are cultures. Some are in wood with the basement in concrete, others are all in concrete, others again are in brick or made from mud paving, or plaster is used for some walls and stone for outside walls, and so on. The same applies to warehouse buildings or to business and office buildings, which can be made of concrete with or without metallic windows or even with metallic girders. The propagation of a signal indoors is likely to be quite different from outdoors, in large open spaces. This is true for all physical measurements,[9] but radio propagation is certainly one of the most disturbed.

The multipath problem and first approaches for its mitigation have already been described in Chapter 8. Unfortunately, the indoor case of multipath is really a lot more difficult than that outdoors, and nothing can be easily done to overcome it. The real difficulty is that indoors, the direct (or LOS for line of sight) path is not always present. In such a case, only the NLOS (non line of sight) path is received by the receiver and directly leads to a wrong pseudo range. This case is identical outdoors, but the potential obstructions are not as numerous as indoors. Figure 9.1 shows a typical indoor configuration of propagation. Simulations are possible in order to evaluate the multipath phenomenon, but require the complete definition of the environment (walls, windows, floor, ceiling, and the materials of all these elements). Thus, it is not possible to imagine making a complete real-time simulation in order to have a real-time evaluation of multipath. This is enhanced by the fact that multipath is also dependent on the location of the receiver (not yet known at time of multipath evaluation).

The other problem specific to indoors relates to the reduced distances to be measured. Thus, when a receiver has to carry out distance measurements, that is, measurements of time, the time "synchronization" needs to be of very good quality.

---

[9]A good example of this is the behavior of a magnetic compass indoors. Because it measures the magnetic field, all metallic elements are bound to disturb the measurements, and of course, this happens. Try approaching a metallic desk with an electronic compass in your hand and you will experience this very obviously.

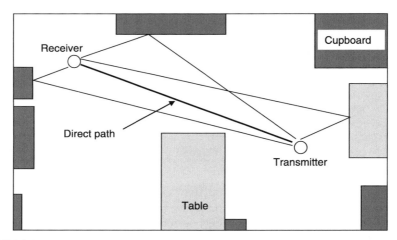

FIGURE 9.1 A typical indoor multipath configuration.

A possible classification of the various techniques used for indoor positioning is proposed in the following subsections, sorted by physical measurements carried out and by technical domains.

Global Classifications of these Techniques by Physical Measurements

1. Time measurements: Bat System (ultrasound), A-GPS, HS-GPS, pseudolites, repeaters, UWB, E-OTD, E-TDOA;
2. Angle measurements: AOA;
3. Power measurements: WiFi (802.11), Bluetooth, RADAR;
4. Non-radio-based measurements: Active Badge (infrared), Cell-Id, gyroscopes, accelerometers, odometers, altimeters, Smart Floor (pressure sensors), camera (image processing).

Global Classifications of these Techniques by Technical Domains

1. Requiring a large specific infrastructure: Active Badge (infrared), Bat System (ultrasound), Smart Floor (pressure sensors), camera, and so on;
2. Using the wireless local telecommunication networks: WiFi (802.11), Bluetooth, RADAR, UWB, and so on;
3. Taking advantage of the existing wireless telecommunication global networks: Cell-Id, E-OTD, AOA, E-TDOA, and so on;
4. Based on satellite navigation constellations: A-GPS, HS-GPS, pseudolites, repeaters, and so on;
5. Complementing the navigation systems: gyroscopes, accelerometers, odometers, altimeters, and so on.

■ 9.3 NETWORK OF SENSORS

Ultrasound

There is a great similarity between the propagation of radio waves in free space and the propagation of sound in air. The frequency of a typical radio wave is about 1×10^9 Hz (1 GHz) with a velocity of 3×10^8 m/s. The frequency of a typical sound wave is around 1 kHz and travels at around 300 m/s. The corresponding wavelengths,[10] which are a characteristic of all propagation signals, are respectively 30 cm and 30 cm: identical! Thus, even if the physical forms of both waves are fundamentally different,[11] their propagations are very similar. Nevertheless, there is a fundamental difference: the velocity of the waves varies from 300,000,000 m/s to 300 m/s, that is, a factor of 10^6. When dealing with time to define distances, it is much easier to achieve accurate measurements when the time constraint is loose. This is the case with sound waves. Thus, ultrasound positioning systems based on time measurements can be quite accurate. To illustrate the comparison in terms of required clock accuracy: 1 ms for an ultra sound system is equivalent to 1 ns for an electromagnetic wave. Furthermore, sound exhibits reflections, but, once again, for a reflected ray that would travel 5 m more, it corresponds to a 16.7 ms delay, which is easily discriminated from the direct path. The reader can now understand the reason for the increased accuracy of ultrasound solutions with quite cheap electronics. An example of such a system has been developed by the Computer Laboratory of the University of Cambridge, UK (Fig. 9.2).

The technical characteristics are summarized in Table 9.6. The upper part of this table is related to the description: the technology, the type of localization achieved (typically giving details on the fact that the localization is done by the terminal, as in GNSS receivers for instance, or by the infrastructure deployed, as in major WLAN based approaches) together with the localization capabilities (the rate at which a location can be calculated by the system). Next is presented whether the positioning is absolute or relative to a given first location that must be known, and finally the types of environments for which it has been designed (indoors, outdoors, or both).[12] The lower part deals with performances in terms of the possible accuracy and range of each beacon. A beacon is considered as being a single element allowing positioning inside the whole infrastructure. Although not absolutely complete, this table gives an acceptable and real overview of a given technique and technology. A similar table will be given for each technique considered in this chapter.

Infrared Radiation (IR)

Infrared radiation is, like radio waves, an electromagnetic wave that is propagated at the speed of light. Time measurements are quite complex to carry out indoors,

---

[10]The wavelength is defined by λ = velocity of the wave/frequency.

[11]A radio wave is an electromagnetic wave made of photons, whereas a sound wave is a mechanical vibration of molecules of air (or also an air pressure wave).

[12]Note that no distinction is made at this stage between different outdoor environments, such as urban canyons or countryside.

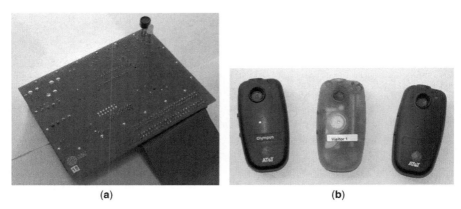

(a)   (b)

FIGURE 9.2   The Bat System: beacon (**a**) and transmitters (**b**). *Source*: Computer laboratory of the University of Cambridge, UK.

unless a precise time distributed clock is deployed, which is still too expensive. Although different techniques could have been implemented, the one most largely used is the detection approach, which consists simply in determining whether a receiver is present in the detection range of the device (example equipment is shown in Fig. 9.3).

The global characteristics of an infrared positioning system are summarized in Table 9.7.

The type of positioning is slightly different from previous ones. "Symbolic" (Table 9.7) means that the positioning is no longer given in terms of spatial coordinates, in an absolute or relative manner, but rather in terms of room numbers or names. One would then know in which office, corridor, or conference room someone is, rather than the absolute positioning.

The accuracy and range figures are related to this concept. The accuracy is the room,[13] the coverage of a single beacon being one room. Thus, there is the need

TABLE 9.6 Ultra sound positioning.

Technical description	
Technology	Ultrasound
Localization	Infrastructure
	Continuous
Positioning	Relative
Environments	Indoors
Performances	
Accuracy	A few cm
Range	A few $10\,m^2$

---

[13] Do not confuse accuracy and reliability. In the present case, even with an accuracy of a few meters (size of a room), the reliability could be excellent in that the right room is given with a very high probability of success.

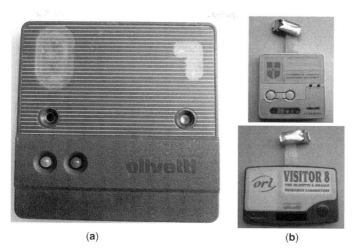

(a) **(b)**

FIGURE 9.3 The "Active Badge" system: beacon (**a**) and transmitters (**b**). *Source*: Computer laboratory of the University of Cambridge, UK.

for as many beacons as there are rooms in the indoor environment where one wants to allow localization.

Pressure Sensors

Another way to locate people is to use pressure sensors. Let us imagine a floor where every floor tile is linked to a pressure measurement element. Then, by observing the value of the pressure on any given tile, it is possible to extract an average number of people that must be on this tile. Of course, the algorithms required to determine the numbers of people present in any given place become more complex as the number of people increases. Some works have shown that 20 people could be located by using neural networks.

The global characteristics of a pressure sensor positioning system are summarized in Table 9.8. The values given in the table should clearly be considered as

TABLE 9.7 Infrared positioning.

Technical description	
Technology	Infrared
Localization	Infrastructure
	0.1 Hz (1 every 10 s)
Positioning	Symbolic
Environments	Indoors
Performances	
Accuracy	A "room"
Range	One beacon per room

TABLE 9.8 **Pressure floor positioning.**

Technical description	
Technology	Pressure sensor
Localization	Infrastructure
	Continuous
Positioning	Relative
Environments	Indoors
Performances	
Accuracy	A few cm
Range	Contact

typical figures and one should understand the underlying reasons for these performances. This remark applies for all the tables in this chapter.

Radio Frequency Identification (RFID)

Combining a radio system (for its ease and low cost of implementation) and symbolic positioning, one can achieve the RFID concept. An RFID is an electronic label that can be associated with any object. The principle is to use a reader that can supply power directly to the label, allowing it not to be active in the sense of electric power. Thus, it is possible to realize so-called "RFID tags" at very low cost and of various sizes and shapes. The tag can be identified because once powered by the reader, it can transmit data, such as an identification number or a location where it is placed. The global characteristics of an RFID-based positioning system are summarized in Table 9.9.

An RFID-based positioning system consists of positioning some tags at various locations. A tag is a very low cost element, so it is possible to place a large number of tags all over a large area. The mobile terminal consists of a reader that can interrogate the tags when passing nearby. Current readers can operate as far away as 1 m (sometimes farther, but this is quite unusual), thus leading to an equivalent accuracy of about 1 m. The positioning method consists in getting the identifier back from the tag and reading a location database to make the link between the identifier and a location. Of course, one can envisage that the tag indicates the location directly to the reader. One can also imagine that the mobile terminal could be the tag rather

TABLE 9.9 **RFID positioning.**

Technical description	
Technology	RFID
Localization	Infrastructure
	Very limited
Positioning	Symbolic
Environments	Indoors, outdoors
Performances	
Accuracy	<1 m
Range	<1 m typical

than the reader; it is less expensive for the terminal, but the new system then requires the location to be transmitted to the tag (this means that the tags have to be programmable).

■ 9.4 LOCAL AREA TELECOMMUNICATION SYSTEMS

Introduction

The wireless local area networks (WLANs) and wireless personal area networks (WPANs) are certain to be widely deployed in the near future. These networks, designed for local telecommunication purposes, are usually installed in indoor environments. As long as global positioning systems (GPS and GLONASS for instance) either do not work indoors, or are not very accurate, there is a need for indoor positioning solutions. The major point is that future applications will certainly be demanding in terms of continuity of service. There is currently no such continuity, mainly because of the poor indoor coverage and performances of localization systems.

The usual methods used for localization purposes were described in the section in this chapter "Introduction to Measurements Used." The one that is usually chosen for WLAN positioning is based on the RSS, which we will focus on. Nevertheless, some WLAN-based systems have been proposed using time of flight of radio signal.[14] For example, the AirLocation® from Hitachi is a network-based positioning system that includes a very precise clock in the base station network, allowing accurate time measurements.

The idea of signal strength measurement is not really a new one as it was considered in the early eighteenth century for measuring the terrestrial magnetic field in order to solve the problem of longitude. The principal difficulty of this approach lies in the fact that there are few locations that show an identical value of power level (in the current case), so the radio visibility of more than one transmitter (base station) is required if an accurate location is required. Nevertheless, the idea is essentially to draw a map of radio signal strength received. A typical result of such a process is represented in Fig. 9.4, where the values are the power levels in dB over a characteristic value (in this case a Bluetooth system using the RSS indicator). From this figure, it is possible to see that many different locations are characterized by, say, an RSSI value of 10. Thus, with only one base station, the accuracy obtained is poor, or indeed the number of possible locations is high. Furthermore, these locations can be scattered throughout the place (due to the non-trivial propagation scheme indoors). If accuracy is not required, this can be a simple positioning method.

When accuracy is required,[15] there is the need for more than one base station. When a few base stations are considered the principle is to find the nearest location

---

[14]It is in fact a time difference of arrival method by comparing the time of arrival at base stations.

[15]In general, as indoor places are smaller than outdoor ones, the typical accuracy required is somewhere between one-third and one-half the floor height. This should then allow the floor level to be determined as well. A typical value of 1 m can then be the tracked value. Note that very few works deal with the full three-dimensional WLAN approach at the moment.

5	9	5	5	6	10,5							
6	10	9	7	5,5	5							
9,5	10	6,5	8,5	6	7	9,5	8	11,5	12,5	7,5	9,5	7,5
7,5	10	9,5	10,5	5	7,5	10	10	10,5	7,5	9,5	9	10,5
5,5	11,5	7	9,5	8	6	11,5	9	7,5	10	12	12,5	8
8	10,5	10,5	8	8	10,5	10,5	8,5	13	11,5	12,5	10	14,5
9,5	6,5	8,5	9,5	10,5	14,5	14,5	13,5	13	10	13,5	12	12,5
11	9,5	12,5	10	11,5	12,5	14	2	19,5	16,5	12,5	10	11
9,5	9	10,5	12	10	13	13	17,5	14,5	13,5	14,5	9	10,5
12	9	8	10	11,5	5,5	11	16	16,5	15,5	13	12	11
9,5	8	10,5	6,5	8	14,5	11	13,5	13	13	14	9,5	12,5
3,5	0,5	1,5	1,5	9,5	11,5	13,5	13,5	14,5	13,5	9	10	9
1,5	3	2	4	0,5	16,5	10,5	9,5	10,5	12	11,5	10,5	7
0,5	3	1,5	5,5	2,5	7,5	11,5	7,5					
0	2,5	1	3,5	2	10	12,5	11					
0	0	1,5	0	0	0,5	7,5	10,5					
0	0	0	0	0	6,5	9,5	11,5					
0	0	0	0	0	2,5	0,5	6,5					
					12	5	2,5					
					0	2,5	1,5					

FIGURE 9.4 A typical RSS map (1 m step in both north and east directions — Bluetooth technology).

within the database containing all the values of all the base stations. It will be the most probable location of the mobile. In fact, many different algorithms have been tried and proposed, all taken from the first phase of establishing the database. This involves a campaign of measurements. Then two methods are possible: either a pattern matching approach consisting of finding the nearest element in the database, or a propagation-based approach, consisting of extracting a model of indoor radio wave propagation, that is, some mathematical relationship between the power level received and the distance from the mobile to the base station in consideration. More sophisticated methods, like defining possible tracks, have been evaluated, with good results. Indeed, if one accepts an increase in the constraints on the "resulting locations," the accuracy can be very good. The main difficulty remains to be able to easily apply the method to a new place.

From a few representative works, it is possible to outline the major trends of a WLAN-based positioning system, as it appears today. The following remarks can be made:

- Pattern matching approaches give quite good results (in terms of accuracy).
- Positioning based on propagation models appears to be less accurate.
- There is quite a large range of accuracy values.
- The distribution of measurements exhibits a rather large error margin (often around 10 dB).

- The orientation of the mobile terminal, relative to the base stations, is a parameter of concern.

- The direct impact of the number of access points that are used in the system is of great influence.

- The use of sophisticated algorithms for a "nearest neighbor" search leads to no improvement.

- Use of privileged tracks provides a tremendous increase in accuracy, when staying on these tracks.

- Dynamic mode, as well as multi-storey considerations, are not yet well described or evaluated.

- More complex infrastructures are likely to provide for both simplification of the calibration phase (almost always required) and efficiency of the location-finding process.

From this first analysis, another approach could be conceived that deals with a "symbolic" positioning, that is, in terms of rooms and corridors instead of accuracy in meters. The principle of a symbolic positioning was used in the Infrared system and could once again be implemented in the present case.

Bluetooth (WPAN)

Within the WPAN domain, Bluetooth was designed in order to get rid of cables in the near vicinity of a terminal. Thus, the ranges afforded by Bluetooth systems are rather small, typically 10 m. In fact, three classes exist, depending on the power level of the signal transmitted, with corresponding ranges of a few meters, a few tens of meters, and finally one hundred meters. The main advantage of Bluetooth compared to RFID, for instance, is the longer range and the radio effect that allows propagation through walls. However, if the same cell identification method is used, then the resulting accuracy is also weakened. New techniques have been imagined, as described in the introduction to local telecommunications, when a few base stations are available at a given point. Then the power level based method can be applied in order to achieve indoor systems with an accuracy of a few meters. The major drawbacks are the need for a large number of Bluetooth modules to cover any given place and the relative complexity in filling the database required to implement a "nearest neighbor" algorithm. The database needs to be modified whenever an indoor element, such as a desk or a cupboard[16] is moved. As this initiation step is really time consuming, a lot of effort has been made in order to find a method so that the database is filled in automatically with permanent continuous measurements, as people are connecting to the Bluetooth network (or WiFi as long as the problem is absolutely identical).

With regard to the possible application and deployment scheme, some proposals have already been made by many authors. A typical example is taken from the Osaka

---

[16]Note also that even normal daily operations are likely to change the environment, such as opening windows or closing doors. The fact of taking these actions into account leads to including error margins on the power levels and, as a direct impact, dramatically decreases the resulting positioning accuracy.

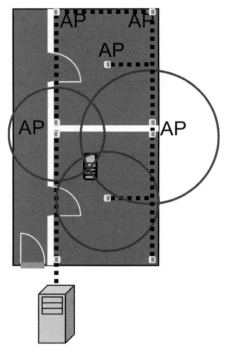

FIGURE 9.5 A typical Cell-Id triangulation approach (AP, access point).

Guide concept, which is a Bluetooth-based positioning system for museum visits. Figure 9.5 shows the way it is proposed to work with a Cell-Id triangulation approach (AP stands for access point). The global characteristics of a Bluetooth-based positioning system are summarized in Table 9.10.

As already discussed, the required network is larger than the telecommunication network, in that the number of access points must be greatly increased. One could argue that telecommunication needs are bound to increase, leading to a natural increase in the number of access points and hence filling the gap described above, but we also know that if increased rate capabilities are really required, manufacturers, helped by technicians, will be able to find solutions with a reduced set of access points in order to stay within acceptable costs. Then, the compromise is, as usual, difficult to be optimized.

WiFi (WLAN)

The majority of work to date has been achieved with WiFi networks. The main difference compared with Bluetooth is the range of the transmitters, which is quite a bit larger in the case of WiFi. The sensitivity of the devices is similar, so the number of WiFi access points could probably be reduced compared to the number of Bluetooth access points used in the same place. Note that the transmitters could be

TABLE 9.10 Bluetooth positioning.

Technical description	
Technology	Radio at 2.4–2.5 GHz
Localization	Infrastructure
	Discontinuous
Positioning	Absolute
Environments	Indoors
Performances	
Accuracy	~3 m (depending on the number of access points)
Range	Network connectivity

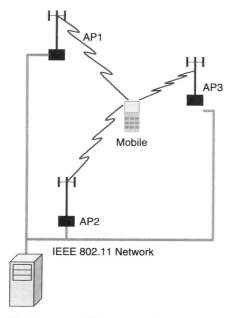

FIGURE 9.6 WiFi positioning system (AP, access point).

either access points or other mobile terminals (this applies to both Bluetooth and WiFi). Although propagation time methods have been reported with WiFi, the RSS approach is preferred. Thus, the technique of positioning is similar to that with Bluetooth (Fig. 9.6). The global characteristics of a WiFi-based positioning system are summarized in Table 9.11.

WiFi networks are primarily designed to allow a mobile terminal to connect to a wider network without wires, and WiFi deployment will be available everywhere indoors. Therefore, the range of any given access point is usually much larger than that of a Bluetooth access point,[17] designed for short-range connection. As the

---

[17]Although Bluetooth was designed for radio links between terminals, some developments have led to network access through the use of access points.

TABLE 9.11 WiFi positioning.

Technical description	
Technology	Radio at 2.4–2.5 GHz
Localization	Infrastructure
	Continuous
Positioning	Absolute
Environments	Indoors (mainly)
Performances	
Accuracy	A few m (depending on the number of access points)
Range	Network connectivity

deployment of a localization system is costly, the number of access points is of prime concern. Thus, WiFi solutions are generally preferred nowadays, certainly because dynamically reconfigurable networks are not yet well developed; indeed, Bluetooth terminals are certain to be more mobile than WiFi access points and thus WiFi positioning is the main subject dealt with in the WLAN positioning literature.

Ultra Wide Band (WPAN)

Within WPANs, the Ultra Wide Band (UWB) approach has a specific status, both because it uses a time-based approach, compared to the classical frequency approach, and because its objective is to provide either low data rates (IEEE 802.15.4a standard planned for June 2007) or very high data rates (IEEE 802.15.3a with typically 480 MB/s in order to achieve a wireless USB link[18]). The interesting feature for positioning is that UWB uses very short pulses. Thus, time is an embedded feature of UWB, unlike other typical radio systems (WiFi, Bluetooth, GSM, UMTS, GPS, and so on.). Moreover, the first application of a UWB system was in radar, because of two factors that will also be of great help for positioning: first, the capacity of these wide-band signals to get past obstacles,[19] and second because of the sharp time discrimination that it is possible to achieve.[20] The UWB approach of positioning is thus based on time measurements from four transmitters (T_1 to T_4 in Fig. 9.7) and is sure to be very accurate because of the very short pulse duration, typically less than 1 ns.

The basic principle of this is to carry out time measurements. One knows that intrinsic measurements can be very accurate, but we are left with the problem of global synchronization,[21] which is achieved through a fifth UWB module (referred

---

[18]Called WUSB! Note that another standard, IEEE 802.15.3c, intends to use the 57–64 GHz band.

[19]Because of the wide frequency band, there is always only a part of the signal that is disturbed by the obstacle, thus allowing the rest of the signal to pass. It obviously gives better results than classical narrow-band signals.

[20]Note that the radar has a specificity that does not exist with other positioning systems considered here: the transmitter and the receiver electronics are driven by the same clock!

[21]This is similar to the GNSS constellations for which the ground segment is in charge of this global synchronization. The delays between the perfect time and the specific times of each satellite are sent to the receivers through the navigation message.

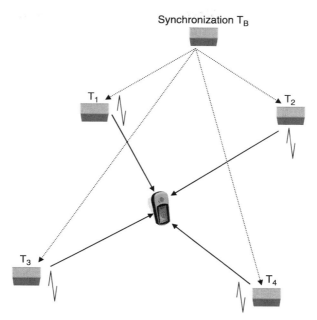

FIGURE 9.7   A typical UWB indoor positioning configuration.

to as "B" in Fig. 9.7). After having made the time measurements, it is possible to calculate either propagation time delays or difference of propagation time delays from a UWB-equipped mobile terminal and the various UWB system modules. Then, a trilateration method is employed, similar to that for GNSS systems, in order to determine the indoor location. Related to accurate time measurements, the resulting accuracy of the positioning is quite good, and results as accurate as a few centimeters have been reported by different teams.

The global characteristics of a UWB-based positioning system are summarized in Table 9.12.

UWB telecommunication systems are facing some difficulties arising from the jungle of wireless standards. The IEEE has decided not to issue a standard yet and to leave the two possibilities (single-band and multiband) open. The bandwidth has also been a subject of discussion; and the GPS community is largely the cause of the bandwidth being moved from $1-10\,$GHz to $3-10\,$GHz. Because of the very low level of GPS signals, GPS people were afraid of this new disturbing signal coming into the GPS band and put some pressure on the standardization organization in order to move the UWB upwards. Unfortunately, others followed and the final band now starts at 3 GHz, with some other potential fears, like, for instance, the WiFi "a" that lies in the 5 GHz band. To try to avoid probable new difficulties, the multiband approach has been proposed: as the official definition of a UWB signal is a signal whose bandwidth is greater than 500 MHz, why not consider channels of, say, 528 MHz and use these channels to fill the entire UWB band? In such a way, 11 bands

TABLE 9.12 **UWB positioning.**

Technical description	
Technology	Radio in the 3.1 – 10 GHz band
Localization	Infrastructure
	Continuous
Positioning	Absolute
Environments	Indoors
Performances	
Accuracy	A few cm
Range	A few m

are available and this method allows the removal of some bands when needed (for example to avoid collision with WiFi "a" or in some countries where other bands might not be available).

Unfortunately, this approach put some constraints on the form of the pulse, which is bound to reduce the positioning accuracy to a few decimeters rather than a few centimeters for subnanosecond pulses. Furthermore, as with WLAN and Bluetooth positioning systems, which do not correspond to the telecommunication network, the UWB positioning capabilities are excellent but are not achieved with the power levels allowed in telecommunication systems. If this were the case, the range would be largely reduced, down to a few meters, compared to a few tens of meters using power levels of a few hundred milliwatts.

WiMax (WMAN)

From WPAN for very small ranges (a few meters) to telecommunication networks for wide ranges (a few tens of kilometers), the wireless systems encountered are, respectively, the WLAN for medium–small ranges (a few hundreds of meters) and the WMAN[22] for medium–large ranges (a few kilometers). The most widely known WMAN is called WiMax. Similar positioning systems can be envisaged, such as WPAN (Bluetooth) and WLAN (WiFi), with an increased range. Unfortunately, a discrepancy between the telecommunication network and the positioning network may once more be highlighted. The former uses transmitters with greater range in order to reduce the number of such transmitters to be deployed, whereas the latter needs far more transmitters. This difficulty has been overcome for Bluetooth and WiFi because of the reduced size of the area being considered (a few hundred square meters). When dealing with a few square kilometers, the foreseen problem is that with only three transmitters the size of the resulting positioning areas could still be quite large, thus requiring a greater number of transmitters (more than three).

---

[22]WMAN stands for Wireless Metropolitan Area Networks.

Radio Modules

All the techniques that have been described in this section relative to the local area telecommunication systems could clearly be implemented by means of radio modules that are not intended to play any telecommunication role. WPAN, WLAN, or WMAN systems are used simply in order to reduce the cost of the positioning system. As already said, this is not necessarily always the case.

Comments

We have seen that for different reasons (number of access points or power levels), the positioning infrastructure is more complex than the telecommunication network. The solution could then be, as suggested in the previous paragraph, to use separate radio systems for telecommunications and positioning. Although a lot of approaches are possible, using a radio link that is already embedded in typical terminals is really much easier. So the optimal solution could well be to deploy two similar WLAN networks: for example, one for telecommunications with no real redundancy, and the second for positioning, with a redundancy of the order of three to five,[23] depending on the accuracy wanted. In such a case, one could envisage that the positioning network could help the telecommunication network in the case of congestion or for marginal actions. Furthermore, unless the positioning is free, one would require a mechanism to allow payments to be collected, and the positioning providers are not necessarily the same as the telecommunication provider (as telecommunications are not really free today). In this case, the fact that the networks are different is a simplification of the problem.

■ 9.5 WIDE-AREA TELECOMMUNICATION SYSTEMS

Following our review of existing radio systems, we will now deal with large-area systems, and typically those used for telecommunication purposes, like the GSM[24] on the one hand and UMTS[25] on the other. It is quite probable that we shall face the same limitations as those discussed in the previous section on local wireless systems and that the wider area covered will be a new feature. In fact, one could now easily understand that methods based on power level measurements are not likely to be really efficient because the equivalent of local system access points are now base stations, which are quite expensive.[26] Thus, redundancy cannot be the prerequisite to positioning. Hence, old methods like the angle of arrival or those based on time measurements

---

[23]Here, the order corresponds to the number of access points that have to be in radio visibility of any given point of the place.
[24]Global System for Mobile.
[25]Universal Mobile Telecommunication System.
[26]Note that the cost of an access point is less than $100, whereas the cost of a UMTS base station is much more. It therefore makes quite a difference in the infrastructure cost.

have been taken out of the array of possible techniques. The following paragraphs will deal with the main solutions proposed, although only the first is widespread.

As a complete description of the main positioning techniques has been given in Chapter 4, we will only discuss the specificities linked to the indoor environment. Typical indoor tables are also given.

GSM

The basic methods used indoors are similar to those used outdoors: Cell-Id, angle of arrival (AOA), time of arrival (TOA), and time difference of arrival (TDOA).

The Cell-Id method is comparable with RSS measurements as it is based on power level estimation, but with only one base station. Note that this is also the basis of the symbolic positioning proposed in the wireless local area domain. The main advantage of Cell-Id is that it works with only one base station in radio visibility. The main advantage of the symbolic approach is to allow increased accuracy in conjunction with the number of access points deployed. The global characteristics of a Cell-Id based positioning system are summarized in Table 9.13.

It is quite obvious that the AOA method is not applicable indoors for positioning purposes. Accuracy is not very good, and this approach is absolutely not applicable if the direct signal (line of sight) is absent. In other words, indoor positioning is not intended really to be obtained with AOA.

Nevertheless, the global characteristics of an AOA-based positioning system are summarized in Table 9.14. Note that this table is only applicable in outdoor environments (and is therefore given for comparison purposes only).

So, power level measurements are not likely to provide good enough accuracy and an AOA method is too complex and really not acceptable for indoors. Quite logically, solutions implementing time measurements have been considered. Different possibilities are open to us, such as direct time measurements or difference of time. The main problem of time-based methods in telecommunication networks is that requirements, in terms of time precision, are once again different for telecommunication purposes and for positioning purposes. Telecommunication exchanges are based on protocols of transmission that include a synchronization feature, usually by way of sending specific data headers prior to the real data transmission, in order to define an identical starting time for both the transmitter and the receiver. For positioning

TABLE 9.13 Cell-Id positioning.

Technical description	
Technology	Radio in the 0.9–2 GHz band
Localization	Infrastructure
	Quasi continuous
Positioning	Symbolic
Environments	Indoors, outdoors
Performances	
Accuracy	Cell size
Range	Cell dependent (100 m to 30 km)

TABLE 9.14 **AOA positioning.**

Technical description	
Technology	Radio in the 0.9–2 GHz band
Localization	Infrastructure
	Quasi continuous
Positioning	Absolute
Environments	Outdoors
Performances	
Accuracy	~100 m
Range	Network coverage

purposes, one needs, as discussed in previous chapters, very good synchronization, because the resulting localization is directly linked to it. Nevertheless, some methods have been proposed, as shown in Chapter 4.

Owing to the poor time accuracy in telecommunication networks, the resulting accuracy is around 100 m outdoors. When multipath occurs, and multipath occurs very often in telecommunication networks, the accuracy drops dramatically to a few hundred meters. Indoor positioning is also affected by this phenomenon, although similar accuracies are often reported for both outdoors and indoors.

The global characteristics of a TOA-based positioning system are summarized in Table 9.15 (typical outdoor performances).

The TDOA method is very similar and results are equivalent. The Matrix system (refer to Chapter 4) reported a typical accuracy of around 100 m. The global characteristics of a TDOA-based positioning system are summarized in Table 9.16. This technique could also be used indoors, but because of the same limitations as for TOA and AOA, the results are not acceptable. An advantage of this technique compared to TOA is the possibility, at no expense, of implementing it directly at the terminal end. This brings it closer to GNSS approaches.

UMTS

The same techniques apply, leading to similar global performance. In some specific cases, the better synchronization available in UMTS networks allows slightly

TABLE 9.15 **TOA positioning.**

Technical description	
Technology	Radio in the 0.9–2 GHz band
Localization	Infrastructure
	Quasi continuous
Positioning	Absolute
Environments	Indoors, outdoors
Performances	
Accuracy	~100 m
Range	Network coverage

TABLE 9.16 **TDOA positioning.**

Technical description	
Technology	Radio in the 0.9–2 GHz band
Localization	Infrastructure or mobile terminal
	Quasi continuous
Positioning	Absolute
Environments	Indoors, outdoors
Performances	
Accuracy	~100 m
Range	Network coverage

increased accuracy. Typical accuracies reported for the Matrix system, for instance, are in the 50 m range.

Hybridization

Let us imagine that no single solution will be found for outdoor and indoor positioning (and also for urban canyon positioning, for in-plane positioning, in-train, or for any other specific environment). In such a case, there will be the need for a multi-technology terminal. Because current systems, in general, are usually an integration of various technologies,[27] there are no *a priori* difficulties. The idea of hybridization is to allow positioning from data coming from different positioning systems. For instance, in relation to GSM and GPS, the idea is to compute a location with two satellites and two base station measurements.

In fact, these approaches, although very interesting in principle, are quite difficult to implement practically, mainly because of the great diversity of physical measurements and their respective errors and noise. In theory, every kind of combination is possible and leads to acceptable results, but in practice one must take into account the time bias between the two (or more) different systems and this can be really impossible because time accuracy is very poor in some systems. Thus, incorporating an unknown in the equations, namely time, leads to adding a new variable in the mathematical system, leading to the need for an additional measurement, removing the theoretical advantage of hybridization.

Nevertheless, there is another kind of hybridization, which consists of aiding the GNSS positioning system, either indoors (but it remains difficult, as described in the next paragraph) or more often outdoors. The way it works is as follows. When obtaining a raw location by means of the GSM (for example) Cell-Id method, this non-accurate positioning can be used by the GNSS receiver in order to decrease the time needed to output the first point.[28] This feature is of great importance in personal localization as the user wants to obtain his location quickly. This cannot

---

[27] As an example, one can note that there are current GSMs that include WiFi, Bluetooth and GPS, or personal device assistants (PDAs) that are available with WiFi, Bluetooth, Infrared, GPS, and a digital camera. So, integration is a daily achievement.

[28] TTFF: Time To First Fix.

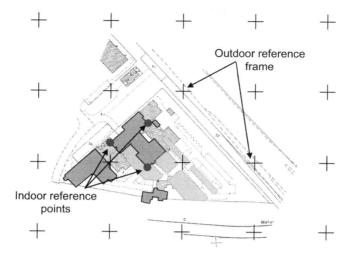

FIGURE 9.8   Building representation and correspondence in WGS84.

be achieved by a standard GNSS receiver because it needs to "download" the satellite ephemeris, which cannot be achieved in less than 30 s (this discussion will be detailed in Chapter 10 when referring to Assisted-GNSS). Furthermore, the GNSS receiver must then compute positioning, considering the last location it output as the starting location of its iterative algorithm (see Chapter 8 for details on the position calculations). Unfortunately for this receiver, very often, in real life, this last location is either far away from the current position or was obtained a long time earlier.[29] In both cases, in addition to the 30 s, there is a need for the receiver to search for the satellites in the sky, to find them, and finally to compute the location. All this can easily take an additional 15–20 s. The total TTFF (Time To First Fix) is then around 45–50 s. This is much too long for a personal user, who needs to obtain the data as quickly as possible. With a Cell-Id positioning, information can be provided to the receiver to help it: the first indication can be the location of the base station to which the receiver is linked.[30] This is not very accurate for personal positioning, but this represents a very good first location for the iterative algorithm. Furthermore, knowing the satellites that are "seen" from the base station, the GSM could give this information to the receiver, which will then know which satellites to search for. Finally, the GSM could also provide a precise enough time.

Note that this second type of hybridization can be achieved with almost all the non-GNSS positioning systems. As an example, let us consider the case of a WLAN indoor positioning system that could be implemented in a building.

The buildings where indoor positioning is implemented are highlighted in Fig. 9.8. In addition to the above explanations, it is obvious that the building is

---

[29]One should understand that the receiver time could be far away from the real time, leading to a misinterpretation of the satellite configuration.
[30]Or any other location obtained through the GSM network.

rather small compared to a GSM cell, leading to a rather accurate first estimation of the location for a GNSS receiver; so, the convergence should be very quick. Of course, to ultimately accelerate the TTFF when the terminal goes outdoors and tracks the satellites, one could easily imagine that ephemeris data could be transferred from a WLAN access point to the receiver.[31]

In such a case, the TTFF could be dramatically reduced due to the fact there will be no need for the 30 s waiting time. As with the GSM hybridization, having such a precise first estimation of the location could also help in finding the satellites[32] very quickly. Furthermore, when considering the different ways to acquire the information for the WLAN network, one can even go a step further. A very simple means could be to have a mass market GNSS receiver located somewhere in the building that has an acceptable sky view, and that is wired to the WLAN network. In this case, the location of the building is available, but also the GNSS time, the ephemeris, and the satellites in view with their positions and power levels. Knowing these data, together with the building's position, it could be another help to the receiver to know the location of the exit doors in order to determine more easily the receiver's location as it is leaving the building.

In case this last solution is not implemented because of the need to deploy a GNSS receiver, another approach would be to consider GNSS-like coordinates for indoor locations. Indeed, it is important to carry out a brief discussion on this matter here. Usually, indoor positioning systems, unlike GNSS, deal with local coordinates; that is, the location is primarily given in a referential that is associated with the building.[33] No absolute references exist either related to North and South or to a global referential. Converting local coordinates into global coordinates was necessary in order that the GNSS receiver was capable of location computation. The way this could be achieved is through the definition of some identified characteristic points of the building in any of the more global coordinate systems (UTM, WGS, Lambert, and so on). Usually, construction plans show such coordinate systems and it is simple to program the transformation between the local coordinates and this wider system. Some implementations of WLAN indoor positioning systems even output the indoor location in WGS84 format.

■ 9.6 INERTIAL SYSTEMS

Inertial systems are described in Chapter 4. Indoor functioning is similar to outdoor, with care needed for specific sensors, such as magnetometers, which are greatly influenced by metallic structures. Remarkable achievements have already

---

[31]The data rate of WLAN connections is currently between 11 Mbit/s and 54 Mbit/s, for a single user. Even with multiple users and considering the real possible data rate, the transmission of the 1500 bits of the ephemeris is achieved in a few fractions of a second.

[32]Note that this is exactly the purpose of the Assisted-GNSS, as described in detail in Chapter 10.

[33]In fact, when moving indoors, almost no one would need to refer to an absolute external referential. This problem is identical to that of GNSS location representation. The real need is a map representation and absolutely not absolute coordinates (except for die-hard sailors).

TABLE 9.17 **Inertial positioning.**

Technical description	
Technology	Inertial: accelerometer, gyroscope and magnetometer
Localization	Mobile
	Continuous
Positioning	Relative displacement
Environments	Indoors/outdoors
Performances	
Accuracy	Time dependent (~ 1 m/mn)
Range	Global

been reported, although the integration to mobile portable devices remains a challenge. The main sensors included in the "inertial" term are

- Accelerometers,
- Gyroscopes,
- Odometers,
- Magnetometers, and
- Barometers — Altimeters.

The coupling with GNSS can be either tight or loose depending on the way positioning is carried out. The loose approach consists of considering either the GNSS positioning or the inertial positioning, depending on confidence indicators (giving preference to one technique). The tight approach consists of embedding the measurements from both techniques in order to produce a resulting positioning.

The global characteristics of an inertial positioning system are summarized in Table 9.17.

■ 9.7 RECAP TABLES AND GLOBAL COMPARISONS

Global synthesis is achieved in Chapter 10, including the GNSS-based indoor approaches.

✓ 9.8 EXERCISES

Exercise 9.1 Carry out a summary of the main specifications required for a positioning system to be used

- By visitors in a museum;
- By passengers in an airport;
- By citizens in an administration center.

The specifications you could consider are, among others, the following:

- Continuous positioning required or on-demand positioning,
- Accuracy,
- Coverage,
- Availability, and so on.

What are the technical solutions you would retain?

Exercise 9.2 Answer the same question as in Exercise 9.1 for the same environments, but considering permanent personnel (technicians, official agents, occasional workers, and so on). Describe the type of applications that could be envisaged and the differences with those examined in Exercise 9.1.

Exercise 9.3 Why are the majority of WLAN-based positioning systems using power level measurements? In order to achieve a few meters of accuracy, what are the major constraints? Can you imagine an approach that would lead to an easier practical implementation (perhaps accepting a less accurate positioning)?

Exercise 9.4 Hitachi has proposed a system carrying out time measurements in a WiFi network (called AirLocation®). Can you imagine the complementary component required to be added to the network in order to allow such a technical approach?

Exercise 9.5 How does UWB differ from classical WiFi or Bluetooth radio systems? Why is this feature of interest for positioning? Give a quantitative description of measurement accuracy and range with current implementations and standards. One needs to access further readings.

Exercise 9.6 Why is the Cell-Id implemented in all mobile networks? What comments can you make concerning the accuracy of this technique for indoor positioning?

Exercise 9.7 What is the main difficulty in implementing either TOA or TDOA approaches within a mobile network? What are the evolutions of UMTS with respect to GSM concerning this matter? Find information and describe the way the Matrix system gives an answer to this difficulty.

Exercise 9.8 Why are AOA measurements not really an acceptable solution for indoor positioning with mobile network transmissions? Give the reasons and how the "direction of transmission" technique has been implemented in mobile networks.

Exercise 9.9 Considering the list of inertial systems given in Section 9.6, try to find the difficulties that indoor positioning will generate.

Exercise 9.10 Answer the same question as Exercise 9.9 for portable devices.

■ BIBLIOGRAPHY

Bahl P, Padmanabhan VN. RADAR: an in-building RF-based user location and tracking system. In: IEEE INFOCOM 2000: Proceedings; 2000. p 775–784.

Bao-Yen Tsui J. Fundamentals of global positioning system receivers — a software approach. John Wiley & Sons; 2000. New York, USA.

Bliesze M, Hupp J. Indoor navigation in DECT networks. Proceedings of the International Symposium on Indoor Localisation and Position Finding; InLoc2002; Bonn, Germany; July 2002.

Eissfeller B, Gänsch D, Müller S, Teuber A. Indoor positioning using wireless LAN radio signals. In: ION GNSS 17th International Technical Meeting of the Satellite Division: Proceedings; September 21–24, 2004, Long Beach (CA).

Fontana RJ. Recent system applications of short-pulse ultra-wideband (UWB) technology. IEEE Trans Microwave Theory Tech 2004;52(9, Pt 1):2087–2104.

Fontana RJ, Gunderson SJ. Ultra-wideband precision asset location system. In: IEEE Conference on UWB Systems and Technologies: Proceedings; May 2002.

Frazer E. Indoor positioning using ultrawideband techniques — analysis and experimental results. 11th IAIN World Congress, October 2003, Berlin, Germany.

Gezici S, Tian Z, Giannakis GB, Kobayashi H, Molisch AF, Poor HV, Sahinoglu Z. Localization via ultra-wideband radios: a look at positioning aspects of future sensor networks. IEEE Signal Processing Magazine, Volume 22, Issue 4, p 70–84.

Gilliéron P-Y, Merminod B. Personal navigation system for indoor applications. 11th IAIN World Congress; Berlin, Germany; 2003.

Heinrichs G. Personal localisation and positioning in the light of 3G wireless communications and beyond. 11th IAIN World Congress, October 2003, Berlin, Germany.

Hightower J, Borriello G. Location systems for ubiquitous computing. Computer, August 2001. Volume 34, Issue 8, p 57–66.

Kaplan ED, Hegarty C. Understanding GPS: principles and applications. 2nd ed. Artech House; 2006. Norwoood, MA, USA.

Kitasuka T, Nakanishi T, Fukuda A. Wireless LAN based indoor positioning system WiPS and its simulation. In: 2003 IEEE Pacific Rim Conference on Communications, Computers and Signal Processing: Processings; August 2003. p 272–275.

Koshima H, Hoshen J. Personal locator services emerge. IEEE Spectrum, February 2000. Volume 37, Issue 2, p 41–48.

Krishnan P, Krishnakumar AS, Ju W-H, Mallows C, Ganu S. A system for LEASE: location estimation assisted by stationary emitters for indoor RF wireless networks. IEEE INFOCOM; 2004.

Mattos PG. Assisted GPS without network cooperation using GPRS and the internet. In: ION GPS/GNSS 2003: Proceedings; September 2003; Portland (OR).

Martone M, Metzler J. Prime time positioning: using broadcast TV signals to fill GPS acquisition gaps. GPS World 2005;16(9):52–59.

Pateli A, Fouskas K, Kourouthanassis P, Tsamakos A. On the potential use of mobile positioning technologies in indoor environments. In: 15th Bled Electronic Commerce Conference: Proceedings; June 2002. Bled, Slovenia.

Randall J, Amft O, Tröster G. Towards LuxTrace: using solar cells to measure distance indoors. In: Location and Context Awareness (LoCA). Proceedings; May 2005; Oberpfaffenhofen, Germany.

Takada Y, Kishimoto M, Kawamura N, Komoda N, Yamazaki T, Oiso H, Masanari T. An information service system using Bluetooth in an exhibition hall. Annales des Telecommunications 2003.3/4;58(3–4):507–530.

Wang Y, Jia X, Rizos C. Two new algorithms for indoor Wireless Positioning System (WPS). In: ION GNSS 17th International Technical Meeting of the Satellite Division: Proceedings; September 21–24, 2004; Long Beach (CA).

Zagami JM, Parl SA, Bussgang JJ, Melillo KD. Providing universal location services using a wireless E911 location network. IEEE Communication Magazine, Volume 36, Issue 4, p 66–71.

Web Link

http://www.cl.cam.ac.uk. Nov 14, 2007.

GNSS-Based Indoor Positioning and a Summary of Indoor Techniques

The advent of positioning over the last few years has clearly been due to the incredible success of GPS. Indoor positioning appears to be one of the most challenging problems for GNSS constellations as it represents their main current limitation. This was identified early on in many works carried out both on the receivers' detection capabilities and the so-called "local elements" or "local area augmentation systems" (LAAS). GNSS-based indoor positioning is not yet fully resolved, although interesting solutions have been proposed and are still under evaluation and development. At the end of the chapter, tables summarize all the indoor techniques dealt with in Chapters 9 and 10. The last part of the chapter is devoted to propositions for future indoor systems using the future diversity of GNSS signals (multiconstellation and multifrequency).

Of course, the first systems whose indoor capabilities were evaluated were the satellite navigation based ones. As will be briefly described in Chapter 11, the history of applications has shown that the success of satellite navigation was anything but planned by the original designers. Being the evolution of the TRANSIT system, designed to fulfil military maritime applications (in order to allow aircraft or terrestrial vehicles a wider range of use) no one could have envisaged, at the beginning, it's indoor application. However, with the advent of modern telecommunications systems for personal users, a real need emerged. Unfortunately for GPS,[1] the power level of the signal received is too low for indoors as the signal margin allowed by the code correlation is about

---

[1] Here, the first system is GPS and not yet GNSS.

Global Positioning: Technologies and Performance. By Nel Samama
Copyright © 2008 John Wiley & Sons, Inc.

10 dB, which is far too low to allow the penetration of walls and other structures. One usually considers that, unless the building is wooden, attenuation generated by structures when a radio signal penetrates, at 1.575 GHz, is between 15 and 30 dB. As it was not possible, at that time, to change the code's length because of the navigation message (see Chapter 7 for more details on the global structure and induced limitations of the C/A code), the only possible direction for GPS was to develop more sensitive receivers. The various techniques are given in the following. For three or four years, from roughly 2000, a lot of effort was made on investigating this approach, which finally appears not to be the definite answer to the indoor problem. Nevertheless, it has enabled GPS manufacturers to propose receivers that nowadays can work inside a car equipped with an athermic windscreen.[2] This was obviously not the first goal but still remains an interesting result.

The Galileo program had to be more efficient with regard to indoor positioning. So, the European Union decided to introduce the concept of "local elements" as a fundamental difference between Galileo and GPS. The timing of this decision coincided with the time when all manufacturers said that high sensitivity receivers would solve the problem. Thus, a universal solution was foreseen at that time. Unfortunately, as is often the case with the "great programs," once someone has a miracle solution, no other idea is investigated. So, when the high sensitivity approach, together with the almost only back-up solution (namely UWB), appeared not to be the ultimate answer, no other acceptable solution was available.[3]

The solutions described below are intended to present the state of the art of the most "promising" approaches based on the use of satellite navigation signals. Of course, as stated in the introduction to Chapter 9, competition exists between solutions that require a local infrastructure and the others. Not having to add more relays or base stations is better, but although a great deal of technical and financial effort has been put into finding a solution, no results have been obtained yet. Thus, no-infrastructure solutions do not resolve the problem, although they do improve indoor performance.

■ 10.1 HS-GNSS

The search domain of a satellite signal is huge, both in frequency and in time. The frequency search is required in order to deal with the Doppler shift, due to the motion of both the satellite and the receiver. As the correlation process is indeed a comparison of a local replica of the satellite code with the incoming satellite code, the replica must take this Doppler shift into account.[4] The time search is required in order to determine the propagation time shift between the transmission time from the satellite to the receiving time at the receiver end. But both searches are rather large: about ± 10 kHz in frequency and as much as 1 ms (the duration of a complete

---

[2]Such a windscreen introduces an attenuation of about 10 dB at 1.575 GHz, making the search for satellites fail in previous receiver technologies.

[3]Only "marginal" solutions, developed by small teams are therefore alternatives. Of course, these solutions are not likely to be applied to a program such as Galileo.

[4]This allows the replica to have the right chip duration and then leads to a good quality correlation.

code) in time. The steps of both searches are also rather small: a few Hertz for the frequency and a fraction of a chip for the time. Thus, the time required to lock on to a satellite is rather long.

Of course, this time is further increased when the receiver tries to find very low power signals, which is bound to occur indoors. Furthermore, with a very low signal, the search process is even more difficult due to the fact that the signal peak is not significantly higher than others. Thus, finding a way to cope with low signals could necessarily help in an indoor situation.

The various methods that have been developed toward this high sensitivity goal include the following:

- Complex electronic systems to allow direct frequency processing in order to find the frequency's peak at once;
- Multiple correlation in order to achieve parallelism;
- Long integration, either coherent or non-coherent, in order to find very low pseudo random noise codes.

The first method can be achieved through the use of a Fourier transform. Unfortunately, this approach consumes a lot in terms of power supply and complexity of the corresponding electronics. As the power consumption of a GNSS receiver is a major concern, other directions were investigated.

The second approach adopts a different philosophy. The idea is to try all the possibilities in the frequency and time domains at once, that is, in parallel. The quasi-immediate electronic architecture would have as many processing channels as there are elementary possibilities. Let us consider a frequency step of 10 Hz for a complete range of 10 kHz (that is, ± 5 kHz); this leads to a thousand possibilities. Let us also consider a GPS C/A code of 1023 chips and an elementary time step of one chip, this leads to another thousand possibilities. The complete search domain then consists of about one million possibilities. If it is possible to build an electronic device including one million parallel channels, then the treatment of all the possible combinations in time and frequency for one satellite can be achieved in one clock time duration. In fact, this should allow, in one clock time, all the correlation values to be output, but the processing of the correct location of the peak still has to be computed. Note that current receivers have typically between 14 and 20 channels, which are usually used by associating one channel with a given satellite. The search process is then conducted in a sequential mode. One of the first industrial realizations incorporated 32,000 parallel correlators. Further products have exhibited more than 200,000 correlators in parallel.

Another way to track very low signals is to use the characteristic of pseudo random noise features, by using the fact that it is not random at all and that if you know what you are searching for, it is possible to "integrate" the energy held in a code by repeating the correlation a few times in a row. Of course, this requires the correlation to be "followed" and "kept" as time goes on. This approach is called the "long integration." There are two kinds of long integration, depending on whether the integration is carried out in a time continuous manner or at some discrete times. Coherent long integration is limited, in GPS, by the global form of the signal.

TABLE 10.1 **High sensitivity GNSS positioning.**

Technical description	
Technology	RF at 1.575 GHz
Localization	Mobile receiver
	Continuous
Positioning	Absolute
Environments	Indoors, outdoors
Performances	
Accuracy	A few m
Range	Global (GNSS meaning)

The navigation message is the main reason (the other one is the time required by the receiver electronics in order to carry out the correlations) — due to the 50 Hz data rate, there is a 20 ms time interval during which the code remains identical (either the code or the inverse of the code, depending on the value of the data bit of the message) to itself. Thus, the receiver can achieve a "coherent" integration within a 20 ms period, at most.

Note that to carry out longer integration, the immediate possibilities would be to have either lower data rates for the navigation message, or even no navigation message at all. Both approaches have been considered in the Galileo program and in the GPS modernization program. Different navigation data rates have been proposed and also so-called "pilot tones," which are signals without navigation data. The purpose of these signals is clearly to help low level signal detection and acquisition.

A typical high sensitivity GNSS (HS-GNSS) positioning system therefore does not require any further infrastructure, other than that of the GNSS. That it theoretically needs nothing in addition to the current constellation is the major advantage of this approach. Unfortunately, although greatly improving the receivers' performances in difficult environments, determination of the correlation peak remains too difficult when the power level received is very low. Thus, false detections are possible and degrade the positioning. Nevertheless, positioning is still possible although imperfect. As the basic principle of the positioning is to carry out classical time measurements, the accuracy cannot be improved compared to an outdoors configuration with good reception conditions.

The global characteristics of a high sensitivity based positioning system are summarized in Table 10.1.

■ 10.2 A-GNSS

A typical Assisted GNSS (A-GNSS) positioning system is shown in Fig. 10.1. It includes a lot of elements: an assisted GNSS server, a special handset that includes the specific "assisted" processing capabilities, and the specific telecommunication protocols for assisted data exchange. Here the basic idea is to "assist" a GNSS receiver to

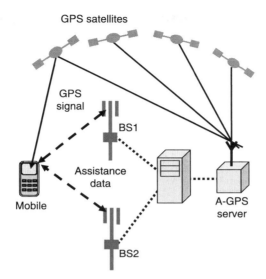

FIGURE 10.1 A typical assisted GNSS configuration (BS stands for Base Station).

allow it both to find a location in difficult environments (in the same sense as for HS-GNSS)ʹ and dramatically reduce the time to first fix (TTFF), which is a major concern for applications such as LBS (Location Based Services) for personal users. Solutions implemented are thus quite similar to those developed for HS-GNSS on the one hand and for hybridization for GSM-like positioning on the other. In addition, using the transmission capabilities of the telecommunication network allows an immediate improvement in the future performances of pilot tones. The A-GNSS server, located at the base station of the telecommunication network, acquires the GNSS constellation navigation message and transmits it to the A-GNSS receiver. Thus, it can simply remove the navigation message from the received signal (coming from the satellites). In such a way, a coherent integration method can be applied and the 20 ms limitation no longer applies. Of course, the use of an HS-GNSS chip is possible and the fact the A-GNSS server has to acquire the constellation makes it possible give both an initial location for the mobile positioning that is rather near the mobile location and a good enough time to facilitate the reduction of the TTFF. As already described in the GSM hybridization approach, another possibility implemented is to transmit information about which satellites to look for first.

The basic principle of the positioning is to carry out time measurements like standard GNSS receivers. The availability of assisted data helps in reducing the TTFF down to a few seconds when leaving an obstructed area, but gives no further answer to indoor positioning, unlike HS-GNSS for example. Furthermore, although not requiring any additional local infrastructure for buildings, this method is only made possible when the assisted server is deployed and the mobile terminal must be compatible with assisted data (that is, is not a current standard receiver). Some A-GNSS providers propose to furnish assisted data on a worldwide basis. In this way, both hardware and complete software suites are available throughout the

TABLE 10.2 Assisted-GNSS positioning.

Technical description	
Technology	RF in the 1.575 GHz and in the mobile network range
Localization	Infrastructure or mobile
	Quasi-continuous
Positioning	Absolute
Environments	Indoors, outdoors
Performances	
Accuracy	From 10 to 100 m
Range	Mobile network coverage

world for rapid deployment. In this competitive world, U.S. companies are the most important ones, both in technological advances and in business development. The global characteristics of a high sensitivity based positioning system are summarized in Table 10.2.

A-GNSS systems are facing the fact that deployment is quite expensive and telecommunication operators want to be sure that potential users will show enough interest in order to cover the investments. The United States situation is very different from that of European countries, for instance. Indeed, the FCC recommendation is a strong invitation to telecommunication operators to implement a location finding solution; A-GNSS is, from this point of view, an interesting approach that is at least currently possible (unlike other techniques, like pseudolites or repeaters). In European countries, regulation is not based on an obligation but on a "best effort" to be carried out by both telecommunication providers and terminal manufacturers. This is apparently not the most efficient way to stimulate industrial development. Furthermore, A-GNSS has shown it provides a real added value compared to "just" GNSS, although this is still not the final answer to indoor positioning.

■ 10.3 HYBRIDIZATION

A-GNSS is already a kind of loose hybridization, but other approaches are already available integrating GSM-like techniques and GNSS-based ones. The only implementation of TDOA in GSM networks is currently the one called Matrix developed by Cambridge Positioning Systems (CPS). For such a TDOA approach, the availability of a good quality clock is valuable. Thus, including a GNSS receiver in the terminal handset, principally for this purpose, is interesting. Once the GNSS receiver is available, one has to consider that when working optimally, it gives much better results than the TDOA method. Thus, the system has been improved in order to include this feature, and is called E-GNSS, for "Enhanced GNSS." The enhancement is mainly due to the fact the TDOA positioning is used when GNSS positioning is of poor quality or not available. In this way, the hybridization is of the loose type. Figure 10.2 shows the architecture of the handset and an illustration of the working mode of E-GNSS (as proposed by CPS).

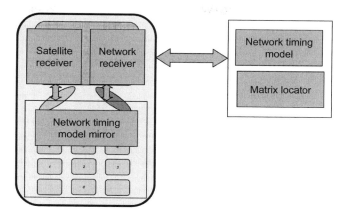

FIGURE 10.2 A typical assisted GNSS proposal.

As both HS-GNSS and A-GNSS are not the indoor ultimate solution, hybridization can be considered identically to previous cases (see Section "Hybridization," Chapter 9). Current approaches are oriented towards the use of A-GNSS and HS-GNSS for outdoor positioning and WLAN for indoor purposes.

■ 10.4 PSEUDOLITES

The GNSS-based indoor positioning systems mentioned above would have been nice solutions as no infrastructure would have been required. Unfortunately there is clearly a need for other solutions. The next two solutions, pseudolites and repeaters, use GNSS-like signals but require a local infrastructure in order to provide indoor positioning. The proposed accuracies are usually better than those proposed by outdoor GNSS, even in very good outdoor conditions (no multipath, good propagation conditions, high received power levels, no obstructions, and a large number of satellites in view). An important point to note at this time is that using pseudolites and repeaters for indoor positioning is still under research (although pseudolites already exist to complement outdoor GNSS systems).

The basic idea underlying the use of a local infrastructure is somehow to reconstruct a constellation associated with the local building. One could call the local infrastructure a "terrestrial satellite constellation." As the problem of indoor reception arises mainly because of low signal levels, using generators can effectively solve the problem by providing higher level signals, that is, those that are powerful enough. The practical implementation of the pseudolite approach consists in building complete signal generators. The shrewdness in the application of pseudolites is their use of similar signals to GNSS. In this way, standard receivers are already enabled to decode such signals without any hardware updates. Only software updates are required in order to tell the receiver to search for and to acquire these new "terrestrial satellites." For the GPS constellation, which is currently the only one that is fully operational, the C/A codes are reserved and it is not possible to use them. Thus,

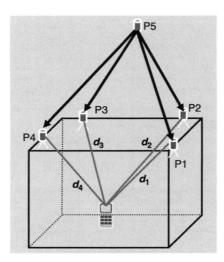

FIGURE 10.3 Illustration of an indoor pseudolite configuration (Pi refers to pseudolite i and d_i to distance between pseudolite i and the indoor receiver).

pseudolites have to choose other pseudo random noise. Fortunately, the 36 reserved codes, although exhibiting the best correlation features, are far from being the only possible ones. There are numerous Gold codes that can be used, with lower correlation quality, but as the power levels are higher for pseudolite generators and the distance from pseudolite to the receiver rather small compared to that between satellite and receiver, this is of no real consequence. Each pseudolite is then considered as a signal generator, as illustrated in Fig. 10.3.

The pseudolite concept is a very old one, originally considered at the beginning of the 1980s. It has been applied, in anticipation, to various types of missions, from augmentation to the current GPS constellation in difficult environments such as open cast mines, or more recently urban canyons, to a Mars exploration positioning system,[5] and a local area augmentation system (LAAS) for plane landing approach. For an indoor system, there is the need for four pseudolites in order to allow three-dimensional positioning (see Chapter 8 for classical GPS equations).

There are two different signal generator approaches, depending on the accuracy required: meter or centimeter. As seen in Chapters 6 and 7, code positioning is less accurate than carrier phase positioning, but requires less complex electronics. Figure 10.4 gives a diagrammatic representation of both techniques.

The code-based technique allows distances to be obtained without ambiguity, given the usual indoor ranges, which are not likely to be greater than one code length, that is, 300 km. The accuracy is then of a few meters, because the running mode is identical to that of GPS, except for the various measurement errors occurring

---

[5]This possibility was envisaged in order to provide planet exploration with a precise deployable positioning system. The other possibility would have been to deploy a complete GPS-like constellation all around Mars...

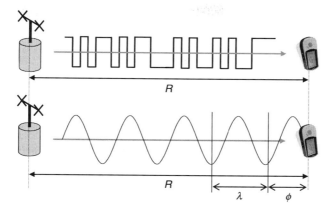

FIGURE 10.4 Code (top) and phase (bottom) techniques for pseudolite generators (R is the distance between the pseudolite and the receiver, λ the wavelength of the signal, and ϕ the measured carrier phase).

while the signal is crossing through the atmosphere. One has to remember that the main errors lie within this part of the propagation of the signals. Of course, to achieve such accuracy, like the standard constellation, the receiver needs to know about the location of the various pseudolites; this can consist either of local or global coordinates. The accuracy of the positioning of the pseudolite is clearly of importance, but with the code technique, this requirement is not too tight: a few tens of centimeters is enough.

The phase-based technique is more stringent for both the pseudolites and the receiver. As it deals with phase measurements, the ambiguities must be solved (see Chapter 8 for details concerning the problem of ambiguities), and the resulting accuracy is a few centimeters. The counterpart is that the location of the pseudolites has to be known to within a few centimeters too.[6] Nevertheless, although the theoretical concept is very clever, some difficulties exist in the practical deployment in the real world of pseudolites.

The first problem is the synchronization required in order to allow the accuracies described above. One has to keep in mind that some satellites have no fewer than four atomic clocks on board in order to supply time accuracy that can lead to a positioning accuracy of a few meters. Such implementation is far too expensive for a locally deployed system for indoor location finding. The way to overcome this problem is through the use of an additional pseudolite that is used, as in the UWB system for instance, as a base station for time reference.

Second is the so-called near–far effect. When indoors, the distances between the pseudolites and the receiver are a lot smaller than those between the satellites and a receiver (a few meters compared to more than 20,000 km). The important point is the relative difference between these distances, that is, the fact that indoors, the

---

[6]Note that the determination of the indoor location of the pseudolites, when the precision required is of a few centimeters, is not so easy. This problem has not been investigated much because it is not the major concern until the technical aspects are settled.

FIGURE 10.5 A typical indoor pseudolite configuration. (*Source*: School of Surveying and Spatial Information Systems, University of New South Wales.)

ratio of one to other such distances can be rather high (as much as 20 or 30). The corresponding ratio for the satellite constellation is always less than $25,600/20,200 = 1.26$. The direct impact is that the power ratio of the received signals from pseudolites can be as much as 20 or 30 to the square (considering a free-space propagation model in d^2). Knowing that a ratio of 16 is equivalent to 24 dB of power attenuation between two signals, one can anticipate the problem. As codes are pseudo random ones, there is a practical limitation in the discrepancy of power levels that can be processed by a receiver, mainly because low power signals are considered as noise. So, for high power pseudolites, that is, those that are the nearest to the receiver, the other lower signals are considered as noise; and they cannot be detected. This problem has an impact on the coverage area or the definition of the positioning of the infrastructure.

Third, the phase-based pseudolites use a differential phase measurement technique that is essentially a relative positioning approach. Then, the determination of the starting point is of prime importance. As there is no real means, as long as the user is indoors, to obtain this first location with the required accuracy (a few centimeters), the pseudolite global system should provide this first point. This step can take quite a lot of time (10–20 minutes is a typical range of values). On the other hand, one could state that this is only needed for high accuracy indoor positioning, which is unlikely to be the typical pedestrian application for instance. Thus, if a user requires centimeter accuracy, there is a time constraint in starting the process. This can, however, still be acceptable.[7]

A typical high pseudolites based positioning system is provided in Fig. 10.5, and positioning results are provided in Fig. 10.6. In this specific case, the pseudolites are installed outdoors, on the roof of the building. One has to remember that this implantation allows one to reduce the discrepancies between the various distances from the pseudolites to the receiver, and thus reduce the near–far effect. The results obtained

---

[7]This constraint existed for the first high accuracy civil engineering systems and has been considered as acceptable.

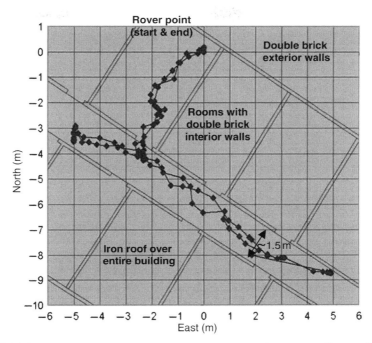

FIGURE 10.6 Typical results of an indoor pseudolite positioning system. (*Source*: School of Surveying and Spatial Information Systems, University of New South Wales.)

from different teams around the world, mainly implementing the phase measurement approach, are coherent and show that problems such as multipath are not so disturbing (although a complete study of indoor multipath for pseudolites still remains to be carried out).

The global characteristics of a pseudolite positioning system are summarized in Table 10.3.

■ 10.5 REPEATERS

Following the same philosophy as for pseudolites, that is, the deployment of a terrestrial local constellation, the repeater approach is a little different in its practical implementation. The basic idea comes from the fact that the need for an infrastructure is clearly a disadvantage, compared to HS-GNSS and A-GNSS solutions. However, it has been seen that there is probably a need for an infrastructure-based solution. Thus, to facilitate an eventual deployment, the infrastructure should be as simple as possible.[8]

---

[8] A pseudolite- or repeater-based system is a specific infrastructure dedicated to positioning, as compared to WLAN or GSM/UMTS infrastructures for which positioning is only a byproduct. The cost of deployment is largely shared with telecommunication purposes (although not totally true for current WLAN indoor positioning systems as discussed in previous sections).

TABLE 10.3 **Pseudolite positioning.**

Technical description	
Technology	RF at 1.575 GHz (continuous wave or pulsed)
Localization	Mobile receiver
	Continuous
Positioning	Absolute
Environments	Indoors, outdoors
Performances	
Accuracy	A few cm
Range	RF pseudolite range

The solution discussed here uses GPS repeaters that transmit signals from outdoors, where the receiving conditions are excellent, to indoors, where a current standard receiver is not able to compute a location. The two hardware configurations described below use repeaters that repeat all the incoming signals without processing (called "RnS," see Fig. 10.7), or that repeat only the signal from one satellite (called "R1S," see Fig. 10.8).

Obviously, the fact of using such a repeater leads to a difficulty when trying to solve the navigation equations with a current standard GPS receiver. The signal propagation path is artificially "curved" through the repeater. Therefore the actual measured propagation time does not give the real distance separating the satellite from the receiver, but the sum of the distance from the satellite to the repeater plus the distance from the repeater to the receiver (plus of course the various additional delays introduced by the repeater hardware). So, the repeater approaches involve designing repeater system hardware and new receiver algorithms to solve the new set of navigation equations.

The Clock Bias Approach

In the RnS approach, a repeater consists only of a receiving antenna, an amplifying chain, and a transmitting antenna. In this case, the signals from all the

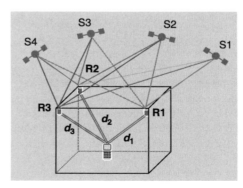

FIGURE 10.7 A typical indoor "RnS" repeater configuration (S refers so satellites, R to repeaters).

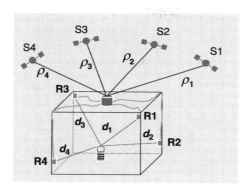

FIGURE 10.8 A typical indoor "R1S" repeater configuration (ρ refers to pseudo ranges).

satellites are transmitted indoors. The computation of the navigation equations at the indoor receiver location (using any technique such as linearization, Kalman filtering, and so on) leads to a solution vector that gives the location of the receiving outdoor antenna together with the fourth component, which is the sum of all the common biases (see Section "Basic Principles of Trilateration" in Chapter 8 for details on clock bias). Thus, it contains the clock bias but also the delays within the electronic structures of the repeater and finally also the free space propagation delay (a time equivalent to the distance d from the repeater to the receiver) from one repeater to the indoor receiver.

As we have no idea of ct_r (assumed to be the real clock bias of the receiver), it is impossible to go back to d. One solution is to carry out such a calculation four times by way of a sequential process, using four repeaters that are located at different points. In such a case, it is possible to obtain two types of information for each calculation: the location of the repeater (the first three components) and the sum of the ct_r and the distance d_i between the repeater "i" and the indoor receiver. Therefore, after one cycle, we have the following vector (taken from the four fourth components of the four solution vectors calculated for each repeater). Refer to Fig. 10.7 and note that the following equations now deal with four repeaters, instead of only the three given in Fig. 10.7):

$$\begin{bmatrix} ct_1 \\ ct_2 \\ ct_3 \\ ct_4 \end{bmatrix} = \begin{bmatrix} ct_r + d_1 \\ ct_r + d_2 \\ ct_r + d_3 \\ ct_r + d_4 \end{bmatrix}, \tag{10.1}$$

where we would like to find d_1, d_2, d_3, and d_4 to determine the actual location of the indoor receiver. This problem is identical to that of a classical outdoor location determination. In fact, the standard method or a "sphere expansion" method can be used. The latter is defined as follows. From the last set of equalities, it is possible to generate a new set by calculating the differences of the coordinates to eliminate

ct_r (which is not easy to determine). One obtains (assuming d_1 is the smallest distance of all the d).

$$\begin{bmatrix} d_2 - d_1 \\ d_3 - d_1 \\ d_4 - d_1 \end{bmatrix} = \begin{bmatrix} ct_2 - ct_1 \\ ct_3 - ct_1 \\ ct_4 - ct_1 \end{bmatrix}. \tag{10.2}$$

From this point, the method consists in choosing the d_1 so that the two largest spheres touch. The three spheres are defined by their radii $(d_{j2} - d_{i1})$, $(d_3 - d_1)$, and $(d_4 - d_1)$. Then, the expansion of the four spheres of radii d_1, d_2, d_3, and d_4 allows the actual intersection point to be determined, which is the position of the indoor receiver.

One needs to define the cycling time as well as the time, the start time, and the identification of the active repeater. Then programs can be run to store the incoming raw data from the mobile receiver. In every case, the basic data required are

1. GPS time;
2. Computed clock bias (obtained from the current GPS location calculation process);
3. Measured clock bias rate (obtained from the Doppler shift).

The mathematical computations are then quite easy to follow, based on the general theory previously described. The complete set of equations is given by

$$\begin{bmatrix} ct_{cal}(t_{R1}) = ct_{osc}(t_{R1}) + \text{del}(R1) + d_1 \\ ct_{cal}(t_{R2}) = ct_{osc}(t_{R2}) + \text{del}(R2) + d_2 \\ ct_{cal}(t_{R3}) = ct_{osc}(t_{R3}) + \text{del}(R3) + d_3 \\ ct_{cal}(t_{R4}) = ct_{osc}(t_{R4}) + \text{del}(R4) + d_4 \end{bmatrix} \tag{10.3}$$

where $t_{cal}(t_{Ri})$ is the computed clock bias at time t_{Ri}, $t_{osc}(t_{Ri})$ is the clock bias rate at time t_{Ri}, del(Ri) is the induced delay of repeater i, and d_i is the distance between the mobile receiver and repeater "i". The unknown variables are the various induced delays (which have to be calibrated in the laboratory), the distances d_i (which represent the x_r, y_r, and z_r coordinates of the mobile antenna), and the three clock bias rates. So the total number of unknowns is now seven (x_r, y_r, and z_r, and the four clock bias rates). In fact, one needs to know about the short-term stability of the receiver's oscillator. If not known, it is still possible to follow it with a rather good accuracy through the Doppler shift. Thus, we can state that

$$ct_{osc}(t_{Rj}) = ct_{osc}(t_{Ri}) + \sum_{k=i+1}^{j} \Delta ct_{osc}(t_{Rk}) \tag{10.4}$$

where $\Delta ct_{osc}(t_{Rk})$ is the measured clock bias rate at time t_{Rk}.

We now have to deal with a set of four equations with four unknown variables: x_r, y_r, z_r, and $ct_{osc}(t_{R1})$. The computation is then absolutely identical to that currently carried out by a GPS receiver, and one can use the linearization method in order to compute a three-dimensional positioning. As one may conclude, we have moved to a local frame of reference for convenience.

In a typical file, GPS time, clock bias, and clock bias rate are stored, and the difference between the evolution of the computed clock bias and the measured clock bias rate is calculated. This demonstrated an interesting feature that caused a lot of trouble during experiments, in that there are sometimes large skips, in one direction or another, where one may consider that there has been a change of transmitting repeater. This is not necessarily always the case: two consecutive skips can be due to a change in the number of satellites used to compute the fix. One has to deal with this problem as one does not compute one's own navigation solution. The induced effect, if not taken into account, is to shift the point well outside the acceptable accuracy range (some computed indoor locations could be 4 or 5 m away from the real point). When removing these results, one obtains some quite good values, in the 1–2 m range.

Unfortunately, this problem of clock bias skips arises frequently. The major reason is that the various filters within the mobile receiver have not been removed. Thus, as one was expecting really raw data, the internal receiver algorithms were trying to smooth these data by either moving from one constellation to another (and generating the skips) or modifying the output data. In this latter case, the additional delay that we were seeking (corresponding to the distance d_i) was greatly modified. Another difficulty of this technique is linked to synchronization.

The Pseudo Ranges Approach

These difficulties generated problems and another scheme was implemented, based on the direct use of the raw measurements. We know that it is not always possible to obtain the real raw data, that is, to get rid of the various filters designed to "improve" the receiver solution. So, it was decided to use so-called "GNSS sensors," designed to deliver a lot of data from the receiver. It is also possible to modify some parameters such as loop smoothness, constellation configuration, and many others.

In fact, it was mainly the Kalman filter parameters, the code loop and the frequency loop, that were tuned. As long as one just wanted to enable the continuity of the location service between outdoors and indoors, there was no interest in

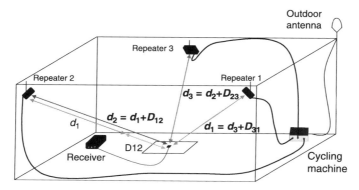

FIGURE 10.9 Indoor repeater-based positioning system.

increasing the accuracy, as long as the goal of a few meters was still achieved (as will be shown below).

The principle of the new technique is simply to "follow" the evolution of the pseudo distances when cycling from one repeater to the next (see Fig. 10.9). One expects skips, of the value of the difference of distances from the transmitting indoor antenna of the first repeater to the indoor receiver, to this same distance from the second repeater to the receiver. In this case, after a complete cycle, we should obtain three differences that will characterize the current location of the receiver.

Theoretical Aspects

A typical curve showing the evolution of the pseudo range over a few cycles is given in Fig. 10.10. To understand this curve, some assumptions have to be made:

- The delays induced by the repeaters, that is, the electronic delays and the cable delays, have been previously calibrated or do not need to be calibrated.

- As long as the curve represents the difference from one time step to the next, values are equivalent to the drift of the pseudo distances (or can also be seen as an acceleration).

- The remaining drift of the receiver's oscillator can be observed (the slight positive slope).

It appears quite clearly that the receiver "sees" the change from one repeater to the next as acceleration. It is also possible to observe that after the transition in the pseudo distance from one repeater to the next, the receiver "comes back" to the natural drift. The time needed for this return depends on the tuning parameters. These parameters are also very important in order to be able to follow the skips from one repeater to the next. Some tunings do not allow the observation and others do.

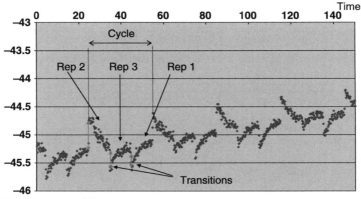

FIGURE 10.10 A typical pseudo-range evolution.

Signal switching from one repeater to the next results in an offset in the phase of the signal received. As in GPS positioning, which needs four satellites to achieve three-dimensional positioning, four repeaters are necessary to have four phase offsets. It is thanks to these values of phase offsets (or phase jumps) that the calculation of the indoor location of the receiver is possible. To understand this, let us write the new indoor pseudo range expression (PR_i) between a satellite and an indoor GPS receiver. One should remember that the signal is received through a repeater R_i:

$$\rho_i(t) = D_{S \to A}(t) + d_i + c \cdot t(t) + \Delta prop(t) + \text{delay}, \tag{10.5}$$

where $D_{S \to A}$ is the line-of-sight distance between the satellite and the outdoor antenna located on the roof, d_i is the distance between repeater R_i and the GPS receiver located indoors, t is the receiver clock bias, $\Delta prop(t)$ is the propagation error delay, and "delay" is the propagation time of the signal received in the cable connecting the outdoor antenna and repeater R_i. We suppose that these cables are identical for the four repeaters. In this case, this delay will be the same for all paths through R_1 to R_4.

Note that the above equation also depends on multipath error delay, which is not considered here. Let us suppose that at time t the signal switches from repeater R_i to repeater R_j. The measured pseudo range at time $t + dt$ through repeater R_j becomes

$$\rho_i(t + dt) = D_{S \to A}(t + dt) + d_i + c \cdot t(t + dt) + \Delta prop(t + dt) + \text{delay}. \tag{10.6}$$

Therefore, the pseudo range difference from time $t + dt$ to time t is given by

$$\begin{aligned}\Delta\rho(t + dt) &= \rho_i(t + dt) - \rho_i(t) \\ &= \Delta D_{S \to A}(t + dt) + \Delta d + c \cdot \Delta[t(t + dt)] + \Delta[\Delta prop(t + dt)],\end{aligned} \tag{10.7}$$

where

$$\begin{aligned}\Delta D_{S \to A}(t + dt) &= D_{S \to A}(t + dt) - D_{S \to A}(t) \\ \Delta d &= d_j - d_i \\ \Delta[t(t + dt)] &= t(t + dt) - t(t) \\ \Delta[\Delta prop(t + dt)] &= \Delta prop(t + dt) - \Delta prop(t).\end{aligned} \tag{10.8}$$

We are interested in Δd, which is the difference between the distances separating the receiver from the two repeaters R_i and R_j. Here, we cannot access its value because it is very small compared with $\Delta D_{S \to A}(t + dt)$ and $\Delta[t(t + dt)]$ values. As Δd appears only at the transition time between successive repeaters, we proceed to a second time difference to detect Δd. In fact, at time t, the signal was still transmitted by R_i, therefore the pseudo range difference between times t and $t - dt$ does not include Δd and it follows that

$$\Delta\rho(t) = \rho_i(t) - \rho_i(t - dt) = \Delta D_{S \to A}(t) + c \cdot \Delta[t(t)] + \Delta[\Delta prop(t)]. \tag{10.9}$$

To extract Δd, we calculate the second variation of the pseudo range. The following equation is obtained:

$$
\begin{aligned}
\Delta\rho(t+dt) - \Delta\rho(t) = {}& [\Delta D_{S\to A}(t+dt) - \Delta D_{S\to A}(t)] \\
& + c \cdot \{\Delta[t(t+dt)] - \Delta[t(t)]\} \\
& + \{\Delta[\Delta\text{prop}(t+dt)] - \Delta[\Delta\text{prop}(t)]\} + \Delta d.
\end{aligned}
\tag{10.10}
$$

If the signal transition between repeater R_i and R_j is very fast (that is, dt is very small), double differences of $D_{S\to A}$, t, and Δprop (first, second, and third terms in the above equation respectively) will be negligible and the double pseudo range difference will finally be reduced to $\Delta_{ji} = \Delta d = d_j - d_i$.

At the end of each cycle we have the following system:

$$
\begin{aligned}
\Delta_{12} &= d_1 - d_2 \\
\Delta_{23} &= d_2 - d_3 \\
\Delta^{34} &= d_3 - d_4 \\
\Delta_{41} &= d_4 - d_1.
\end{aligned}
\tag{10.11}
$$

Note that the sum of first three equations gives the last one. This is the reason only three differences are retained from the four: for example the first three. Expressing distances d_i by the coordinates of receiver (x_r, y_r, z_r) and those of repeaters R_i $(x_{Ri}, y_{Ri}, z_{Ri}, i = 1, \ldots, 4)$, it follows that

$$
\begin{aligned}
\Delta_{12} &= d_1 - d_2 \\
&= \sqrt{(x_{R_1} - x_r)^2 + (y_{R_1} - y_r)^2 + (z_{R_1} - z_r)^2} \\
&\quad - \sqrt{(x_{R_2} - x_r)^2 + (y_{R_2} - y_r)^2 + (z_{R_2} - z_r)^2} \\
\Delta_{23} &= d_2 - d_3 \\
&= \sqrt{(x_{R_2} - x_r)^2 + (y_{R_2} - y_r)^2 + (z_{R_2} - z_r)^2} \\
&\quad - \sqrt{(x_{R_3} - x_r)^2 + (y_{R_3} - y_r)^2 + (z_{R_3} - z_r)^2} \\
\Delta_{34} &= d_3 - d_4 \\
&= \sqrt{(x_{R_3} - x_r)^2 + (y_{R_3} - y_r)^2 + (z_{R_3} - z_r)^2} \\
&\quad - \sqrt{(x_{R_4} - x_r)^2 + (y_{R_4} - y_r)^2 + (z_{R_4} - z_r)^2} \\
\Delta_{41} &= d_4 - d_1 \\
&= \sqrt{(x_{R_4} - x_r)^2 + (y_{R_4} - y_r)^2 + (z_{R_4} - z_r)^2} \\
&\quad - \sqrt{(x_{R_1} - x_r)^2 + (y_{R_1} - y_r)^2 + (z_{R_1} - z_r)^2}
\end{aligned}
\tag{10.12}
$$

TABLE 10.4 Repeater positioning.

Technical description	
Technology	RF at 1.575 GHz (continuous wave or pulsed)
Localization	Mobile receiver
	Continuous
Positioning	Absolute
Environments	Indoors
Performances	
Accuracy	$\sim 1-2$ m
Range	RF repeater range

which is a typical hyperbolic system of positioning equations. In this system the unknowns are (x_r, y_r, z_r), the position of the receiver. The resolution is carried out either by linearization or using a hyperbolic solving algorithm (see Chapter 8 for details). The problem of indoor repeater based positioning thus consists in measuring the differences Δ_{ji}. These differences correspond to the code phase jumps while switching from one repeater to the next. The main goal is then to detect and measure these code phase jumps at the time of signal transition between two successive repeaters.

The global characteristics of a repeater positioning system are summarized in Table 10.4. Repeater systems present some advantages in comparison with pseudolites, in particular in the electronics of the repeater itself, which is quite simple, and the absence of the near–far effect, as there is always only one repeater that transmits at a given time, thus removing the problem of discrepancies in power levels from different transmitters. Another very important advantage of this approach is the "time differential" method that is implemented. The accuracy reported is quite a lot better than that allowed by the correlation function of the Gold codes in use in the GPS system, and the reduction in both distances (indoors) and propagation errors (as differences are calculated) are not enough to explain these good results. Indeed, the time differential approach highlights the performances of the code loop. Thus one is facing the behavior of a feedback loop of first, second, or third order. This behavior is very good when one is interested in the discontinuities, as they are highlighted by these loops. In classical GNSS receivers, there is the need to follow the pseudo range and such skips as those generated by repeater cycling are not searched for. Nevertheless, with the indoor concept, these skips are the only values that interest us and loops are inherently designed to give these peaks. This is the reason the accuracy that can be achieved with the repeater approach depends on the behavior of the feedback side of the loop rather than on the correlation side of it.

The disadvantage of repeaters compared to pseudolites is the fact that there are clearly different modes for outdoor and indoor environments and the receiver would need to switch from one mode to the other. This is not a great problem. For instance, this is a similar idea to using GNSS for outdoors and WLAN for indoors, but it should be taken into account. Furthermore, using similar frequencies to the outdoor constellation is likely to generate some interference, which is the terror of

GNSS signals. Once again, this is not a great problem as many possible solutions exist, but nevertheless the problem also has to be solved.

■ 10.6 RECAP TABLES AND COMPARISONS

First, one has to associate the positioning techniques with their maturity levels. For instance, HS-GNSS, A-GNSS, and Cell-Id are solutions that already exist. TDOA is implemented by some companies but is not well developed to date. Other techniques, such as the WLAN-based one, pseudolites, or repeaters are the subject of research programs[9] and have only been implemented on a very small scale, thus only showing feasibility concepts. This point has to be considered when drawing comparisons, but data does not appear in purely technical tables.

The various techniques described so far all have advantages and drawbacks and the final choice is largely driven by the application. As there is no single technique that appears clearly to be sufficiently versatile, we present three tables in an attempt to make a classification:

- The accuracy table, which includes the type of positioning, the achievable accuracy, and the TTFF (time to first fix) to allow GNSS-based comparisons;
- The coverage table, which shows the availability and the coverage areas of the techniques considered;
- The complexity table, which shows indicators of cost for terminal and infrastructure complexities in order to evaluate the overall cost of any solution.

Furthermore, footnotes are included in order to provide some information regarding the reasons for the choices that have been made in filling in the tables.

Table 10.5 is presented to allow the reader to make his/her own choice from the whole range of technical solutions. It is possible to choose personal coefficients for accuracy, infrastructure complexity, as well as terminal complexity.

The following terms are used:

- For type, "A" indicates absolute positioning, "R" relative positioning or displacement, and "S" symbolic representation and/or positioning.
- For accuracy, when applicable, a value in meters is given. Otherwise, the best estimated figure is used.
- For time to first fix, this figure is intended to allow GNSS-based comparisons and is sometimes not relevant.

Remark: the values indicated in Table 10.5 are typical indoor figures (for instance, the magnetometer exhibits a much better accuracy outdoors, but due to metallic structures, it can be a lot worse than the $10°$ mentioned).

Table 10.6 summarizes the areas covered by each approach. The various cases considered for availability are indoors (In) and outdoors (Out), which have to be

---

[9]However, some commercial systems are available for sale implementing WLAN pattern matching or even time-based measurement approaches. UWB is also a technique that can be purchased for positioning.

TABLE 10.5 Accuracy table.[a]

Technique	Type	Accuracy	TTFF
Sensor network			
Infrared[b]	S	Room	<1 s
Ultrasound	A/R	A few cm	<1 s
RFID	R	<1 m	<1 s
Physical	A/R/S	A few cm	A few s
WLAN			
Wi-Fi	R/S	A few m	<1 s
Bluetooth	R/S	A few m	<1 s
UWB	A	$\sim$10 cm	<1 s
WiMax	A	$\sim$10 m[c]	<1 s
GSM/UMTS			
Cell-Id	A	100 m to 20 km[d]	A few s
TOA	A	>100 m	A few s
TDOA/OTD	A	>100 m	A few s
AOA	A	>100 m	A few s
GNSS			
A-GNSS	A	>10 m	<1 s
HS-GNSS	A	>10 m	<1 s
Pseudolite	A	$\sim$10 cm[e]	<1 s
Repeater	A	$\sim$1 m	<1 s
Inertial			
Gyroscope	R	$\sim$1 m/mn[f]	<1 s
Accelerometer	R	$\sim$1 m/mn[g]	<1 s
Magnetometer	R	$\sim$10°	<1 s
Odometer	R	A few %	<1 s
Other			
Camera	A/R	<1 m	A few s
Laser	A/R	$\sim$1 cm	<1 s
TV	A	$\sim$200 m	<1 s

TTTF, time to first fix. Positioning: A, absolute; R, relative; S, symbolic.

[a]Note that this table refers to indoor positioning accuracy.

[b]The infrared system considered here is similar to "Active Badge." One could have taken into account a purely optical positioning system that is indeed the "Laser" line in the table.

[c]Estimation (to be confirmed).

[d]The highest figure corresponds to a rural situation where only a few base stations are deployed. A typical value is rather in the 100–500 m range.

[e]For phase-based measurements.

[f]The unit used corresponds to a cumulative error of 1 m after 1 min of measurements, typically.

[g]Idem footnote "f".

seen as the two typical cases a positioning system has to face (one could add urban canyons to be complete). This vision is thus mainly oriented towards the location-based services domain. Note that the order of appearance is important in the table: "In/Out" means that the main availability of the technique is indoors, although it can be used outdoors. "Out/In" means the opposite and "Out & In" means that there is no real difference between the performances in both environments.

TABLE 10.6 Coverage table.

Technique	Availability	Coverage
Sensor network		
Infrared	In	Spot location
Ultrasound	In	Spot location
RFID	In/Out	Spot location
Physical	In	Spot location
WLAN		
Wi-Fi	In	Building
Bluetooth	In	Building
UWB	In	Building
WiMax	Out/In	Local/Wide
GSM/UMTS		
Cell-Id	Out/In	Wide
TOA	Out	Wide
TDOA/OTD	Out/In	Wide
AOA	Out	Local/Wide
GNSS		
A-GNSS	Out/In	Global
HS-GNSS	Out/In	Global
Pseudolite	Out & In	Building
Repeater	In/Out	Building
Inertial		
Gyroscope	Out & In	Local
Accelerometer	Out & In	Local
Magnetometer	Out/In	Local
Odometer	Out/In	Wide
Other		
Camera	In	Building
Laser	In	Building
TV	Out/In	Wide

See main text for explanation of availability.

The purpose of the coverage column is to be more precise about the idea of availability. This should allow the reader to have a means to evaluate the total requirements in terms of complexity of the infrastructure for a typical deployment. The various cases considered are

- "Spot location" for a restricted area positioning (typically a few square meters);
- "Building" for a limited indoor coverage;
- "Local" for a local area range (typically a few acres);
- "Wide" for a large area coverage of a few hundred square kilometers (to thousands of square kilometers);
- "Global" for a real coverage of the whole Earth.

Table 10.7 summarizes the complexity of each approach. For both the infrastructure and the terminal, the costs are given in terms of Low, Medium, or High; this is not

TABLE 10.7 **Complexity table.**

Technique	Infrastructure		Terminal	
	Deployment	Cost	Complexity	Cost
Sensor network				
Infrared	New	High	New	Low
Ultrasound	New	High	New	Low
RFID	New	High	Existing	Low
Physical	New	High	N/A	N/A
WLAN				
Wi-Fi	Existing +	Medium	Existing	Low
Bluetooth	Existing +	Medium	Existing	Low
UWB	Existing +	Medium	Existing	Low
WiMax	Existing +	Medium	Existing	Low
GSM/UMTS				
Cell-Id	Existing	Low	Existing	Low
TOA	Existing	Low	Existing	Low
TDOA/OTD	Existing	Low	Existing	Low
AOA	Existing	Low	Existing	Low
GNSS				
A-GNSS	Existing +	Medium	Soft Dev	Low
HS-GNSS	Existing	Low	Hard Dev	Low
Pseudolite	New	Medium	Soft Dev	Low
Repeater	New	Medium	Soft Dev	Low
Inertial				
Gyroscope	Without	Nil	Integration	High
Accelerometer	Without	Nil	Integration	High
Magnetometer	Without	Nil	Integration	Low
Odometer	Without	Nil	Integration	Low
Other				
Camera	Without	Nil	Integration	Low
Laser	Without	Nil	New	Low
TV	Existing	Low	Integration	Low

See main text for an explanation of the terms used.

very accurate, but technological developments are so rapid that this granularity appears to be sufficient for our present purpose.

Concerning the deployment of the infrastructure, the table proposes to make a difference between already existing structures, such as GNSS ones or a standard GSM network, with those requiring an update of existing systems (this is the case for WLAN indoor positioning for instance, where an increased number of access points is generally required). Of course, the possibility of a requirement for a totally new infrastructure is also indicated.

The same approach applies to the terminal complexity field. The possible values are

- "Existing" if no further development is required ("Existing +" in the case where only a very limited update of the current deployment is required);
- "Soft Dev" when software developments are needed;

TABLE 10.8 Overall indoor decision table.

Technique	Accuracy		Infrastructure		Terminal		Result
	R_a	C_a	R_i	C_i	R_t	C_t	
Infrared	5		2		4		
Ultrasound	10		2		6		
RFID	8		0		3		
Physical	8		2		6		
Wi-Fi	5		6		7		
Bluetooth	5		6		7		
UWB	8		5		7		
WiMax	5		6		7		
Cell-Id	0		10		10		
TOA	2		8		9		
TDOA/OTD	2		8		9		
AOA	2		5		9		
A-GNSS	7		10		10		
HS-GNSS	7		10		10		
Pseudolite	10		5		5		
Repeater	9		7		9		
Gyroscope	6		10		5		
Accelerometer	6		10		5		
Magnetometer	2		10		7		
Odometer	6		10		5		
Camera	7		10		5		
Laser	9		10		2		
TV	2		10		4		

See main text for an explanation of the parameter rotation.

- "Hard Dev" if hardware developments are required to existing systems;
- "Integration" when only integration developments would be enough to propose the technique from existing systems;
- "New" when totally new developments, both hardware and software are needed.

Finally, the idea of Table 10.8 is to allocate a rating between 0 and 10 to all the techniques for each of the three parameters Accuracy[10] (R_a), Infrastructure[11] (R_i) and Terminal[12] (R_t). These values result from a partial analysis of Chapters 9 and 10.[13]

---

[10]"0" is allocated to Cell-Id with accuracy in the 100 m to 10 km range, and "10" is allocated to the pseudolite positioning based on phase measurements or to the ultrasound technique, which can achieve a few centimeters.

[11]"0" is given when an important new infrastructure is required and "10" when the required infrastructure is already available.

[12]From "10" when current terminals are sufficient down to "2" when they require complex new electronics.

[13]Note that a synthesis cannot only be drawn from these three parameters, thus leading to questionable values. The values reflect mainly my own point of view and are subject to discussion. Therefore, they have to be considered only as an example of the way to fill in the table.

The coefficients C_a, C_i, and C_t (respectively for accuracy, infrastructure, and terminal) allow the reader to apply the criteria in order to find the most suitable solution for a specific case. Applying the coefficient to the values gives a result for all the techniques. It is then quite easy to find the best technique for any given application.

The interesting point to note is that applying this approach to different applications, and hence different coefficients, leads to different techniques. Moreover, it is quite difficult to find one technique that matches a few sets of coefficients, that is, several applications. The conclusion is certainly that some techniques should be put aside before the start of the game or conversely, some should be pointed out for further development to address the difficult problem of indoor positioning.

■ 10.7 POSSIBLE EVOLUTIONS WITH AVAILABILITY OF THE FUTURE SIGNALS

HS-GNSS and A-GNSS

The availability of satellites and therefore the coverage of urban canyons, for example, will be largely improved in the future. One could also consider that indoor coverage will benefit from future signals. This is true in many ways. First, the increased number of satellites will improve the availability of more than one or two signals coming through the windows for instance. With more than 100 satellites (as compared to only some 30 GPS satellites today) providing more than 10 different signals (as compared to the only L1 C/A really available today), one can easily imagine the potential for improvements. Second, the increased power levels of the new signals (a few dB) will also help in the detection process in difficult environments. In addition, the availability of pilot tones is also important for the longer integration process that we know is a very powerful way to detect weak signals.

Unfortunately, although real improvements will be made, all the abovementioned features are not likely to solve the indoor problem. A few dB of increased power is not enough, the long integration process has been tested for some years now, and the availability of several satellite signals through the windows will provide the positioning with very high DOPs, leading to a poor accuracy location.

Pseudolites

Current pseudolite indoor systems are based on phase measurements in order to provide high accuracy. Perhaps some work could be carried out to develop further approaches based on codes. In such a case, the diversity of future codes and frequencies is interesting. For example, the availability of pilot tones allows increased accuracy indoors where there is no need for navigation data, other than knowing the pseudolite locations (data that do not change). Pseudolites are simple compared to repeaters is that these are simultaneous transmissions from all the pseudolites,[14]

---

[14]However, the simultaneity does not allow time-based differential measurements, which is the major advantage of repeaters...

although this does cause the near−far[15] effect. A combined approach using pseudo-lites and repeaters could therefore be developed, taking advantage of the positives of each method (see the next section for details).

Repeaters

Perhaps the best solution for infrastructure-based indoor positioning could be a mix of pseudolites and repeaters, using the benefits of each. From pseudolites we could use the simplicity of the receiver hardware, which remains identical to a standard receiver. Also from pseudolite could be kept the possibility of transmitting many different signals simultaneously, when required. From repeaters the main advantage seems to be related to the cycling method, a sort of time differential method that allows much more accurate distance measurements. This solution could take advantage of the evolution of the GNSS constellation with the numerous future signals (see Chapters 3 and 6 for details).

In this section, we describe how to use the multifrequency capabilities of the current constellations of GPS and GLONASS. These two constellations currently offer a second civil signal in the L2 band, called L2C, for a few satellites. The specific case of the Galileo constellation is also described below.

Two-Band Approach

Let us consider the typical repeater infrastructure of Fig. 10.11. In this figure, the repeaters are represented by R_i and are located indoors. The antenna in the middle of the figure is the outdoor receiving antenna, which receives the satellite signals.

The variables used are the following:

- d_k^i is the distance between the repeater R_i and the GPS (or GLONASS) receiver k.
- $D_{SA}^i(t)$ is the distance, at instant t, between the satellite i and the outdoor receiving antenna, located on the roof.
- $t_k(t)$ is the clock bias, at instant t, of the receiver k with reference to GPS (or GLONASS) time.
- $t^i(t)$ is the clock bias drift, at instant t, of satellite i with reference to GPS (or GLONASS) time.
- $I_i^1(t)$ (respectively $I_i^2(t)$) is the ionosphere delay for the propagation path from satellite i and the outdoor receiving antenna at L1 frequency (respectively L2C), at instant t.
- $T_i(t)$ is the troposphere delay, at instant t, for the propagation path from satellite i and the outdoor receiving antenna.
- $\rho_{k,j}^i(t)$ is the pseudo distance between satellite i and receiver k, measured through the signal transmitted by repeater R_j.

The electronic delays induced by cables between the outdoor receiving antenna and the transmitting repeater should be zero (this is obviously not true but of no impact in

---

[15]See Section 0.4 for details.

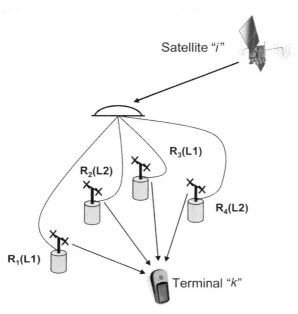

Satellite "*i*"

R₃(L1)

R₂(L2)

R₄(L2)

R₁(L1)

Terminal "*k*"

FIGURE 10.11 Two-band indoor architecture for the case of GPS and GLONASS.

the calculations as long as the differences between such delays for the different repeaters are negligible — this is the real assumption that is made here).

In the present case, we consider that R_1 and R_3 are transmitting at L1, R_2 and R_4 at L2. The first transmission is achieved by both R_1 and R_2, at instant t. Thus, the measurements of pseudo distances obtained from R_1 and R_2 are the following:

$$\begin{cases} \rho_{k,1}^i(t) = D_{SA}^i(t) + d_k^1 + ct_k(t) - ct^i(t) + I_i^1(t) + T_i(t) \\ \rho_{k,2}^i(t) = D_{SA}^i(t) + d_k^2 + ct_k(t) - ct^i(t) + I_i^2(t) + T_i(t). \end{cases} \tag{10.13}$$

Then R_1 and R_2 are stopped and R_3 and R_4 start to transmit, at instant t'. The pseudo distances obtained from R_3 and R_4 are then as follows:

$$\begin{cases} \rho_{k,3}^i(t') = D_{SA}^i(t') + d_k^3 + ct_k(t') - ct^i(t') + I_i^1(t') + T_i(t') \\ \rho_{k,4}^i(t') = D_{SA}^i(t') + d_k^4 + ct_k(t') - ct^i(t') + I_i^2(t') + T_i(t'). \end{cases} \tag{10.14}$$

Then, one can compute differences:

$$\begin{cases} \rho_{k,3}^i(t') - \rho_{k,1}^i(t) = D_{SA}^i(t') - D_{SA}^i(t) + d_k^3 - d_k^1 + ct_k(t') - ct_k(t) - ct^i(t') \\ \qquad\qquad + ct^i(t) + I_i^1(t') - I_i^1(t) + T_i(t') - T_i(t) \\ \rho_{k,4}^i(t') - \rho_{k,2}^i(t) = D_{SA}^i(t') - D_{SA}^i(t) + d_k^4 - d_k^2 + ct_k(t') - ct_k(t) - ct^i(t') \\ \qquad\qquad + ct^i(t) + I_i^2(t') - I_i^2(t) + T_i(t') - T_i(t). \end{cases}$$

$$\tag{10.15}$$

The cycling is carried out with the successive transmission of R_1 and R_2, followed by that of R_3 and R_4, and so on. As with the standard repeater approach, this sequence is achieved through the use of a specific electronic command board.

Thus, it is theoretically possible to obtain $d_k^i - d_k^j$. Nevertheless, our choice is clearly to detect the variations of these distances at transitions from R_1 and R_3 to R_2 and R_4. It is indeed the observations at these transitions that allow the quite good accuracy achieved with the repeater approach (mainly because the code loop can detect very small variations while they are "instantaneous," which is exactly the case we are dealing with). At the transition instant, one can observe a skip in the pseudo distance that is the sum of

- "$d_k^i - d_k^j$";
- The variation of the satellite to outdoor receiving antenna distance;
- The variation of the clock biases;
- The propagation errors.

Concerning the clock biases, the evolution over time of these parameters is almost linear. Thus, when computing the "double difference" (the variation of $d_k^i - d_k^j$), this parameter is almost totally eliminated. The same applies to the difference of the satellite to outdoor-receiving-antenna distances, which is also negligible. The variation of the atmosphere delays, from one instant to the next, is neglected as well.

With this technique, it is possible to obtain quite accurate two-dimensional positioning (in this case, the z_r coordinate is considered as known):

$$\begin{cases} d_k^3 - d_k^1 = \Delta[\rho_{k,3}^i(t') - \rho_{k,1}^i(t)] \\ d_k^4 - d_k^2 = \Delta[\rho_{k,4}^i(t') - \rho_{k,2}^i(t)]. \end{cases} \tag{10.16}$$

This dual-frequency repeater approach is slightly different from the repeater approach described in previous sections, in that it is possible to output a complete two-dimensional computed location at each transition. In the previous mode, one has to wait for three transitions, unless one uses a specific calculation method called "the sliding mode" that uses the last three transitions to achieve the calculation. In such a way, a positioning is available at each transition, but in a dynamic mode, it introduces an error related to the fact that "old" data are used. This is no longer the case for the proposed dual-frequency approach.

Note that once again, only one satellite is needed: the d_k^i distances are identical for all the satellite signals that go through a given repeater. Nevertheless, the availability of many satellite signals allows useful redundancy.

Four-Band Approach

In this four-band case, we keep the same infrastructure, but instead of using two frequencies, each repeater is assigned with a specific signal. Then, the cycling method is once again applied in order to achieve better accuracy (as described above). This is quite similar to the pseudolite approach in the sense that an indoor local constellation with a variety of signals is thus designed, but the main specificity of the repeater approach is preserved. Each repeater transmits all the incoming signals on a given frequency. This allows both a very simple synchronization between repeaters (that are synchronized because of the use of the space constellation) and a very simple implementation of the hardware of the repeaters. A first implementation of this approach could be as follows:

- Repeater 1: GPS-L1 signals;
- Repeater 2: GLONASS-L1 signals;
- Repeater 3: GPS-L2 signals;
- Repeater 4: GLONASS-L2 signals.

With two such measurements, it is possible to compute a two-dimensional positioning (by the way of the classical linearization algorithm applied in standard GPS for instance). Three repeaters are required in order to allow two transition measurements, then leading to a two-dimensional positioning. Four repeaters are required for a three-dimensional positioning; the repeaters have the role of the satellite for the space constellation. The deployment is similar to that shown in Fig. 10.12.

Please keep in mind that the accuracy goal is equivalent to that achievable outdoors, that is, a few meters in order to allow real and complete continuity of the localization service. Furthermore, we are using modernized civil signals of the current constellations that will make available the so-called L2C for all satellites. Of course, extrapolations to any radio signal are possible, and specifically to those of the future Galileo constellation (see below for a special application of this approach for Galileo).

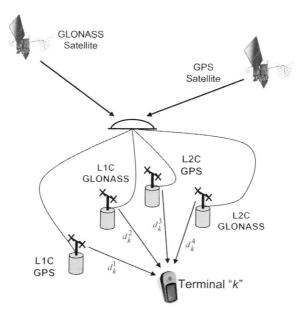

FIGURE 10.12 Four-band indoor architecture for the case of GPS and GLONASS.

TABLE 10.9 Example of multiconstellation and multifrequency cycling.

	t_1	t_2	t_3	t_4	$t'1$
Repeater 1	L1 GPS	L1 GLONASS	L2 GPS	L2 GLONASS	...
Repeater 2	L1 GLONASS	L2 GPS	L2 GLONASS	L1 GPS	...
Repeater 3	L2 GPS	L2 GLONASS	L1 GPS	L1 GLONASS	...
Repeater 4	L2 GLONASS	L1 GPS	L1 GLONASS	L2 GPS	...

Note that the main advantage of cycling is in the reduction of errors in pseudo range measurements. Indeed, using the described system without any cycling is possible indoors, but leads to similar errors of positioning as outdoors, that is, roughly a few meters in the best cases. The cycling concept allows time differential calculation, decreasing errors to differential errors that are very small when small time intervals are considered (typically less than 1 s). The global principle is thus built on three concepts, summarized in Table 10.9, which shows a complete four-repeater cycling (allowing three-dimensional positioning). The three concepts are respectively the multiconstellation one, the multifrequency one, and finally the cycling concept. Table 10.9 presents the case for the GPS and GLONASS constellations.

In such a configuration, it is possible to measure the variations of distances, at transition instants, for the four repeaters. The location computation may then be carried out at each transition. To understand this feature, let us consider the behavior of the receiver at time of transition from t_1 to t_2 (imagine that t_1 is just before the transition and t_2 just after). Considering that d_i is the distance between repeater i and the indoor receiver, one obtains

- $d_2 - d_1$ from the L1 GLONASS signal;
- $d_3 - d_2$ from the L2 GPS signal;
- $d_4 - d_3$ from the L2 GLONASS signal;
- $d_1 - d_4$ from the L1 GPS signal.

This multiconstellation multifrequency cycling method allows full three-dimensional positioning indoors, with, in theory, very good accuracy of about 1 m, obtained with a rate equal to the cycling time. As this cycling time can be as small as a fraction of second, this means that the positioning rate could be greatly increased if required for future indoor applications. Furthermore, the infrastructure, although still required, is composed of simple hardware structures.

The Galileo Case

The multifrequency concept can advantageously be applied in the case of Galileo, because of the great embedded diversity of frequencies available. Let us consider Fig. 10.13. In such a configuration as the one described in this figure, four signals are being used, namely L1, E5a, E5b, and E6. Note once again that in the present concept, all these signals come from a single satellite. The four repeaters form the architecture of the local terrestrial constellation for indoor purposes. When one decides to define the system of equations that drives the positioning

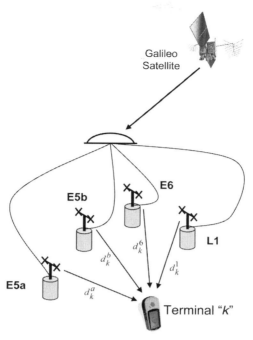

FIGURE 10.13 Multi-frequency approach with Galileo signals.

calculation, it is easy to see that it is absolutely identical to that which drives standard current GLS L1 receivers.

This system is given by

$$
\begin{cases}
\rho^i_{k,\text{E5a}}(t) = D^i_{\text{SA}}(t) + d^a_k + ct_k(t) - ct^i(t) \\
\rho^i_{k,\text{E5b}}(t) = D^i_{\text{SA}}(t) + d^b_k + ct_k(t) - ct^i(t) \\
\rho^i_{k,\text{L1}}(t) = D^i_{\text{SA}}(t) + d^1_k + ct_k(t) - ct^i(t) \\
\rho^i_{k,\text{E6}}(t) = D^i_{\text{SA}}(t) + d^6_k + ct_k(t) - ct^i(t),
\end{cases}
\tag{10.17}
$$

where all the variables are identical to those used in previous paragraphs. It is also possible to use another form for this system where (x_r, y_r, z_r, ct_r) are the coordinates of the searched-for location and (x_i, y_i, z_i) the coordinates of the repeaters in the local reference frame. Then Δt is the difference between the clock bias of the satellite and that of the receiver:

$$
\begin{cases}
\rho^i_{k,\text{E5a}}(t) = \sqrt{(x_{\text{E5a}} - x_r)^2 + (y_{\text{E5a}} - y_r)^2 + (z_{\text{E5a}} - z_r)^2} + ct_r \\
\rho^i_{k,\text{E5b}}(t) = \sqrt{(x_{\text{E5b}} - x_r)^2 + (y_{\text{E5b}} - y_r)^2 + (z_{\text{E5b}} - z_r)^2} + ct_r \\
\rho^i_{k,\text{L1}}(t) = \sqrt{(x_{\text{L1}} - x_r)^2 + (y_{\text{L1}} - y_r)^2 + (z_{\text{L1}} - z_r)^2} + ct_r \\
\rho^i_{k,\text{E6}}(t) = \sqrt{(x_{\text{E6}} - x_r)^2 + (y_{\text{E6}} - y_r)^2 + (z_{\text{E6}} - z_r)^2} + ct_r
\end{cases}
\tag{10.18}
$$

Such an approach allows three-dimensional indoor positioning, with a similar accuracy to outdoors. This is achieved using only one Galileo satellite and an infrastructure that requires only repeaters including filters and amplifiers. Nevertheless, as the cycling is an interesting way to achieve better accuracy, it is possible to imagine the same concept but including the cycling. Figure 10.14 shows the proposed method.

The repeaters are represented and the transmitted frequency is given at the initial time. In this scheme, the frequency transmitted by any given repeater changes at each transition. The new set of navigation equations is now as follows (the notations used are also those of Section 10.5):

$$
\begin{cases}
\Delta\rho_{k,\text{E5a}}^{i}(t) = D_{\text{SA}}^{i}(t+dt) - D_{\text{SA}}^{i}(t) + d_2 - d_1 \\
\qquad + ct_k(t+dt) - ct_k(t) - ct^i(t+dt) + ct^i(t) \\
\Delta\rho_{k,\text{E5b}}^{i}(t) = D_{\text{SA}}^{i}(t+dt) - D_{\text{SA}}^{i}(t) + d_3 - d_2 \\
\qquad + ct_k(t+dt) - ct_k(t) - ct^i(t+dt) + ct^i(t) \\
\Delta\rho_{k,\text{L1}}^{i}(t) = D_{\text{SA}}^{i}(t+dt) - D_{\text{SA}}^{i}(t) + d_4 - d_3 \\
\qquad + ct_k(t+dt) - ct_k(t) - ct^i(t+dt) + ct^i(t) \\
\Delta\rho_{k,\text{E6}}^{i}(t) = D_{\text{SA}}^{i}(t+dt) - D_{\text{SA}}^{i}(t) + d_1 - d_4 \\
\qquad + ct_k(t+dt) - ct_k(t) - ct^i(t+dt) + ct^i(t).
\end{cases}
\tag{10.19}
$$

The method is now similar to that developed in previous paragraphs and includes the calculation of the "time difference" of differences of pseudo ranges. The resulting accuracy should be improved down to 1 m, thanks to an increased complexity of the cycling process. And here again it is possible to output a three-dimensional positioning at each cycling transition, with signals from only one satellite.

For the case of Galileo, it is also possible to imagine a scheme where both the approaches described above (with and without cycling) are used simultaneously.

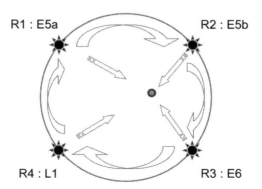

FIGURE 10.14 Implementation of the cycling concept with Galileo signals.

Continuity of service can be achieved using both approaches. Without cycling, the accuracy remains at a comparable level to outdoors, but with the cycling, the accuracy should certainly be increased, if required. As long as both schemes give a location at each transition, it is also possible to design a global system that can switch from one scheme to the other. For example, this could be used when accurate positioning is not required continuously. The cycling could then be designed on demand, depending on the application's specifications.

✓ 10.8 EXERCISES

Exercise 10.1 Why are GNSS signals not available indoors? What are the main issues and describe all the possible solutions. Your comments must be quantitative and expose the main implications of the proposed technical solutions; for instance, an increased transmitted power must be linked to the increase of satellite power and corresponding life-time, as well as the impact on other satellite signals, and so on.

Exercise 10.2 Describe the principles of HS-GNSS and achievements to date. What are the current limitations on the increase of sensitivity? What are the major drawbacks of an HS-receiver? (Note, further reading is required.)

Exercise 10.3 Give details on the way Assisted-GNSS allows for

1. Reduction of time to first fix (TTFF);
2. Ease of implementation of a high sensitivity approach;
3. Easier acquisition of satellites (quite similar to (2));
4. Aid for position calculation.

Exercise 10.4 Comment on the possibility of achieving indoor positioning with Assisted-GNSS. Is the indoor reception of mobile network signals of any help in achieving indoor GNSS positioning? What is the operational mode when the mobile network is not available? Does Assisted-GNSS provide a better accuracy than non-assisted GNSS?

Exercise 10.5 What are the objectives of so-called "Long-Term Ephemeris" (LTE) and what are the issues (one needs further readings to fully answer this question)?

Exercise 10.6 What are the meanings of tight and loose hybridizations?

Exercise 10.7 What is the classical architecture of a pseudolite-based indoor positioning system? What main differences exist between a pseudolite signal and one transmitted by a satellite?

Exercise 10.8 Explain the near–far effect and the reason it occurs. What kind of solution could you propose in order to reduce the impact of this problem in a typical indoor environment?

Exercise 10.9 What are the reasons why most work relating to indoor positioning with pseudolites uses phase measurements? How is the initial point then calculated? Would it be possible to cope with code measurements? What kind of performance could then be obtained?

Exercise 10.10 Explain the principle of a repeater-based positioning system. Such an approach uses a kind of time differential method: describe it in detail and explain how this effect allows better accuracy than outdoor GNSS positioning.

Exercise 10.11 Describe the way in which one has to consider the near–far effect in the case of the repeater-based approach. How is this effect largely attenuated compared with when using pseudolites?

Exercise 10.12 How is the difficult problem of synchronization dealt with in both the pseudolite and repeater methods? Comment.

Exercise 10.13 What could be the advantages of future signals and the availability of a large number of GNSS signals for indoor positioning? Please widen your reading and give a personal and a positively critical point of view.

Exercise 10.14 How could the combination of the pseudolite and repeater approaches improve the performance of an indoor positioning system?

■ BIBLIOGRAPHY

Avila-Rodriguez JA, Wallner S, Hein GW. How to optimize GNSS signals and codes for indoor positioning. In: ION GNSS 2006: Proceedings; September 2006. Forth Worth, (TX).

Bartone C, Van Graas F. Ranging airport pseudolite for local area augmentation. IEEE Trans Aerosp Electron Syst 2000;36(1);278–286.

Carver C. Myths and realities of anywhere GPS — high sensitivity versus assisted techniques. GPS World, September 2005.

Eissfeller B. In-door positioning with GNSS — dream or reality in Europe. In: International Symposium European Radio Navigation Systems and Services: Proceedings. Munich, Germany; 2004.

Francois M, Samama N, Vervisch-Picois A. 3D indoor velocity vector determination using GNSS based repeaters. In: ION GNSS 2005: Proceedings; September, 2005. Long Beach (CA).

Im S-H, Jee G-I, Cho YB. An indoor positioning system using time-delayed GPS repeater. In: ION GNSS 2006: Proceedings; September 2006. Forth Worth, (TX).

Jee GI, Choi JH, Bu SC. Indoor positioning using TDOA measurements from switched GPS repeater. In: ION GNSS 2004: Proceedings; (USA) September 2004. Long Beach (CA).

Kaplan ED, Hegarty C. Understanding GPS: principles and applications. 2nd ed. Artech House; 2006. Norwood, MA, USA.

Kee C, Yun D, Jun H, Parkinson B, Pullen S, Lagenstein T. Centimeter-accuracy indoor navigation using GPS-like pseudolites. GPS World 2001.

Kiran S. A wideband airport pseudolite architecture for the local area augmentation system. Ph.D. Dissertation. School of Electrical and Computer Engineering; Ohio University; Athens; 2003.

Parkinson BW, Spilker Jr JJ. Global positioning system: theory and applications. American Institute of Aeronautics and Astronautics; 1996.

Progri IF, Ortiz W, Michalson WR, Wang J. The performance and simulation of an OFDMA pseudolite indoor geolocation system. In: ION GNSS 2006: Proceedings; September 2006. Forth Worth, (TX).

Rizos C, Barnes J, Wang J, Small D, Voigt G, Gambale N. LocataNet: intelligent time-synchronised pseudolite transceivers for cm-level stand-alone positioning. In: 11th IAIN World Congress: Proceedings; October 2003. Berlin, Germany.

Samama N, Vervisch-Picois A. Current status of GNSS indoor positioning using GNSS repeaters. In: ENC GNSS 2005: Proceedings; July 2005. Munich, Germany.

Suh Y-C, Konish Y, Shibasaki R. Assessing the improvement of positioning accuracy using a GPS and pseudolites signal in urban area. Available at: www.chikatsu-lab.g.dendai.ac.jp/s_forum/pdf/2002/10_suh.pdf.

Sun G, Chen J, Guo W, Ray Liu KJ. Signal processing techniques in network-aided positioning: a survey of state-of-the-art positioning designs. IEEE Signal Process Magazine, July 2005; Volume 22, Issue 4, pp 12–23.

Syrjärinne J, Wirola L. Setting a new standard – assisted GNSS receivers that use wireless networks. InsideGNSS 2006;1(7):26–31.

Van Diggelen F, Abraham C. Indoor GPS technology. Global Locate Inc. Available at: http://www.globallocate.com/SEMICONDUCTORS/Semi_Libr_Piece/IndoorGPSTechnology.pdf.

Web Link

http://www.gmat.unsw.edu.au/snap. Nov 15, 2007.

Applications of Modern Geographical Positioning Systems

This chapter is devoted to describing some applications of positioning systems. They are mainly associated with the huge development, far beyond the hopes of the first designers, of the satellite-based navigation constellations. The description given does not detail the requirements but rather the major achievements. Thus, many positioning techniques could be used instead of GNSS in some cases, although the ease of using and deploying such receivers is often a key issue. The limitation of availability and coverage, together with the great interest in integrity, is also highlighted.

Satellite navigation is being used in many different application areas, from purely commercial to highly scientific. In between, there are many professional domains that have found a great interest in using GNSS, mainly in order to reduce the time and complexity of deployment of previously used systems. This is notably the case in the civil engineering and construction industries, where high accuracy receivers have been in use for years now.

The classification of applications can be carried out in many ways: mass market versus professional, commercial versus scientific, or by main domains such as surveying, environment, and agriculture. We propose to review them in chronological order, from the oldest maritime use to the latest location based services (LBS) domain. Some possible classifications are then provided and discussed.

■ 11.1 INTRODUCTION

This chapter is mainly based on current satellite navigation positioning systems; that is, the applications described are those enabled when using a GNSS receiver.

Global Positioning: Technologies and Performance. By Nel Samama

Thus, it is very restricted to specific conditions. For instance, no applications are indoors. Some are in quite difficult environments for current GNSS, like urban canyons. We decided to propose two chapters (the current one and Chapter 12) in order to make a clear distinction between what could appear as promoting GNSS technology by describing all the possible applications, and what could be done, inconsideration of current limitations, in the future.

◼ 11.2 A CHRONOLOGICAL REVIEW OF THE PAST EVOLUTION OF APPLICATIONS

The first applications planned were clearly military, and more precisely concerned the ability to provide a good departure location for intercontinental missiles launched at sea. Nevertheless, a civilian availability of early systems was also planned and implemented, mainly for commercial maritime purposes. At that time the maritime domain was the only one interested because receivers were expensive and performances not sufficiently good to develop in any other market. As accuracy improved and costs decreased, other fields, like terrestrial transport, have shown some interest in positioning systems.

Besides large public applications that are bound to produce high revenues due to the large number of units sold, some professional domains, such as geodesy or civil engineering, rapidly saw the huge advantages such a system could produce, both in terms of accuracy and ease of carrying out measurements. Unfortunately, GPS signals were altered by the application of Selective Availability (SA), thus reducing the accuracy of the system.[1] On the other hand, this could be considered as a lucky event. Furthermore, two frequencies were not available, except for military purposes, as no civil signal was broadcast on L2 (only the P code). Therefore, some small groups decided to find a way to overcome both SA and P code. This led to techniques such as differential approaches or codeless treatments, still widely in use today for professional purposes.

Global navigation satellite systems are now reaching a maturity phase, although much improvement is still possible. Thus, the cost of the positioning function (based on GNSS) is now less than a few dollars for mass-market products and the latest ongoing developments[2] will lead to a further spread of GNSS.

TRANSIT and Military Maritime Applications

We have already described the origin of the satellite-based systems for positioning and navigation (see Chapter 1); and the military need for an accurate enough location as the starting point of inertial intercontinental missiles launched at sea. Terrestrial systems, such as LORAN, were already in use and allowed such a process, but the coverage was limited, mainly because of the limited range of the LORAN beacons, although it is some tens of hundreds of kilometers, especially

---

[1]On the other hand, this could be considered as a lucky event.
[2]Such as high sensitivity receivers or SBAS wide availability.

over the sea. But of course, it is quite difficult to deploy such a beacon network all over the world.

Furthermore, there was no need, in the late 1960s, to cope with altitude at sea, as by definition the ship was "on the geoid." Thus, a two-dimensional positioning was sufficient.

The First Commercial Maritime Applications

Very soon after the military availability of TRANSIT (three years later indeed), the U.S. administration made the system available for civilian use, and particularly to commercial maritime vessels. The main advantages of having an automatic positioning device are to allow optimized routes over the oceans and also for increased security when facing bad weather.

Taking this fact into account, one could easily state that the satellite-based positioning systems developed by the United States have always been available without interruption, at no cost, to the civilian community from 1967. Unfortunately, the rapidly growing application market and the economic importance of the sectors involved have led to the fact that a military-driven system is no longer totally satisfactory.

Maritime Navigation

The GPS system was also available to civilian users from the very beginning. Naturally, the first community interested was the maritime one, for both professional and recreational purposes. The level of security provided by the availability of a much more accurate positioning system, compared to previous calculated locations, and using receivers of low price allowed the wide distribution of GPS technology on board recreational boats. Once again, the fact that GPS accuracy is less in vertical coordinates than in the horizontal plane is really of no importance for such applications.

Furthermore, the Selective Availability, switched on during the first years of GPS, was not a real handicap as it still remained more accurate than previously used techniques — it was so much easier to obtain, and only had real limitations in very specific cases. A typical example was when one wanted to enter the harbor just by using the automatic pilot driven directly by the output of a GPS receiver. Initially, there were at least two difficulties: the first was related to GPS accuracy, and the second due to the accuracy of maritime maps and data available. In order to achieve such a performance, one needed to have data such as the channel dimensions and exact location, relative to the GPS referential. Neither was available at these early stages.

Nevertheless, this first "obvious" application field was certainly almost the only one that could have been imagined at the start of the development of satellite-based positioning systems. The full story is far from this reduced set of applications, as described in the following sections. Typical maritime applications (see Fig. 11.1) include rescue and replenishment of off-shore platforms, cruising positioning, digging waterways, or positioning and monitoring of off-shore platforms.

(a)   (b)

FIGURE 11.1 Typical maritime applications: Rescue (**a**) and commercial activities (**b**). (© ESA-S., CORVAJA.)

Other typical applications consist in coupling GNSS receivers with dedicated sensors such as radar, echo-sounders, fish-finders, and so on. At sea, the receivers are usually quite crude receivers as there is no possibility to implement techniques such as map matching, as one might be situated anywhere. An additional system is the AIS (Automatic Identification System). All ships of more than 300 tonnes, as well as all passenger ships, must be equipped with an AIS transponder, which continuously transmits data concerning the ship's identity, route, position, and speed. The so-called AIS radar aimed at yachts displays the information received from the ships in the vicinity. A typical screen is displayed in Fig. 11.2.

By clicking on any given boats on the screen, it is possible to access all the corresponding data for any given ships. As this system can also show the route followed by the ship, it is quite easy for enforcement bodies to draw a parallel between the route and any suspicious oil slicks in the wake of the boat.

Time-Related Applications

The first specifications of the GPS in the early 1970s encompassed the time requirement. Of course, given the technical approach in order to obtain accurate distance estimations, a good time distribution over all the satellites and ground stations is required. Furthermore, the manner in which a fix is calculated gives rise

USER'S POSITION

SELECTED TARGET   SELECTED TARGET'S DATA

FIGURE 11.2   Example of an AIS display screen. (*Source*: Nasa Marine.)

to the possibility to obtain, from the location solution vector,[3] a time indication exhibiting an accuracy of comparable value to the distance error evaluation, namely a few tens of nanoseconds.[4] This leads quite directly to being able to consider a GPS receiver as a fantastic tool for synchronization purposes.

Some very demanding fields, in term of synchronization, exist in different domains: seismology, communication networks such as the Internet or the wireless telephone, and also banking or finance. As will be described in Chapter 12, we believe that this tight relationship between time and positioning is the essence of the future revolution of our daily life, not only because of the necessity to deal with a precise time in order to achieve accurate location finding, but first due to the main goal of our modern lives, that is, optimizing time.

For telecommunication systems for instance, the transmission of the transmitter time to the receiver[5] is currently achieved through the use of so-called "synchronization bits" or "sequence." This approach requires a given amount of resources to be devoted to this task and also needs a sufficient bandwidth to be available. In the case of microsecond synchronization, this is considerable, but not really in the nanosecond domain. Let us now imagine that every mobile phone is equipped with a GNSS receiver for navigation purposes. The synchronization function will then be de facto available, with better performances than today's, enabling optimization of the wireless link as well.

---

[3]The fourth coordinates related to the receiver clock bias with reference to the GPS time.
[4]This applies from when the Selective Availability was switched off. Note that really observable values are even better than these figures, being often less than 10 ns.
[5]This is of prime importance. Without such a synchronization, messages would not be understood by the receiver.

Geodesy

Once the problem of reference frames was solved, GNSS receivers could offer a simple way to follow the evolution of an earthquake. It is also an interesting way for geodesy cartography or for topographical measurements (surveying) to be carried out. Very accurate receivers are then required, such as dual-frequency and carrier phase measurements (in order to achieve centimeter accuracy, or even better, as required in these cases).

Note that these kinds of applications are terrestrial-based ones, unlike all the first domains. The development of GPS was based on the desire to increase the number of potential users, the first being air transport (military aircraft as the first target). This is the reason altitude became so important, together with taking into account the receiver's velocity. However, it appears also that terrestrial users had potential needs,[6] partially covered with GPS. Furthermore, user communities were able to propose technical innovative solutions in order to meet their own requirements, such as high accuracy for instance.

Civil Engineering

Civil engineering was another community that saw a great advantage in using high accuracy positioning receivers. In fact, nothing was impossible without such equipment, but GPS receivers have highly simplified certain phases for the initial positioning of concrete masts (see Fig. 11.3). High accuracy GNSS receivers (down to the centimeter level) are now used for many different tasks from absolute positioning of a road as well as the heights of the various layers (gravel, bitumen, and so on). The limitations of the current GPS coverage are only highlighted when the upper levels are in place — at that point no more positioning is possible at the lower levels.

Positioning is also interesting for those running open cast mines (see Fig. 11.4). It can help in digging with a uniform height over a whole site, allowing geological real-time analysis of the shape of the mine. In some cases, when the digging is deep, a problem of satellite visibility can occur; that is, not enough satellites are in view to provide a good positioning, either because the minimum number of satellites is not reached or the Dilution of Precision values are too high. In such cases, the use of pseudolites has been reported.

Other Terrestrial Applications

One of the first application fields of high performance mono-frequency receivers is agriculture (and also the fishing industry, which has similar needs in terms of accuracy requirements). The main uses consist of having an analysis tool in order to optimize the spraying of fertilizers and other herbicides and insecticides, and the

---

[6]Note that it sounds like history is going backwards. Do not forget that when the problem of longitude was first solved by the extrapolation of Galileo's observations, it was first used in order to solve land claims between local landowners.

FIGURE 11.3 Civil engineering's use of high accuracy GNSS receivers (© ESA-BARBARA NSC). (*Source*: Leica Geosystems, Switzerland.)

management of set-aside lands. The installation is achieved in the best possible conditions: tractors and other agricultural machines move slowly, have enough electric power to supply the receiver, and the typical accuracy needed is 1 m. Specific software allows one to have a graphical representation of the farm work, together with automatic time alerts for cultivation purposes.

With the increased importance of ecological and environmental matters, this approach can also be used in order to demonstrate and enhance the changing agricultural practices in this field. This is certainly a good motivation to develop this market.

FIGURE 11.4 Open cast mines. (*Source*: Leica Geosystems, Switzerland.)

FIGURE 11.5   Transportation applications. (© ESA-S. CORVAJA; ESA-P. SEBIROT.)

Up to this point,[7] the applications described were never really dynamic — they were either static (civil engineering, time-related, and so on) or very slow moving (maritime or agriculture for instance). When switching to the road sector, or more generally to the transportation sector (see Fig. 11.5), the speed of the mobile increases and this piece of data has to be taken into account.

The first terrestrial application of GNSS technology to the car industry was the navigation system — this will be described in the following sections. Another way to use GNSS receivers has been in fleet management systems, still in use today. The idea is to integrate a GNSS (GPS for the first trials) receiver and a telecommunication system; the satellite navigation part allows individual locations of the equipped trucks or vehicles to be obtained, and the telecommunication part is designed to get all the positioning data from the mobiles back to a central management unit. Thus, the central controller has a real-time visibility of the status and positioning reports of the fleet.

The first implementations concerned truck fleets used for various delivery tasks, soon followed by taxi companies or bus fleets. In the case of trucks, the main objective was to follow the goods and to check the route for both surveillance purposes and retrieval after theft. As presented later in this chapter, these first trials led to well developed auto systems. In the case of a taxi fleet, the goal was to optimize the service to customers, as what they want is to optimize their time and to be sure the taxi will arrive quickly. Knowing the location of all the taxis in your fleet is bound to help in getting one to the customer's place as quickly as possible. Once again,

---

[7]Typically mid-1990s.

the fundamental features of current in-car navigation systems were incorporated in these first professional deployments. Of course, it is easy to understand that many other pieces of data, in addition to the location and time of location of the taxi, are required. Is the taxi free? Are the relevant roads jammed? Among all the data of interest, one can state that traffic information in real time[8] is certainly one of the most important.

In the case of bus fleets, the purpose was different. The difficulty with buses, compared to the metro or train, is the lack of centralized information for users, or conversely the difficulty in planning a route using several different buses. Also important is the fact that when traveling on dedicated lanes (which is more and more usual nowadays in France, for instance), it is quite easy to give an accurate time for all the stops, but not when traveling on "open" lanes. Then, when a user arrives at the bus stop, he can never be sure if the bus has already left. Thus, an interesting approach is to give information to the user, such as when the next bus is going to arrive or how to reach a destination, together with a time estimation of the journey. Some experimentation has been carried out recently and operational systems are available.

When dealing with the data transfer required in order to send all the data to the central managing unit, specifically for position data which are of reduced size but repetitive (to achieve real time), a problem of communication links appears. The encapsulation of the message is so huge that one needs to transmit from five to ten times as much data as real useful data. The telecommunication system reaches its limits mainly because there is a need for a destination node address, sender's node address, synchronization needs, integrity check, and so on. Many of these headers can be removed for some applications where the availability of a common reference clock could be turned to a huge advantage, considering a fixed allocation of time slots to every user participating in the exchanges. In such a system, as every node is synchronized through the GNSS clock distribution technique, the header is reduced to a minimum. The identification of the transmitter is deduced from its time slot, as it is uniquely attributed to one user. Furthermore, in the case of transmissions towards the central managing unit, the destination address is the same for all the mobiles. Last, but not least, a transmitter can enter the system without any prior exchange; in other words it will automatically synchronize itself using its own time reference (GNSS one). Some features of such a system have been implemented and demonstrated.

■ 11.3 INDIVIDUAL APPLICATIONS

It is noticeable that the first applications described so far are mainly professional ones. This is due to the fact that receivers were, initially, quite expensive and required accuracy not available to civilian users. Thus, only professionals or small highly technical groups had enough money, interest, and skills to drop into this new positioning

---

[8]Maybe real time is not enough in certain cases such as emergencies or full optimization. One certainly would like to have access to predictive traffic information. This is not yet widespread, but first elements are available (see following sections).

world. By the end of the 1990s, industrialists found that the mass market was attainable through two personal applications: walking/riding and car navigation. The first one requires no specific development except for a user-friendly interface that allows fixes to be stored and the walker's displacement to be shown. Thus, it was possible to get it on the market quite rapidly. For the second application, car navigation, the requirements are a little bit more stringent as there is a great need for maps and databases to store addresses. Nevertheless, because this application had been under development since the early days of the availability of GPS signals, this was reaching maturity. In the meantime, some more specific applications appeared (tourist information systems, Bluetooth-based GPS receivers that can be run on a Personal Digital Assistant (PDA), integration of GPS receivers into mobile phones, and so on), together with the advent of Galileo, which represented a new boost. Galileo officials tried for years to find new applications in order to justify the Public–Private Partnership: this has stirred up various potential user communities.[9]

Let us have a quick description of some applications and give explanations of the main issues.

Automobile Navigation (Guidance and Services)

A car navigation system is a lot more than a positioning device (and represents only a very small part of the total cost of the system), but the quality of this determines the performance of the system. The main elements are:

- A digital map database;
- A positioning module;
- A route guidance module;
- A human–machine interface;
- A route-planning module;
- A map-matching module.

The first four are required as the basis of the system; the last two are additional features. Note that the map-matching feature is a very powerful approach to overcoming inaccuracies in the positioning module (and also in the map itself!) but is only applicable to car navigation, and not at all adaptable to the pedestrian world.

Digital map updating is a very heavy charge and is currently achieved by a few large companies like NavTeq or TeleAtlas. How much databases cost depends on the size of the maps (that is, one country only or many different countries, such as Europe), but it is still a high percentage of the total cost. Updated maps are available

---

[9]Remember that the Galileo system tried mainly to find the so-called "killer application" that would be a source of high revenues and that could be achieved through the use of Galileo and not GPS or GLONASS. Many projects were funded in order to find it (or them), but it has been very difficult to find such an application. Furthermore, all applications proposed were compatible with GPS capabilities and this contributed to promoting satellite-based positioning (which was exclusively GPS-based at that time... and still is).

from providers every three months and the industry upgrades their maps every six months. Although mass-market users are not really used to upgrading so often... they are not happy when the map is out of date. Another way to overcome this drawback is given in the LBS section.

The positioning module is far from being just a GNSS receiver, especially when embedded in the car (and not added as an afterthought). These in-built systems are usually more expensive than portable devices, but offer a wider range of features including a larger color screen, additional sensors in order to allow dead reckoning (using non-radio-based positioning approaches, see Chapter 4, Section 4.5, for details) or even additional services, like traffic information. In the case of this automobile application, a simplification applies: the location is basically a two-dimensional vector.[10] This also simplifies the map databases quite a lot. Note that map databases are still two-dimensional. This has to change for future applications, such as optimizing a train's power consumption depending on the route profile, but this is another story. Coming back to dead reckoning, the major sensors that are in use today are odometers, magnetometers, gyroscopes, and accelerometers. Having access to part or all of these data, it is possible to carry out very valuable position estimations. The major drawback of these sensors is the fact that instantaneous values do not lead to an instantaneous calculation of an absolute location, but only allow a relative displacement to be estimated. Thus, measurement errors accumulate with time and lead to an important cumulative error after a while. For indoor applications, it is considered very difficult to stay within a 1 m radius error after 10–15 minutes of navigation.

An odometer allows measurements of the distance traveled and also direction of movement if a differential system is deployed.[11] A magnetometer allows one to determine the orientation in a discrete manner.[12] An accelerometer gives the instantaneous acceleration vector, usually in three dimensions by using three sensors. Gyroscopes allow the variation of orientation of the movement of the car to be determined, here again usually using three sensors in order to achieve three dimension vectors.

The route guidance and the route planning modules are based on complex algorithms that allow the best way from a departure location to a destination location to be determined, given certain optimization factors. The main current optimization goals are "the shortest route, in distance" or "the shortest route, in time." Note that distance and time are the two opposite main goals. The main parameters the user can usually adjust are the average speed on various types of roads and the fact there are preferences concerning the type of roads. For example, a user can specify he does not want to use small or toll roads, and so on. Then, roads are represented by a graph and the determination of a route is somehow a graph optimization problem. Each

---

[10]Note that this is once again no longer the case for pedestrians and indoor environments where altitude, or height, is of prime importance.

[11]For this, an odometer is allocated to each front wheel. In such a case, it can be shown that depending on the direction of the car, the differential value output by both odometers can give an indication of the direction.

[12]The measurement is valid at the time of measurement only. One can increase the refresh rate of measurement, but the integration process will still accumulate errors.

road is an edge and, depending on the final goal, each edge is assigned a given weight. The best route is then the one that minimizes the overall cost function.

As the map database includes information such as the direction of the traffic and various elements such as tunnels for cars only, the height of bridges, and so on, more features are available depending on the user's means of transport. On a bicycle, motorways and other restricted roads are forbidden. When the user is on foot, then one-way roads no longer exit. These are specific examples of different optimization goals and the way they impact on the parameterization of the system.

Another very interesting feature implemented in all car navigation systems, whether embedded or transportable, is the map-matching approach. Very quickly, manufacturers found that GPS accuracy was not good enough to give the user any certainty about his exact location, especially if he could see that the system was wrong, which can generate frustration and bad press. Furthermore, a typical accuracy of 10–20 m leads to some difficulties where streets are close together and parallel. A powerful approach to this kind of problem has been to implement so-called map-matching. The main assumption that is made is that the car (or rather the positioning module) lies somewhere on a street, and not elsewhere — when driving a car, this seems reasonable. Then, processing is carried out on the location provided by the positioning module, in order to make some comparisons with previous locations, direction of movement, compatibility with car dynamics, and so on. Then, this can be compared to the actual map. Once again, the system tries to find the nearest possible true location considering the location given by the positioning module and the cartography. This approach is so powerful that it is often referred to as a new positioning sensor.

The last feature that has not yet been mentioned is clearly "access to services."[13] For a car driver, there are obviously potential needs such as the location of the nearest service stations, restaurants, or hotels. All these could be included in the databases, together with complementary information such as phone numbers, availability, opening hours, and so on. In the case of the availability of a telecommunication link, an automatic reservation could also be of interest. This link to the telecom world is of prime importance[14] when one wants to access services, although some of them are accessible via the radio. One of the most interesting services when navigating from one location to another by car is traffic information. It requires a huge infrastructure but is already in place over Europe for the major roads. It is broadcast through the Radio Data System (RDS) and has been encoded uniformly all over Europe; the service is called TMC for Traffic Message Channel. The optimal car navigation system is thus the one that takes into account the traffic messages, calculates alternative routes (to the ongoing one), compares the global cost functions of the various possible routes, and proposes the best ones to the driver, allowing him to bypass traffic jams.[15] Such products already exist. The problem is now clearly the reliability of the traffic information and the fact that a user will certainly be more interested in predictive rather than real-time information (we always want more than what

---

[13]This is clearly the philosophy of Galileo in order to provide the program with revenues.
[14]See note 13.
[15]Users are nowadays usually interested in "saving" time.

we have). Future possible approaches are described in Chapter 12. Finally, concerning traffic information, it is noticeable that this feature changes the usage of navigation systems: from occasional use in unknown places only and not for routine displacements, to a daily use, where the optimization of the time spent in traveling is really an added-value service.

Tourist Information Systems

The combination of a system that has the ability to drive you from one location to another and potentially huge databases, leads quite naturally to tourist information systems. The main advantage is that tourists are usually moving in unknown environments, where navigation exhibits highly satisfactory levels, and they typically want to access geographical sites efficiently. They do not have the same time constraints as workers, but it is more relaxing to know this kind of help is available. Such a guidance system could also propose different features such as a quick one-day visit of an optimized number of sites or a thorough "inspection" of a single site. Of course, you are free to follow these instructions or not, knowing that at any time your "traveling friend" can get you back on track. The availability of a car park, hotel, for example, plus detailed information, provides the user with a wide range of configurations. The rapid development of memory capacity does not really limit any needs imaginable today; the real difficulties lie rather in the collection of information than in the storage.

Local Guidance Applications

As an extension to the tourist information system proposed in the previous section, one can easily imagine a guide for archeological sites or museums. Some are outdoors and already exist, while some are indoors. No technology similar to GNSS is yet widely available for such purposes. Nevertheless, initial trials have been reported for a museum guidance system using a Bluetooth positioning module for indoor location findings. Features such as precise positioning and orientation are generally required, although not yet provided by any deployable solutions. One has to go back to automation and robotics to find such systems, which require a large infrastructure. Users and future providers are clearly seeking simpler solutions, together with increased accuracy, because indoor places are usually smaller than outdoor ones. See Chapters 9 and 10 for more details on achievable performance of indoor positioning techniques. Perhaps one could imagine a complex system in the future that integrates GNSS capabilities together with indoor positioning, achieved for example through the use of scattered UWB modules throughout most of the building and also including reliable orientation and velocity sensors. In this case, it is possible to think of such applications as an automatic indoor virtual guide, making appropriate comments while the visitor is standing in front of a work of art.

Answering the question "Who would be interested in knowing his location?" suggests the case of blind people, who could take advantage of being helped in their movements, in the same way that guide dogs help now. Systems are already available,[16] but further questions have to be addressed. What level of detail should there be in the databases? What levels of detail do user communities require? For example, in the case of blind people, is it pertinent to add information concerning the width of the sidewalk or its height at the crossing place? And then, will it be pertinent too to add the location of signposts or traffic lights? To what degree of accuracy?

The next generation of systems will certainly incorporate a lot more data than those of today, but this will require both a huge effort in collecting and keeping these data up-to-date, but also in processing them, in order not to be snowed under with too many vocal or displayed messages. First attempts have been reported in this direction by implementing tactile actuators on white sticks for blind people.

Location-Based Services (LBS)

It is forecast that the two major domains of applications of the GNSS will be transport and Location-Based Services (LBS), which will represent more than 70% of the total market. The term "LBS" is used to designate all services that have to use the location of the user as input data. Usually, the mobile terminal is considered to be somehow a communication device, such as a mobile phone. Thus, the interesting part of LBS lies in the fact that the user could access services, and providers could sell services, thus generating revenues, potentially on a wide scale. It is also the LBS sector that will force the mobile telecom industry to enter the game, whereas so far they have played a waiting game.

Large companies have decided to push forward the LBS sector and it appears that take up is slower than expected. Nevertheless, important projects are currently being carried out in the Galileo program in order to define clearly the potential of the LBS sector, together with its major trends.

The first real difficulty is to cope with the problem of personal navigation only by using existing technologies of positioning. We have already seen (Chapter 10) that indoor positioning is not yet provided by the GNSS constellations,[17] or by any other means currently in place.[18] This observation leads directly to the current efforts, which are either conservative or visionary. "Conservative," in this context, means those services based on very classical applications such as "finding a friend," "having access to a point of interest (POI)," "looking for a weather forecast," or such like. By "visionary," the reader should understand if to mean applications such as when firefighters intervene in a burning building with many people involved and whose locations should be transmitted to a central unit for crisis management in real time in extreme conditions.[19] The same could apply to the security forces. In these cases, positioning should be possible within 1 m both outdoors and indoors. Some techniques allow this achievement outdoors, but certainly not indoors. As an

---

[16] These systems are indeed products that "include" various technologies.

[17] However, the latest high sensitivity GPS chips, like the SiRF III, push detection and tracking to the ultimate limits thus allowing "light indoor" positioning.

[18] UWB or WLAN approaches are commercially available, but really not widespread at the time of writing.

[19] Such as collapse of walls for example.

example of this one can consider the result of a Galileo project called AGILE that carried out this kind of work and ended up with a classification of applications as shown in Fig. 11.6.

Thus to date, LBS are mainly directed at the professional sector that wants to find industrial solutions with guarantees of services, and also recreational activities where the abovementioned limitations are acceptable. Of course, this constitutes a real brake to the development of the LBS. The main applications are then mainly due to the integration of telecommunication services with the positioning capabilities of a mobile device. Thus, when available, the terminal's location is used to reduce the search domain for weather information, points of interest selection, or to help to find friends. In other cases, the availability of the location can also modify the human–machine interface by skipping some pop-up screens (the assumption is then that the request concerns the immediate geographical environment of the user).

As a first conclusion on LBS, one could state that today's LBS are largely oriented towards the integration of location into telecom services and are not

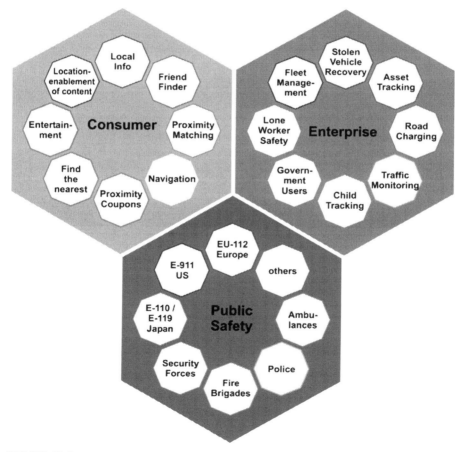

FIGURE 11.6 Classification of potential LBS applications. (*Source*: Hanley et al. 2006.)

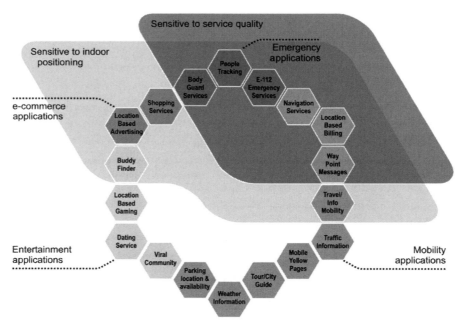

FIGURE 11.7 Classification of potential LBS applications with associated environments. (*Source*: Abwerzger et al. 2007.)

proposing real innovations. There are maybe two reasons for this: the technical limitations of existing positioning systems and the lack of emulation in this field. One can be almost certain that once the technologies are available, many applications will pop up (see Chapter 12 for first thoughts). In order to illustrate the abovementioned limitations and to make a link with the indoor positioning described in Chapters 9 and 10, the AGILE project drew another classification of Fig. 11.6 taking into account the environment (Fig. 11.7). This clearly shows that many applications require indoor positioning.

Emergency Calls: E911, E112

Section 11.8 deals with the privacy issues that are bound to decrease the interest of potential users in LBS. No one is going to be really happy to be "followed" or "tracked" in real time... unless the advantages outweigh this (foreseen) drawback. There is one application that is certain to provide the positioning industry, in general, with an enabling feature: the emergency[20] call. Such calls, E911 in the United States and E112 in Europe, are reserved numbers anyone can dial in the case of an emergency. Local relays are then activated and specific operators will

---

[20]Note that there is a real difference between emergency and assistance, based on the notion of urgency. In the case of assistance, time is not vital, whereas it is in the case of on emergency.

route the call as quickly as possible to the appropriate service: police, fire brigade, medical center, and so on.

The basic idea of an emergency call is the possibility for an emergency call center to locate the caller precisely. Indeed, the first difficulties arose a long time ago, with the first analog telephone centers. When a user called the fire brigade, he often forgot to tell the operator his location because he was too busy coping with his home fire. There was then no possibility for the operator, once the caller had hung up, to find the location where he should send the brigade. With the advent of electronic centers, the problem disappeared as it was very easy, by searching the database, to go back to the address for the telephone number, which was visible to the operator. The problem arose once again with the advent of mobile phones, where the location of the user was once again unknown. However, in the case of accidents or emergencies, this kind of service appears to be a real need. Thus, the United States Federal Communication Commission (FCC) decided to oblige telecom operators to provide a user's location in the case of an emergency call. The introduction of this regulation has been delayed quite a lot, mainly because of the technical difficulties in achieving the required performances. Nevertheless it encouraged the navigation industry to find new approaches and urged the telecom industry to take this problem into account.

In the United States, some companies decided to move into the Assisted-GPS domain and to develop a complete set of products. Due to the compulsory part of the FCC regulation, it was really efficient in "shaking up the industry." In Europe, things are a little bit different with E112. Only "best effort" is required by the regulatory authorities. The direct impact is the speed of progression of positioning features in the mobile phone industry, which is much lower in Europe than in the United States. Furthermore, Northern American companies are gaining expertise in these domains. The relative delay in Europe could also be considered as a "wait and see" approach, which can lead to an optimization of efforts. In fact, one can hardly state that Assisted-GPS and High Sensitivity GPS are the ultimate answers to satellite navigation positioning being available everywhere.

Security

Knowing the location of people or goods can certainly help in protecting them, both against others or even against themselves. Thus security applications are one aim of positioning systems. Different sectors are addressed; for example, relating to the following situations:

- Protection of assets;
- Keeping track of children;
- Management of groups;
- Surveillance of prisoners;
- Locating people in mountains.

Among these sectors, certain solutions are already available. Knowing about the real limitations of the various possible techniques, let us discuss some real implementations. As an example, people management systems exist and are typically provided by telecom operators. Collaborators are then tracked and their location reported to a central management unit. This is interesting when one needs to access a specific person but do not want to create disturbance. It is possible to determine a disturbance indicator depending on the geographical positioning of the people.

For instance, if it appears clearly that someone has left his meeting (because he is now walking down "main street" for instance), one can easily call him.[21] Unfortunately, the coverage of GNSS is not yet good enough, although it is very accurate in available locations. The technique deployed is simply using the Cell-Id of the mobile network. The location is very inaccurate but there is still a location. Of course, the use of the location is quite different from what it would have been in case of availability of an accurate positioning. Nevertheless, it is already on sale commercially.

This application can be transposed into the personal market, for security or recreational purposes: it would be nice to have an idea of the location of family members: either children coming back from school or parents leaving their office.[22] This is also the first step for "find a friend" like applications.

Games

Besides the fact that specific games can be developed in the light of positioning capabilities, where the players are localized and the evolution of the game depends on these locations,[23] another activity is developing "geo caching." This is based on a user "hiding" an object, whatever it is, in a given location. The idea is then to allow other players to find this object. Treasure hunting then becomes the logical extension of this.

■ 11.4 SCIENTIFIC APPLICATIONS

Many other applications exist, and specifically in scientific domains because the high achievable accuracy (a few millimeters[24]) allows numerous interests, in seismology, climatology, and geophysics for instance.

Atmospheric Sciences

GNSS signals are affected while traveling through the atmosphere because the refraction index varies in relation to a vacuum. It is thus possible, through GNSS measurements, to carry out an analysis of the various levels of the atmosphere — troposphere and stratosphere for the lower layers, but also the ionosphere for the upper one. Of course, regarding the importance of ionosphere propagation-induced

---

[21] He is maybe speaking with the "big chief" who would not appreciate being interrupted by a phone call, but he is at least no longer in the meeting room: the next step is to implement personal cameras...

[22] In such a way, it works both ways; parents can "keep an eye" on their children, but children can also "check" how close their parents are.

[23] Geolocation games include Virtual Maze, Nibbons, Geko™ Smak, and Memory Race.

[24] When using multifrequency phase measurements and accepting very long static runs.

error in GNSS positioning, such analyses are also likely to improve mass-market receiver accuracy.

Troposphere Analysis

The troposphere is the lowest layer of the atmosphere, typically located between the Earth's surface and an altitude of 8–15 km, depending on the latitude and season. This layer contains around 90% of the total atmosphere mass and is where most meteorological phenomena take place. The water cycle develops in this layer and its importance is clear for many domains, including GNSS signal propagation modeling.

An analysis of the troposphere can be carried out with permanent GNSS stations. From continuous phase measurements of a static GNSS dual-frequency receiver, it is possible to estimate the evolution of the correction value for the wet part of the troposphere. This is acceptable at least at the vertical to the station. The accuracy of such measurements can be as small as 1–2 mm. The dry fraction is estimated by local pressure measurements. If the number of stations is sufficient and if simultaneous observations are carried out from many different satellites, it is also possible to estimate the dissymmetry of the atmosphere, notably when clouds are passing in the sky.

From the high number of stations throughout the world, the data gathered allow the precise estimation of water vapor content and its evolution. Delay in obtaining these data should be reduced to a minimum in order to be valid and to improve the various meteorological models and propagation modeling. Internet sites allow rapid use of these data and also provide very precise orbital parameters that augment scientific applications.

The main advantages of GNSS-based measurements rely on accuracy, availability, and continuity, leading to high reliability. However, these measurements give local, not global data, and the latter require further modeling.

Stratosphere Analysis

The stratosphere lies above the troposphere and has a depth of about 15 km. Data concerning this layer (temperature for instance) are used in meteorology and also long-term climate studies. The main idea is to use satellites located in low Earth orbits (typically 750 km) in order to analyze occultation phases, which occur when the GPS satellite moves around the other side of the Earth. It is then possible, through Doppler measurements, to determine the pressure and temperature gradients, in three dimensions, of the stratosphere layer.

Ionosphere Analysis

The ionosphere is the upper layer of the atmosphere, typically located at between 60 and 800 km of altitude. It is an ionized layer and thus leads to delay due to collisions between the GNSS signal and the ionized particles. This is a real disturbance as the principle of GNSS positioning relies on time measurements converted into distances. If the velocity of the signal is mis-estimated, the positioning will exhibit corresponding errors.

A correction of the propagation model is possible through the use of different frequencies, which also allows determination of the Total Electron Content (TEC), an important parameter that could be used for modeling improvements. These measurements can be carried out either by static ground receivers or by low Earth orbiting satellites, once again using the occultation method. Then, ionosphere maps can be drawn on local or even global scales. The main uses for these are GNSS ionosphere propagation modeling for GNSS purposes, but also for satellite telecommunication systems.

Tectonics and Seismology

The use of very accurate positioning (a few millimeters) allows observation of slow-motion movements over long periods — this is typical of Earth surface displacements. Continents are moving at a few centimeters per year; and a long-term analysis of static receivers allows the movement of continents to be precisely determined. The same applies to studies concerning, for example, the impact of tides on the Earth's surface.

The analysis of earthquakes, and more generally seismology, means dating the various observations in order to allow correlations. Precision dating is possible with GNSS due to the fine time management required for these systems. Thus it helps in determining whether two distant phenomena are linked to each other or not, using an underground wave propagation model to estimate the time bias that would have occurred in the case of related events.

The ionosphere observation can also provide prediction[25] of terrestrial events. It has been shown that earthquakes or tsunamis induce changes in the constitution of the ionosphere that can be observed through GNSS readings. The augmentation of the number of available satellites with the combination of GPS and Galileo constellations should increase observation capabilities.

Natural Sciences

Animal Surveillance

Positioning systems could enable animal management in different types of applications. At first sight, it could help in defining migration movements of wild animals. This has already been achieved through the installation of miniaturized GNSS receivers coupled to transmitting devices; which allows animals to be followed continuously in real-time. It is also helpful in the case of protected species. By permanent monitoring, any harm done to the animal can be precisely dated and located, allowing optimized pursuits. This can also help human populations located near dangerous wild animals, in detecting their presence and coping with sharing the same environment. This locating feature can be used to study very specific wild

---

[25]Prediction is not really the right term, because the GNSS-based observations happen at best in real time. But observations can be used to send alerts concerning meteorological or seismological events to the population in order to prevent the dramatic effects of tsunamis, for instance.

behavior such as the sense of orientation developed by traveling pigeons. Equipped with miniaturized recording receivers, it has been possible to know the route followed by pigeons. Of course, even with this information the mystery has not yet been solved, but this is an appreciable tool for analysis.

Environment Surveillance

The double feature of time and location could help in precisely determining the degradation of certain environments. Satellite-based observation allows this in different ways, but local measurements provide accuracy. The Amazon deforestation is one example. Meteorological observation and prediction could also be used in order to affirm trends and to play a role in political decisions concerning the Earth's climate. Furthermore, maritime buoys are used to determine major oceanic currents. Glacier evolution can be measured, and ice-floe displacements analyzed.

11.5 APPLICATIONS FOR PUBLIC REGULATORY FORCES

Both GPS, with its Precise Positioning Service (PPS), initially designed for military applications, and Galileo with its Public Regulation Services (PRS) signals, have designed signals that can exhibit a high protection level against radio electric perturbation. This is typically achieved through the use of both very long codes and cryptography. Such signals have obviously a real interest for military applications, but also for crisis management.

Safety

One of the prime examples of an application requiring high reliability (that is, withstanding perturbations) is the guidance of mass transportation systems. Transporting people by plane, train, or boat needs constant accuracy as well as being able to be alerted immediately in case of the failure of any component of the positioning system being used. This last aspect is commonly called "integrity." Indeed, this is the confidence one can have in the measurements being carried out. EGNOS, WAAS, and MSAS are intended to provide GPS with this feature and Galileo will also provide integrity. Such applications are thus possible as long as the accuracy and availability requirements are met. This is clearly the case for the sea but still needs some upgrading of availability for rail transportation. The integrity concept, compatible with aviation demands, has still to be shown to be reliable in city centers where multipath is of prime concern.

Prisoners

Some prisoners, when at the end of their prison term, are freed on condition that they report regularly to the police. In recent years, a new method has been developed in which an ex-prisoner has to be present at a given location, at the prison, police station, or even at home, at pre-determined times. At home, the police could check the presence of the ex-prisoner by way of an electronic device that carries out proximity detection. If, at a given time, the presence of the device is not detected, an alert is sent. Many disadvantages of this method are obvious, and in particular the fact that once the alert has been given, the police forces have to find the ex-prisoner.

A new method is currently being implemented, using both GPS for location and GSM for data transmission. The complete system is included in an electronic bracelet and allows restricted location-based displacement of the prisoner. It can also be used the other way round: to eliminate suspicion in the case of theft or aggression. A commercial system exists and is also available to keep track of elderly people.

This bracelet can also be used in other applications where there is a real need to be able to follow, to find, or to locate people. The Columba device is intended to provide peace for those dealing with people suffering from Alzheimer's: patients, relatives, and doctors. A patient can "stray" and forget where he or she is; and in such a case, even very near the hospital or house, it is possible he will not be able to find his way back, with dramatic results in some cases. The management of medical centers may be held responsible and relations deeply affected. The bracelet is likely to provide tranquility as it will be possible to locate the patient or to install voice dialog with him eventually. Although the total cost of this equipment is still quite high (about €400 without the communication charges), this certainly is the way forward.

■ 11.6 SYSTEMS UNDER DEVELOPMENT

Besides the various applications already mentioned, and often well implemented, based on GPS signals, there are some research projects that aim at finding new applications and new ways of using satellite navigation signals. This section is devoted to some of them, especially in the automobile domain. The first is called ADAS (Advanced Driver Assistance Systems) and is aimed at applications concerning

- Automatic cruise control;
- Automatic control of headlamps;
- Warning of dangerous bends;
- Aid for night vision;
- Reduction of fuel consumption;
- Braking assistance and trajectory correction;
- Collision avoidance at intersections;
- Lane crossing prevention (in case of falling asleep);
- Warning of speed limit.

An example is the adaptation of headlamps to road conditions. In bad conditions, that is, fog, heavy rain, and so on the headlamps can automatically, depending on the road's profiles, be adapted to help the driver. For example, at an intersection the beam can be enlarged, on the highway the range can be increased, and under a bridge the beam narrowed. The data required to achieve such features are either in the map database or are transmitted from the infrastructure to the navigation system of the car.

Another example, also taken from the automobile domain, concerns the road services available through future telecommunication links. It could include, for instance;

- Communication between the car and an information center;
- Communication between vehicles and the infrastructure; or
- Communication from car to car.

◼ 11.7 CLASSIFICATIONS OF APPLICATIONS

There are many ways to classify applications, depending on the criteria one wants to choose: accuracy required, coverage needed, technique employed (dual frequency, autonomous, phase or code measurements, one or multisystem), autonomy, and so on. The classification that is often suggested by the professionals in the GPS sector is aimed at helping the user to be directed to the right product, whereas the classification obtained from the official leaflets of the Galileo program is clearly aimed at describing all the possibilities of the future GNSS in order to create revenues.

Another classification could be made according to the type of entities that are equipped, and it appears that the main classes are humans and transportation vehicles. The only exceptions to this are related to time-dependent applications that are a little specific. A domain interested in the "instrumentation" of objects is the transportation of goods. It is already possible to follow a parcel from sending to reception and to know its location[26] by Internet access. The technique currently used is based on RFID. In the near future, if there is enough demand, it will be possible to know any object's location. Projects are currently dealing with dangerous goods, but an extension to all sorts of objects is easy to imagine, if not easy to implement.

Still another way to classify relies on the number of frequency bands that are being used in the receiver and whether it is implementing a differential approach or not (although autonomous receivers with EGNOS or WAAS are differential). As described in Chapter 3, two frequency bands are common to both GPS and Galileo (E5a/L5 and L1). Mass-market receivers, which are certain to equip objects on a large scale, will certainly, at least until the advent of complete software receivers, remain single frequency on L1, as this frequency band includes the highest number of signals. Professional receivers will, on the other hand, test the various constellations and frequency band combinations; there are quite a few possibilities if one includes the three frequencies of GPS, the three frequencies of GLONASS, and the four frequencies of Galileo.

---

[26]In fact, it is rather the last known location.

■ 11.8 PRIVACY ISSUES

Besides technical issues, the localization of people and goods is bound to raise some questions relating to privacy. Although it is already possible to follow someone through his mobile phone or credit card, using a mobile terminal able to locate a person with an accuracy of a few meters is a difficult idea to accept. In addition, the terminal will have been designed for this purpose, as opposed to mobile phones or credit cards whose primary functions are not localization.

Well aware of this potential difficulty of usage, together with the possible wrong use associated with such personal data, the European Union has proposed, in its 2002/58/CE directive dated July 12, 2002, an article (n°9) in relation to this concern. It drew up some points relative to the use of localization data not directly linked to telecommunication networks (that are needed by the mobile networks in order to transport calls). The substance is as follows. The use of localization data should be based on

- Anonymity or acceptation;
- Limited use only for the associated service and for the right duration corresponding to this service;
- Need of the service provider to inform the user how localization data should be used (in case a third party is involved, for example);
- Possibility for the user to temporarily withdraw his acceptation, freely and easily, for each transaction;
- Access to these data should be restricted to people acting under the provider's authority.

From these guidelines, it is easy to formulate a few rules based on "common sense," as follows:

- No one should be localized without their knowledge, ... unless in case of emergency!
- The localization data should be destroyed once the corresponding service has been delivered.
- Any transmission to a third party should be clearly stated to the user.
- In the case where this third party is not directly involved in the service, the acceptation of the user should be gained in advance.

It is quite easy to imagine situations where all these guidelines are not really applicable. The case of breakdown service subscription is not one of them, where being asked for acknowledgment will be acceptable and will also help you to be confident that your request has been taken into account. The case of a geographical based mobile game is totally different. If this means that you have to accept pop-up messages every 10 s, it will largely decrease the interest of the game. Studies on acceptance levels and the fears of various types of people are certainly needed in order to understand the possible usages and inherent limitations of positioning systems.

FIGURE 11.8 Handheld receivers. From left to right: Magellan eXplorist series 100 and 600 (used with permission, © 2007 Magellan Navigation, Inc.); Silva Multinavigator (© Silva Sweden AB), and Garmin rino (Garmin International).

■ 11.9 CURRENT RECEIVERS AND SYSTEMS

As for application classifications, it is possible to present current receivers in many ways. We have decided to choose the following plan (the major companies providing these receivers are also listed):

- *Mass-market handheld receivers.* (Garmin, Magellan, TomTom, ViaMichelin, Mio, and so on);
- *Application-specific mass-market receivers.* (Garmin, Magellan, TomTom, and so on);
- *Professional receivers.* (Trimble, Javad, Septentrio, u-blox, NovAtel, NordNav, Ashtech, Thales, Leica, and so on);
- *OEM receivers.* (u-blox, Garmin, Trimble, and so on);
- *Chips.* (Trimble, Global Locate, Alcatel, Infineon, and so on).

Mass-Market Handheld Receivers

These are general purpose, autonomous receivers (see Fig. 11.8). Some include additional sensors, such as micro barometers in order to provide both a meteorological forecast[27] based on pressure change as well as a local altitude. It can be used indoors, for example, to help in defining the floor level as the typical accuracy is around 1 m in the vertical.[28]

---

[27] This feature is useful when hiking.

[28] Note that this is only valid for local measurements once a reference altitude has been defined, at the bottom of a building for example, and gives acceptable accuracy while atmospheric pressure does not change (over a limited amount of time).

The receiver's performances are equivalent for all the systems although some technological advances sometimes occur from one generation to the next, as happened with the SiRF III chipset. The difference in prices, which can go from one to ten, is mainly due to the packaging: cartography of one or many countries, the size of the screen, black-and-white or color screen, software suite available, additional sensors, access to services, and so on. The Garmin receiver in Fig. 11.8 has an interesting feature that integrates GPS capabilities with communication (walkie-talkie mode), thus allowing typical "find a friend" behavior.

As a first demonstration of possible integration, some years ago, Casio sold a wristwatch including a GPS receiver. The main problem, besides the integration performance, was the actual autonomy of the system while running GPS: about 45 min.

In 2004, Garmin (see Fig. 11.9) proposed a new type of wristwatch designed for runners: it includes both navigation capabilities and an assistant to improve or just check your performance in real time. It is based on the knowledge of your location and time of running; thus it can tell you if you are ahead or behind compared with a previous run.

The first non-GPS devices (except PC) that allow navigation were PDAs either linked to a GPS receiver through special card drivers (MMC, SDIO, and so on) or Bluetooth. The following generation incorporates the GPS antenna directly into the PDA (see Fig. 11.10 left) and the latest one (to date) also integrates the GPS antenna into a mobile phone (see Fig. 11.10, right).

The current trend is clearly to push forward the integration of all functions into a single device that is indeed a mobile phone. Thus, an HP6550, which runs under Windows® Mobile 5, includes a mobile phone, a camera, an IR port, Bluetooth, WiFi, a complete set of PDA functions, and an Assisted-GPS chipset. These platforms are nevertheless quite difficult to handle for developers as the rate of renewal is extremely rapid. Operating systems are changing regularly and chipsets too. Thus, the efforts made are not necessarily usable for the next generation, available

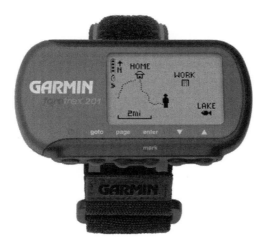

FIGURE 11.9 Wristwatch receiver. (Garmin International.)

(a)   (b)

FIGURE 11.10   Mobile phone and (**a**) PDA receiver (**b**). (Garmin International.)

a few months later. An example of this is given by the Imco'sys mobile terminal, which is oriented towards mobility by allowing a smooth transition from mobile network to WiFi-based communication and integrates a GPS, all this under linux (in order to facilitate compatibility with next generations).

Application-Specific Mass-Market Receivers

The automobile market is one of the two major domains that are considered as being the future of GNSS revenues, either through receiver sales or through specific road tolling applications, for example. The current market offers a wide range of products including sophisticated maps, two- or three-dimensional views,[29] complex routing algorithms, turn-by-turn navigation, both with the possibility to specify the means of transport, including traffic information, and so on. Well-known manufacturers are TomTom, Magellan, Garmin, and so on. Figure 11.11 gives the picture of some commercial products.

Some products are also designed for specific applications, like the speed limit controller that tells the driver when he drives over the maximum allowed speed, and also can warn of changing speeds zones. In such a case, to lower the cost,

---

[29]Note that this is currently not real three-dimensional visualization as no digital ground representation is used. Thus, this is only an aerial view of the map. Japanese companies are already working towards a real three-dimensional representation, including buildings and service stations, for example. For buildings, it is currently only the outside shape that is of interest.

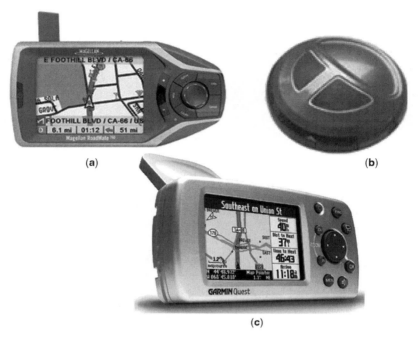

FIGURE 11.11 Automotive receivers. (**a**) A Magellan MRM760 (used with permission, © 2007 Magellan Navigation, Inc.); (**b**) an Inforad radar detector (Inforad Ltd); (**c**) a Garmin iQuest (Garmin International).

which is mainly due to maps, displays, and routing and navigation algorithms, some have no display at all apart from LEDs and digital display. This includes cartography of roads together with the associated speed limits. It can also help in telling the user where fixed speed traps are located.[30]

Maritime receivers have the specificity to operate in "good" environments (although navigation on inland waterways is sometimes not that good). The major concerns were then associated with maritime maps and with integration with other sensors, like radar for detection purposes or sonar for water depth evaluation or fish-finding. A typical screen is shown in Fig. 11.12.

The aviation sector can be seen as a two-part domain: civil aviation with big planes and tourism aviation with small planes. In the case of the second category, a few specific receivers have been designed, mainly with a focus on ease of use while flying and providing dedicated screens.

Professional Receivers

There is a real gap between mass-market receivers and professional ones. The number of professional units sold is much smaller and this explains the higher

---

[30]Especially in France, where more than 2000 such devices have been recently deployed!

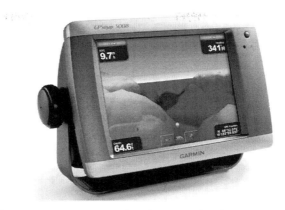

FIGURE 11.12 A maritime receiver. (Garmin International.)

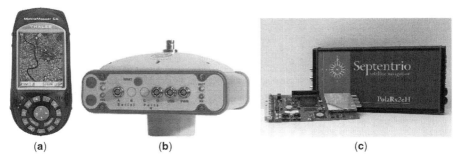

(a) (b) (c)

FIGURE 11.13 Professional receivers. (**a**) A Mobile Mapper CE (used with permission, © 2007 Magellan Navigation, Inc.); (**b**) a Maxor GGDT (Javad Navigation Systems); and (**c**) a PolaRx2eH (Septentrio).

prices, but the features are really different. They are either associated with robustness or specific environmental conditions or linked to the achievable performances in terms of accuracy. Some examples are given in Fig. 11.13, which shows respectively a Geographical Information System oriented receiver from Magellan, a dual-constellation dual-frequency receiver from Javad, and a dual-frequency accurate to the decimeter from Septentrio.

In addition to these high performance receivers, there is an increased availability concerning the software receivers whose superiority relies on the Galileo ready feature for the L1 common band of GPS and Galileo. As the main processing is achieved through PC-based software, only this part of the receiver will have to be updated once the Galileo signals are fully defined and available. In this sense, as happened with EGNOS, where EGNOS-compatible receivers were available long before operational validation, it is obvious that Galileo receivers will be available at an early stage.[31] An example of such a receiver is the NordNav R series, shown in Fig. 11.14.

---

[31]There are already full L1 GNSS enabled software receivers.

FIGURE 11.14   The NordNav R-30 software receiver. (NordNav.)

The really interesting feature of the software approach is the flexibility to upgrade the receiver. Furthermore, the cost of the receiver is expected to drop compared to classical FPGA (Field-Programmable Gate Array) architectures. In addition, the increased power of mobile terminals is a predictive indicator of the growing achievable performances of such receivers.

Original Equipment Manufacturer (OEM) Receivers

Original Equipment Manufacturer (OEM) is the term used to define equipment that is designed to be integrated in more complete systems. There are currently two types of such receivers: cards and embedded modules. They are able to output typical GNSS data streams (messages) that allow developers to design specific applications. Typical messages are under the NMEA (National Marine Electronics Association) norm, or are proprietary ones. The current level of integration is given by a typical total size of $1 \times 1 \times 0.2$ inches, to which one must add the antenna (see Fig. 11.15 left). Larger cards are usually available for sophisticated dual-frequency receivers.

(a)   (b)

FIGURE 11.15   The Antaris module (**a**) and associated chipset (**b**). (u-blox.)

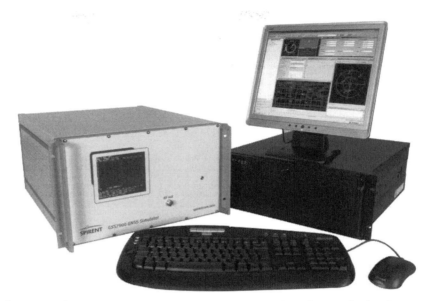

FIGURE 11.16   A Spirent GSS7900 GNSS simulator. (Spirent Communications.)

Chipsets

The final category is the one used for integration in specific terminals such as mobile phones of car navigation systems. These chipsets exhibit the same performances as other receivers and are currently usually composed of two chips: the radio part and the digital processing part (see Fig. 11.15).

Constellation Simulators

The last category of instruments specific to the satellite navigation sector includes equipment that has been designed for receiver or application development purposes: constellation simulators. The most well-known company providing such simulators is Spirent Communication, which proposes a large range of products, from one satellite generator to a complete GPS and Galileo 24-channel simulator (see Fig. 11.16).

The aim of these systems is to provide developers and engineers with realistic and repeatable signals in order to facilitate the perfection of receivers' algorithms or new architecture design. It allows the definition of scenarios that can then be reproduced identically as many times as the user wants. It is then possible to establish a normalized specification procedure for various environment types and to evaluate the receiver's performances. This can also be used for comparison of receivers.

■ 11.10 CONCLUSION AND DISCUSSION

The real questions concerning the scope of the applications are dependent on the industry and providers rather than users. In fact, there is no risk in asserting that users will take advantage of the development of positioning systems whatever happens.

Although not yet mature, the technology will evolve in one way or another and application fields will cope with the actual limitations. On the other hand, telecom operators are equally not in trouble, because when available, it is clear that services associated with localization will require data transfer; and they will be deeply involved at that time. Of course, increasing the rate of use of their networks is important (and directly linked to increased revenues), but the benefit of location-based services is not obvious for them, thus it is not currently a priority.[32]

These questions are in fact related to the GNSS industry and GNSS officials that want to show both users and telecom operators the potential of this domain. In the case of Galileo, it is even more critical, because the Public–Private Partnership and the Galileo Operating Company (GOC) are based on forecast revenues.

As a conclusion, the efforts spent in trying to find new applications and the competition thus created both increase the benefit to the user communities, which will see lives change in the near future. Chapter 12 is aimed at discussing some of the first elements of these possible changes.

A few parameters have to be considered and classified when designing a positioning system; accuracy, reliability, availability, and coverage are certainly some of the most important, but a complete system is much more complicated than only these features. In terms of cost, the terminal and the infrastructure required, if any, are a fundamental concern. The same applies to ease of deployment or ease of use, or even ease of handling and tuning. The following sections are aimed as discussing accuracy, availability, and integrity.

Accuracy Needed

Table 9.4 shows a summary of achievable performances of various techniques but does not deal with other system parameters. Looking at this table, some techniques appear to be really very powerful and accurate. The cost is often not even correlated and some very cheap approaches, like RFID for instance, may be the perfect solution. This is true unless the real needs of the application are considered. Accuracy cannot be considered alone. Which types of environments are needed? With at coverage? In real time? The technical solution implemented will vary considerably, with the same accuracy required, but considering also these complementary aspects. In some cases, although all ranges of accuracy between 1 cm and 100 m are achievable, no real implementation would be possible as these side parameters are indeed the real constraints.

Without going back over the discussions of Chapter 9, although dependent on the applications, the accuracy is also a function of the environment. Ideally, it should be better in "difficult" environments where visual landmarks are more numerous and more complex to apprehend — the guidance could then be more judicious. Unfortunately, with GNSS, this is exactly the opposite way round, leading sometimes to unexpected performances.[33] This is the case indoors too, where current

---

[32]It will certainly be totally different once one operator provides its customers with an attractive location-based application. Then, all the others will have to follow quickly.
[33]From the user's point of view.

technologies available, such as Assisted-GNSS or High-Sensitivity GNSS, allow "light indoor" positioning, but exhibit worse performances than outdoors. However, the opposite would have been required as usual indoor dimensions are of a few meters. An accuracy that is not compatible with the automatic determination of floor level is not acceptable for many indoor applications.

Availability and Coverage

Availability and coverage are not the same thing: the first aims at assessing the percentage of time, in one way or another, the positioning is possible for a given system and the second aims at defining the conditions to be followed in order for the system to output a location. Typical features are, for instance, the fact that GNSS signals are 100% available but the location calculation is impossible indoors, thus leading to a coverage limited to places where at least three or four signals are visible. For a typical mass-market user, availability and coverage are almost identical and result in limited coverage.

This limited coverage is one of the disadvantages that make application development very difficult. The lack of localization continuity is a major drawback and has led to a lot of research on hybridization of techniques, namely GNSS for outdoors and GSM/UMTS for non-covered areas or WLAN, UWB, and so on. This too is not yet fully acceptable because of the real discrepancy between the positioning performances achievable by technology. It is quite difficult for a user to cope with the very different accuracy figures relevant depending on the techniques used, which is furthermore transparent (or should be) to the final user. In that sense, a coupling between GNSS and UWB has a real meaning as it provides coherent accuracies for both outdoors and indoors and can be almost transparent. The coverage thus realized is of good quality. The last point, which brings us back to the previous section, is the complexity of deployment of a full UWB indoor positioning system.

Integrity

The following comments have to be considered with the current limitations described in Chapter 8. Integrity is of prime importance for applications of life and death, but is difficult to integrate for mass-market users. In fact, the major point of interest to a mass-market user is the accuracy provided when the positioning is calculated by the receiver. The reliability indicator, which is the basic primary function of integrity, is not yet a major concern. It is bound to change a little bit with the combination of Galileo partners and emergency applications.

✔️ 11.11 EXERCISES

Exercise 11.1 Make a parallel between the evolution of techniques and the evolution of applications. For instance, concerning GPS, comment on the link between technical improvements and new applications. For example, what are the latest technical improvements of GPS receivers for car navigation? What was, for instance, the real purpose of the high sensitivity receivers now available?

Exercise 11.2 Describe the different GNSS positioning techniques with respect to their main user characteristics: accuracy, stand-alone, differential, time to first fix, coverage, availability, limitation, and so on. Once you have defined a set of applications (that you can freely choose), make the link between the technical approach to be implemented and the applications.

Exercise 11.3 From the first classification developed in Exercise 11.2, can you see an interest in providing a higher level of accuracy than today's, for mass-market receivers and applications? For what kind of applications and environments?

Exercise 11.4 What are the applications that would need GNSS positioning? What are the applications that could not work with GNSS positioning? In this latter case, what kind of system may be used?

Exercise 11.5 Can you imagine application requirements that are not met by any techniques proposed in this book? Specify thoroughly such requirements and describe whether achieving them would require the development of current techniques or a totally new approach. Give your vision of the technical method needed to resolve the problem.

Exercise 11.6 What is the main interest of a GNSS simulator? What are the main technical issues in achieving such a simulator? Describe, by analyzing further reading, the way it actually works.

Exercise 11.7 Let us consider a future L1 single-frequency receiver. Give your analysis of the performance of such a positioning device for personal applications (assuming the receiver is included in a mobile phone for instance). In addition, imagine that Bluetooth, WiFi, and UWB are also available. Would it be enough? Why? How would the positioning engine work?

Exercise 11.8 Propose some innovative applications that use positioning, whatever the positioning technique required. Compare with existing applications described in Chapter 11 and conclude regarding the necessary features of the positioning system. In which direction would you point an industrial team if this option was available to you? Answer the same question for a research team.

■ BIBLIOGRAPHY

Abwerzger G, Wasle E, Fridh M, Lem O, Hanley J, Jeannot M, Claverotte L. Hope beyond the hype — location based services and GNSS. InsideGNSS 2007; 2(4):54–63.

Ashkenazi V. Geodesy and satellite navigation. InsideGNSS 2006; 1(3):44–49.

Business in satellite navigation. GALILEO Joint Undertaking; Brussels; 2003.

Coordination Group on Access to Location Information for Emergency Services (CGALIES). Report on implementation issues related to access to location information by emergency services (E112) in the European Union. Available at: http://europa.eu.int. Nov 15, 2007.

Hanley J, Scarda S, Wasle E. Application of Galileo in the LBS environment (AGILE). In: ENC-GNSS 2006: Proceedings 2006; Manchester, UK.

Kaplan ED, Hegarty C. Understanding GPS: principles and applications. 2nd ed. Artech House; 2006. Norwood, MA, USA.

Liu J, Hasegawa K, Wakamori M, Ogawara K, Ronning M, Fay L, Hayashi N, Osborn B, Duguid D. E911 implementation for automotive application. In: ION GNSS 2005: Proceedings; September 2005; Long Beach (CA).

Onidi O. Directions 2004. GPS World 2004; 15.

Rowell J, Sabel H. ITS applications using GNSS–digital mapping industry opportunities. In: The European Navigation Conference, GNSS: Proceedings; 2003; Austria. Graz; 2003.

Schleppe JB, Lachapelle G. GPS tracking performance under avalanche deposited snow. In: ION GNSS 2006: Proceedings; September 2006. Forth Worth (TX).

Swann J, Chatre E, Ludwig D. Galileo: benefits for location-based services. In: ION GPS/ GNSS 2003: Proceedings; September 2003. Portland (OR).

Web Links

http://www.magellangps.com. Nov 15, 2007.

http://www.septentrio.com. Nov 15, 2007.

http://www.spirentcom.com. Nov 15, 2007.

http://galileo-in-lbs.com. Nov 15, 2007 should be closed end of 2008.

http://www.nasamarine.com. Nov 15, 2007.

http://www.silva.se. Nov 15, 2007.

http://www.garmin.com. Nov 15, 2007.

http://www.javad.com. Nov 15, 2007.

http://www.u-blox.com. Nov 15, 2007.

http://www.leica-geosystems.com. Nov 15, 2007.

The Forthcoming Revolution

The eleven preceding chapters have made a review of numerous possible positioning techniques and the associated performances and limitations. In this final chapter we will consider the positioning techniques that exist and that are widely deployed, and we will try to answer the following question: How will our daily life be changed? At the end of the chapter, some possible technical directions are proposed that will have to be developed in order to make the positioning available on a large scale, in a lot of environments. Note that these techniques are not necessarily based on worldwide systems but rather on the use of telecommunication exchanges on a wide scale.

This chapter deals with the changes in the organization of our daily lives that could be induced by the availability of positioning. No hypothesis is made concerning the technical solution (or solutions) that could be used in order to provide the positioning; assumptions are made that it is available in every kind of environment for both people and objects and that it is also possible to have this information transferred to any place or to any person (via a telecommunication network). The first part of the chapter is oriented towards reflections concerning the evolutions of the perception of both time and space, and the possible link between these evolutions.[1] In fact, there are similarities between the evolution of time and what could be the evolution of the use of positioning. The conclusion of this first part is that people often try to optimize their time and positioning can certainly be a valuable aid in that aim. The second part deals with the evolutions that can already be sensed in today's applications, such as transport or Location Based Services (LBS). The third part deals with the possible revolution induced in our daily life by positioning — cases of a student and a district nurse are taken as illustrations. Finally, a brief description is given, as a conclusion

---

[1]Note that having an individual time has been available for many years, whereas having the knowledge of an individual position is in its infancy.

and glance forward, to what could be the approaches of positioning that one could imagine for the future.

■ 12.1 TIME AND SPACE

The perception of time has changed a lot over the centuries, to the current omnipresent availability of a precise time that can thus be shared by everybody. By briefly analyzing the evolution of the effects of this availability of time on people's lives, some parallels are drawn concerning possible changes induced by the availability of positioning.

A Brief History of the Evolution of the Perception of Time

At the very beginning, time and space were notions that people felt — the number of days of walking needed to reach a given place and drawing simple maps of places. This was achieved long before writing was available.

With the increasing diversity of his activities, a human being has increased both his living space and the need to measure time in order, for example, to better organize commercial activities. The lunar calendar appeared to help in this task, in that the observation of the phases of the moon was enough to give a date. Unfortunately, this was limited for activities such as agriculture, which relies more on annual cycles. Then solar calendars appeared that allowed the collective organization of the activities of society. The notion of a year and months was already present. Furthermore, it was quite precise enough for seasonal activities. Further improvements were rapidly required in order to divide the day into time units to organize activities within a given day. The initial approaches were based on the sundial, but the obvious problem was that the duration of a unit of time is not the same in every season; thus, a day time unit lasted longer in summer than in winter. Ingenious water clocks (clepsydras) were devised to solve this problem, which also had the advantage of making the time available at night. Time became available, and the next steps were to make it both transportable and synchronized from place to place.

The monks were the first to develop "clocks" in order to synchronize their religious practices. The first achievements were based on rings and gongs. Here, the interesting point is the fact that it allowed for synchronization for a whole group of people (those that heard the bells), but knowing the precise time was absolutely not required.[2] Universal time was nevertheless not yet a worry as life revolved around local affairs. Furthermore, the night remained "another world," but it was acceptable to use the Sun for time. The evolution was to develop clocks that were able to "ring" various time of the day, even without a dial and hands. The most advanced such clocks were also able to ring at night in order to organize the whole life of the village.

---

[2]Note that this notion could be interesting in the case of positioning: there is no permanent need for knowing precise positioning, as long as it is possible to know the path to be followed and maybe the time it will take to reach the next stopping place.

The next step in the management of time measurement and restitution was the advent of mechanical dials that allowed people to "locate" themselves within the day. Representations are used (often based on religious or astronomical symbols) in such a way that even those that did not read were able to understand the time. All the mechanisms used at that time were based on gravitational effects, meaning that it was not possible to use them at sea (this leads us back to Chapter 1).

Meanwhile, western countries started to expand their influence around the world, where difficulties appeared for commercial activities and synchronization. The first trains were in operations, but clocks were still synchronized on the Sun mid-day and time drifts were "visible." Trains raised the need for a coordinated universal time and this was the starting point for time zones.

The industrial revolutions brought about a change in attitudes towards time. Work was no longer related to the task but to a given amount of time. New relationships were created between employers and employees. New claims arose concerning the rights of workers, who sometimes organized strikes in support of their claims. Industry realized that "time is money" and life itself became defined in relation to time. In addition, time became a global notion, shared worldwide. With this globalization appeared the (paradoxical?) need for an individual time-keeper; everybody needed to be synchronized with the rest of the world, or at least with his professional and personal neighborhood.

Over the last few centuries, time has clearly increased its ascendancy in men's activities. Financial transactions are nowadays fundamentally based on time, and the Internet and all telecommunication networks must be synchronized.[3] Almost every action is quantified in time (and hence in money). At work this is clear, but also for travel, either professional or personal, leisure, entertainment, and so on. In the development of time measurement, these has also been the disappearance of the mechanisms that were the visual demonstration of time passing. Some displays no longer have hands but give digital values.

Comparison with the Possible Change in our Perception of Space

As described in Chapters 1 and 2, the representation of the Earth has changed a lot over the centuries. As time was being synchronized around the world, there was also a need for more accurate representation of the world in terms of maps, routes, and so on. Note that although many different needs are at the origin of this requirement, time is certainly one of the most important. The world's activity is largely based on time, so it is very important to be able to evaluate the time needed for any given trip, either of people or of goods.

If we try to make a comparison between the evolution of time measurement and the evolution of positioning systems, it is certainly possible to say that positioning is today in the situation that time faced more than 150 years ago with the advent of portable clocks. This was the technical feat that allowed the appropriation of time by everybody. The equivalent in positioning is now available with satellite-based

---

[3]Even if global synchronization is not required in the case of GSM networks, for example.

positioning systems (thanks to the pioneer, GPS). A few features of the first portable watches and basic GPS receivers of today have parallels: the similar approach of needing an identical referential worldwide, the availability of a personal local measurement, and the possibility to synchronize[4] with anybody else using a similar device. In addition, time and position are closely linked in GNSS and this feature will help bring together the two aspects.

There is another technical achievement that is fundamental for the dissemination of portable positioning devices and their incorporation into everybody's lives: telecommunications. When someone uses the time read on a wristwatch, it is automatically shared with others as the uniqueness of the common referential is enough. This is absolutely not the case for positioning. Even when considering a shared geographical referential, the position is a specificity for one person. In order to share this information with others, there is the need to communicate this information. This is why the advent of both positioning and telecommunications is sure to provide a wide development in positioning (maybe on a similar scale to what happened for time).

In the scope of this evolution, it is possible to consider that positioning could be profitable in domains such as ubiquity, or in other words the automatic discovery of anyone's environment, or also in group management. For ubiquity, it is clear that if the positions of all people and objects were easily available, in all possible environments, and at almost no expense, the environment of everybody could be discovered. The telecommunications required are almost available today, although not positioning (but this is another story). An extension of this could be that people would need to define themselves with some criteria that would lead to belonging to a group of like-minded people. The abovementioned discovery of environments could then be to find, from a geographical point of view, people or objects that belong to your group (or any other group). This is currently being implemented in the LBS community, with applications such as "find a friend" or "find a point of interest." The idea is to extend these first applications to everything.

When compared to the evolution of time and its impact on society, it is even possible to imagine that positioning could be used in many other ways (considering positioning as the combination of positioning and telecommunications). Knowing how people are moving around in the city,[5] it is possible to organize the "waves" of movement and then to define the policies to be followed by the town council in terms of roads, infrastructure, and public transport, for example. This could also be an interesting tool in forecasting increases in pollution and probably in the reduction of these and their consequences. This leads us to transportation. The health and safety authorities could also use positioning-related devices. Emergency calls are already in use, but one can imagine that the abovementioned group management approach could be part of the management of any emergency call. For example, if somebody fell ill in the street, an alert could be transmitted to people who are geographically close and who have been identified as competent in this medical field. This raises the

---

[4]"Synchronize" either relates to time or to position.
[5]The proposed concept can easily be applied to a country or even to the world, as well as to smaller structures like a district or inside a company.

problem of definition and access to the corresponding information files, and also to privacy issues, but could be one direction of future developments.

Of course, transportation will certainly remain a predominant domain as it is in essence highly demanding of positioning. Nevertheless, great modifications could occur in the coming years due to environmental considerations. In this field positioning could also be a central component. Applications currently under development that will be available in the near future are briefly described in the section on "transportation" that follows, but let us imagine future possible uses of positioning for transportation. Current worries worldwide are about ecological matters and transportation is of major concern. Positioning could be used to help reduce the energy consumption of the western world. A first approach would be to allow a complete analysis of the traffic and the behavior of users in order to reorient public policies in terms of public systems, either locally, regionally, or on a nationwide scale.[6] Furthermore, in order to give a tool to every citizen, positioning systems could be used to calculate the real energy consumption for any given activity. As transportation is used for numerous reasons, this approach could be spread to all activities, from business to leisure or to shopping. This could also be used to rank, from an environmental point of view, the various events, from international conferences to Olympic Games.[7] A first implementation could be to add up how much energy every person has used to come to the event. Knowing the distances traveled, the time taken, and the various means of transport used, each person could evaluate his own energy equivalent value. Adding up all individual values will allow the final evaluation of the energy cost of the event. This could then be used in order to reduce this cost for future events. Of course, the ideal would be to be able to make a predictive calculation to help define the best place to organize the event, the best associated transport policy, and so on. But prediction is another domain (a further discussion is given in the section "first synthesis," which follows, on this point).

In the same direction, one could imagine that companies also could be evaluated with a comparable "carbon footprint." This would be the addition of the energy consumption of its activities, but also the consumption of its employees' energy equivalent values while coming to work, moving around for their work, and so on. This looks like the "permit to pollute" allocated to European companies in recent years, but this could be used in two steps. For the first stage, this can help the company to evaluate its practices and to take positive steps. For the second stage, some kind of regulation could be envisaged to reduce the energy equivalent value. This could be achieved on a company scale or on a local, regional, or nationwide scale.

First Synthesis

From this very rapid analysis of the changes in time and space, some interesting comments can be made, as follows.

---

[6]It is also possible to think on a worldwide scale but it would probably take a little bit more time...

[7]It is on purpose that large international events are considered, to highlight the great implications of such approaches. Of course, it is certainly easier to implement similar solutions on a reduced scale (city, district, or even inside a company or a block of houses).

Humans, at least in developed countries, are mainly interested in optimizing their time. The advent of portable clocks was a considerable move in this direction, but the advent of positioning is certain to provide the second revolution towards this same goal.

There are close similarities between time and position, in that they are concepts that can be shared with others (although positioning needs telecommunications in order to achieve this), that they are both global (that is, synchronized in time and using a unique referential for all users, either a coordinated time or a geographical representation of the Earth), and that they are bound to modify our daily organization.

It is not necessary to "understand" the physical measurements, or even to know about them. As for modern digital clocks that have no hands or dial, positioning devices are not always required to display the position. As a matter of fact, the display is required only when you need to see your location, but will disappear as soon as the whole system is clever enough to give you only the correct relevant information. For example, you do not need to know your position and those of your friends as long as you know they are about ten minutes away from you, although this information can be obtained through a positioning system.

Time and positioning are certainly required for environmental applications. These measurements can help in evaluating, precisely measuring, and implementing real-time solutions to this difficult problem to come. In addition, they could help establish the reality of the worldwide situation.

Although positioning systems allow real-time measurements, the demand will certainly be for predictive methods. It is interesting to measure in real time that somewhere there is a problem, but what we really need is a tool that also gives some help to solve the problem. In the case of positioning systems, this could be achieved, in the future, by having a close link between positioning and telecommunication systems in order to allow the definition of behavioral models that could be used for prediction. This kind of model could be obtained from large measurement campaigns.

■ 12.2 DEVELOPMENT OF CURRENT APPLICATIONS

Based on positioning technologies currently under development, it is possible to give a brief summary of a few domains where applications are likely to use new features in the near future. The technical part is mainly based on GNSS and hybridization, with inertial systems, RFID, or WLAN to overcome the limitations of GNSS (that is, indoors).

Transportation

An interesting point concerning the evolution of the availability of time was that, at the beginning, it was available on a collective level (the ringing of church bells). Only with the advent of portable devices did it become individual. Positioning currently acts the other way round — portable GPS are available and provide valuable applications on a personal level (autonomous navigation). Some already foreseen applications will expand the use of positioning to a collective level, thanks to the telecommunication capabilities of modern systems.

In the scope of "individual" applications in the automobile domain, ITS (Intelligent Transport Systems) will provide a set of aids to the driver, based on the position of the car. For example, car headlights might change automatically as a car approaches an intersection, or depending on the speed of the car and the type of the road (motorways, inside cities, or in the countryside). It is also possible to have this positioning information coupled with rain or night sensors (which already exist on many cars) in order to take into account additional information. Among others, the information on the profile of the road can be efficiently used to prevent drivers from taking a bend too quickly or to help the automatic cruise control to adapt the speed and reduce the petrol consumption when climbing a hill.

Another aspect of ITS relies on the possibility, in the near future, to set up communications between vehicles. In such a case, a driver would be alerted in the case of potentially dangerous road configuration (for example, another car arriving at too high a speed at the next intersection or a car parked ahead on the side of the road at night). The concept of "ubiquity" is once again considered when combining positioning and telecommunications. The possibility to develop systems such as collision avoidance or a sensor for a driver falling asleep could be envisaged. Automatic "intelligent" cruise control depending both on the positioning (the type of road, the current speed limit, and so on) and the environment (rain, night, other cars, and so on) can then be developed.

Another individual aspect of car navigation is that everybody has his own habits, based on experience, "natural" learning, and personal preferences. This means that one driver will make a detour with ten bends to avoid a traffic light while another one will prefer to go in a straightforward manner. This cannot easily be taken into account in current navigation systems (lack of cartography information, complexity of the displacement preparation, and so on). The approach currently under development is to have systems that will "learn" the habits of the driver and incorporate them into the algorithms.

On the collective level, positioning data are used in order to define traffic conditions. This means accessing individual data such as position, speed, and direction of movement, and maybe also the route being followed. Once gathered, these data could be sent to a central server that will then, assuming the amount of data is comprehensive enough, compute and evaluate the real conditions: heavy traffic, accident, fire brigades in operation, police intervention, road works, and so on. This requires that a sufficient number of cars are equipped and that the telecommunication infrastructure is in place. Furthermore, the downlink from the server to all the users has to be defined.

Positioning systems could also be of assistance to buses, trains, and planes. Concerning buses, an important aid of positioning systems would be to give information to passengers. The bus stop will certainly be a central point for information, preparing the journey and schedules. Interconnection with other public transport systems could also certainly be valuable (metro, tramways, and so on). Concerning trains, a lot has already been achieved, but the optimization of power consumption depending on the profile of the track, tourist information to passengers depending on the location of the train, direction of where to go in a railway station, and so on, are probably the next steps. Air traffic is a target of GNSS systems due to the high reliability level required for planes. This would be valuable for any positioning system. Of course, it could help reduce the bottleneck in airports by providing a better organization of flights, but it could also be interesting for passengers to optimize their movements inside the airport and to manage their transfer time. Telecommunications, if organized, could facilitate the whole organization of stopovers, and so on.

Cartography

There are two aspects that will be improved in the near future (a few years from now) concerning cartography: the representation of the environment and the estimation of the time of arrival of a planned trip.

Everybody knows of the difficulty for a user to locate himself from a two-dimensional representation of his environment (a map). This comes from the need to both locate himself on the map and to represent the map in the real environment. A first approach could be to display a representation in three dimensions of the buildings, the roads (together with traffic lights and road signs), and probably notable places (monuments, parks, and so on). This would be of great help for the attractiveness of navigation systems. Such representations are given in Fig. 12.1.

A first step could be a diagrammatic representation of buildings, then a more sophisticated one using virtual reality. A second step, already available on a limited scale, consists of having photographs of the front views of building and pasting them on to the virtual representation; this gives a very realistic display of buildings. The final stage, in the longer term, would be to have real photographs or movies of the reality, in real time. What you see is where you are!

Of course, this could (and should) apply to all possible environments where positioning is intended to be available, that is, outdoors but also indoors.

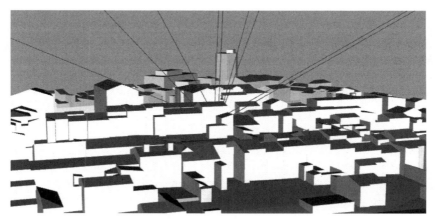

FIGURE 12.1 Example of three-dimensional displays for navigation. (*Source*: Ergospace.)

For indoor navigation, only preliminary studies have been carried out, mainly by research teams or for architectural purposes. The transcription to navigation is a huge job; as many attributes are very different from outdoors:

- Pedestrians are likely to be located everywhere indoors;
- Map-matching algorithms are certainly quite different (although probably very interesting in order to improve the reliability of positioning);
- The floor level must be considered (current outdoor maps are primarily two-dimensional projections); and so on.

Location-Based Services

Location Based Services (LBS) are considered to be the second major domain of applications, after transportation. All the foreseen possible activities are described in Chapter 11: find a friend, follow colleagues, know the position of relatives (or dependents), emergency calls, geo-advertising, and so on. If one considers that positioning is available, all these services are bound to find interested customers, ready to pay for them. The fact these services are not already widely deployed is probably because the corresponding positioning techniques are not so easy to implement with acceptable performance (refer to Chapters 9 and 10 for details).

Considering that positioning is available can really broaden the scope of applications. Once again, telecommunications are involved and allow some exchanges to be made automatically. The availability of positioning could certainly be included in this loop in order to carry out certain repetitive tasks. For example, one could think of an automatic opening of the door of the garage when your vehicle approaches it. For home applications, one could also imagine that the central heating automatically switches from the idle mode to the "comfort" mode when you leave your office (considering your trip back home will take 45 minutes). Of course, if you are not planning to go home directly from work, you could easily change this automatic scheduling by a simple phone call. This principle can be applied to numerous daily activities.

One could also imagine that your position could be obtained by other means (not necessarily your mobile phone). In such a case, it could be imagined that if your mobile is down or you have left it at home, the mobile network, knowing your location, could enable you to be reached on a phone other than yours. This means that your location device is also equipped with transmission capabilities. In a similar way, you could be warned if one of your family or friends is in your vicinity. But this is a little bit more intrusive to your private life!

■ 12.3 THE POSSIBLE REVOLUTION OF EVERYBODY'S DAILY LIVES

Let us imagine that the technical side is available and that positioning of people and objects is as easy as knowing the time. What could the changes be in our daily life? How could this have an impact on our behavior, every day? Take two situations, respectively days in the lives of a student and a district nurse, and let us try to imagine such modifications.

A Student's Day

From Waking Up to Departure: Morning at Home

Regarding breakfast, no real fundamental change is likely to occur due to positioning systems.[8] Things are quite different concerning the preparation of the journey from home to university. It would be quite easy to display on a large flat screen all the positions of the people, transport systems, and objects we will have to deal with in the morning. Thus one could display buses that go to the railway station; and one could then synchronize oneself with the appropriate bus. Traffic jams are also displayed and alternative traveling modes are proposed. In addition, a friend could appear on the display. He has taken his car this morning and is ready to come and pick you up. Of course, today this is particularly interesting for you because you have to carry heavy documents. Note that the weather forecast is also information that could be taken into account.

From Home to University: Transport Optimization and Scheduling

An accident occurred on the usual route, so you are informed that you will have to make a detour; this hitch will be profitable to another friend, who is now on your new route and you can contact him and offer to pick him up. He accepts but is in no hurry as his first course only starts in two hours. This will be the chance to carry out some research at the library. Moreover, he can already book a seat in lecture room A from 9 to 10 a.m.

Arriving at the university, you are automatically guided to the best available car park corresponding to your morning lectures. Of course, the system can also discuss with you the situation where your car has three people with three different courses at three different locations. In addition, you may choose to leave your car at the central car park and use the public transportation system (which is less polluting); an information system can help you in your journey, depending on your current location. The major achievement of this global displacement system is to be able to cope with multimodality; you can use a car, a bicycle, or be on foot and you will be given ideas concerning your trip. You are, of course, free to switch the device off.

The Morning Session at the University

The class is planned to start at 9:00 a.m.: the lecturer is ready but the video projector of the amphitheater is down and the logistics people are not available because they are all busy organizing the welcome meeting for an international conference. This is in fact not a problem because the lecturer can display on his PC screen the location of all available video projectors for this morning. Furthermore, knowing the amphitheater's location on the campus, the system reduces the search to the

---

[8]Unless the breakfast is delivered by some modern milkman who is late...

nearest video projectors. Thus, the lecturer is able to know that a projector is available, book it and make this information available to everybody.

All the students are present at 9:00 a.m., as usual (!), but a colleague of the lecturer invited some industrialists and the mayor to this lecture, the subject of which is important for city activities. All are informed of the 10 minute delay and can carry on with their discussions.

During the lecture, many slides are used, but also very interesting drawings on a white board as well as valuable comments, both from the lecturer and the audience. Everything is recorded through cameras and microphones and only people that are present in the amphitheater are allowed to download the corresponding files in their electronic devices. The security of these transactions is ensured through the location of the users. Similarly, one can also think of the possibility of asking the lecturer a question that he could record on his terminal and organize for future answers.[9]

At Lunch Time

During the lecture, the outside world has not stopped and the dynamic organization of your lunch-time session progresses. The friend you were supposed to eat with has left the campus for personal reasons and was unable to tell you; you are informed through your location alert system that detects your friend's position is not compatible with your agenda. This is achieved through messages that are available only on-demand in order not to disturb you during the lecture.

Thus you can know about the progression of the people you are interested in and can decide to organize your lunch with them. Of course, privacy issues are of prime importance. The notion of belonging to groups could be a way to cope with privacy: one person could be a "member" of a few groups, or all the groups (and hence be "visible" for everybody), or none of the groups (and hence not be "visible").

For leisure activities, meteorological conditions could be taken into account, in order to determine whether the sailing is confirmed or not. Thus, knowing the locations of others and their movements, dynamic appointments could be managed. In the case of a typical lunch, restaurant availability could also be taken into account for the choice, as well as the time required to go to the restaurant, depending on the transportation means to be used. The "synchronization," both in space and time, of all the participants could also be considered.

In the Evening and at Night

For the organization of the evening, the way back home, and the night, similar approaches could be imagined, once again based on the knowledge of the location of people and objects, associated with additional information; such as the weather or the traffic. Evening work could certainly also take advantage of location information; knowing that such and such a group of friends is gathering together to revise physics or history is potentially interesting.

---

[9]Note that research teams are working on such scenarios. Currently, location is considered to be available either with GNSS, WLAN, or sensor networks. The real implementation is not the major concern of these works.

Comments

Note that some of the concepts described above are widely expandable to many situations. Keeping track of friends or colleagues can easily be transcribed to children or, the other way round, to parents or grandparents. Hospitals could also be interested in such a feature, as would tour operators: this could allow greater independence and also increased security. The notion of belonging to groups appears to be a wide concept, applicable in many various situations.

It is interesting to understand that the main feature of location-based approaches is the dynamically reconfigurable environment. It is also remarkable that the same question (for example the best way from home to university or from university to home) can give totally different answers depending on many parameters: weather, traffic, public transport systems, friends, luggage, other activities planned during the day, and so on. This is clearly possible through the use of both positioning and telecommunications.

A District Nurse's Day

Preparation of the "Journey"

Our district nurse has a hard day: many patients to visit, specific medicine to pick up, one of which should only be prepared one hour before being administered. Thanks to the availability of the positioning of people and objects, the organization of the journey is quite different from what it used to be. There are no more fixed appointments that had to be made one week in advance and many late changes that lead to a waste of time for patients. The way it is organized is the following: all the patients that our nurse planned to visit today have been put in a single group. Usually, all patients can have access to the nurse's position but not to the other patients' positions. Nevertheless, in some cases, as in this example, all patients can access the other patients' positions.

The order of the journey is not defined before the early morning because too many outside constraints are likely to induce modifications (traffic, weather, and the availability of medicine primarily). Once the information is gathered by the nurse, the journey is settled with the possibility to plan dynamic meeting places. For example, some medical acts can be carried out in the nurse's van, which includes a specific place for this purpose. The availability of the positioning of all the patients to the nurse is the fundamental piece of information that allows the dynamic meeting feature. Of course, time synchronization is carried out with the pharmacist too.

A first version of the planned day is sent to all the patients. However, they know that this is not set in stone and they have the possibility either to tell the nurse of a change in their own scheduling or decide to live their life quietly knowing that dynamic modifications are possible. Everybody is a little bit freer.

Displacements and Automatic Re-Scheduling

All the features that have already been described in the case of the displacement of the student are also available here. If the van is not required for the displacement, the nurse could choose to use public transport, and once again, meeting points could be arranged with patients using these transportation systems.

An interesting point concerns the dynamic meetings; optimization techniques and prediction models are used in order to give travel time information to both the nurse and her patient. This is like a missile and target approach, where the missile has to go to the predicted position of its target at time of impact, and not to the position of the target at current time. In our present situation, this is similar. Dynamic appointments have to reduce waiting times (hence "wasted" time) for both the nurse and her patient. Of course, one can choose to spend some time in the library or at the museum waiting for the nurse, but this is a parameter of the meeting that one can easily configure.

Finally, new patients could be added to the nurse's journey in the case of an emergency, even if it leads to the cancelling of a planned visit. This could be put off till the next day. Nobody is frustrated as this is the usual way it works and no one has waited for nothing.

Reports

Positioning could also be useful for the end of day reports of our nurse. She has to send her conclusions and the results of some analyses of her patients to the hospital or to the doctor. This transmission could also be achieved more efficiently using positioning; when driving back home and passing in front of the hospital, a secured transmission link is established that allows the downloading of her reports. Similarly, when passing in the vicinity of some doctors' offices, she can automatically send the corresponding reports. Moreover, if she passes close to someone who is a member of the hospital administration that is in charge of receiving the reports, such a transfer can be carried out. Of course, this requires secured transmission and verifications in order to avoid loss of reports or multiple transmissions.

Objects

Positioning is nowadays thought of very often in terms of people, but locations of objects can also be interesting. The video projector was an example, but mobile equipment in a hospital could be located and its use optimized. Following parcels is already a feature available with large delivery companies, for which positioning relies on restricted-area RFID-based detection of the parcel. This is a valuable service provided to customers, but remains quite limited to specific areas. Improvement is possible through permanent localization. This could certainly additionally allow dynamic pick up of parcels. If yours remains for some time in a warehouse close to you (or close to one of your displacements), you could be invited to collect it. The result is an economy of energy and time. Let us come back to the projector (or any other piece of equipment) — knowing the place it will have to be available next time, one could reduce the energy needed for its displacement, and there is no longer the need to take it back to the central office to put it in storage.

Pushing the fiction one step further, it is possible to imagine vehicle displacement in cities in a completely different way from today. Imagine that pollution is at such a high level that cars are no longer allowed in cities. Everybody is obliged to use small electric cars that can be rented in a suburban car park. They are the only means to enter the city, apart from public transport systems. These cars are rented in a very different way from today, in that the rates depend on both the time one keeps the car and the distance traveled. Once your displacement is finished, you can either keep the car at your disposal (and pay for it) or leave it wherever you want. Then, another person, knowing the exact locations of free electric cars, could choose to rent it for a new displacement. Reduced fees are applied when you take the car back to a charging center. Free parking places are displayed on a screen in the car, and information is available concerning all the services available in the vicinity of your route (charging centers as well). By programming your trip to a given destination in town, you could additionally make it available to any person who would like to share the cost of the car on the basis of a shared trip. Dynamic appointments could thus be organized in order to further decrease the total energy consumption of the global displacement of people in the city. This can of course be applied to movements of all professionals and parcels (or objects).

Ideas in Development

The applications described above can be classified in the following domains:

- Belonging to groups of like-minded people;
- Sharing the availability of objects;
- Ubiquity;
- Energy saving;
- Dynamic time management and sharing;
- Real-time and predictive models.

For almost all these approaches, it would already be possible to implement telecom-based systems, without positioning, but the availability of positioning provides a tremendous simplification of the deployment.

Some applications could probably mix the different domains mentioned above. For example, dynamic groups are easily imaginable — you walk along the street and, due to the notion of ubiquity, that is, the ability to discover your environment in relation to your position,[10] and the notion of groups, you are linked to other people with similar interests. Imagine you are close to a museum and an announcement is made through the wireless network (whatever it is) that there is a 30-minute guided visit on Ancient Egypt. You can choose to join the newly created group that will attend the seminar. The same could apply for a cinema or

---

[10]Ubiquity is a concept in computer science and telecommunications that is based on the ability to discover your environment while moving. This discovery is mainly based, today, on the possibility to create a wireless radio link between transmitters and receivers. Note that this is a kind of relative positioning, not the GNSS absolute positioning dealt with in this book.

the city hall, with specific movies or municipal meetings. The main idea is thus to be open to your geographically close environment.[11] Of course, groups are private or public, free or profitable, personal or professional, and so on.

The real-time capabilities of organizing (or rather re-organizing) a day are once again something we are not used to. The possibility to share resources in order to reduce energy, pollution, time, and money is worth thinking of in the years to come. Furthermore, the availability of positioning for people and objects could help in both managing the "tides" and developing predictive models that could then be used to reduce congestion and its impact. This could also be used in order to orient public (or private) policies in numerous domains (transport, energy, construction, social affairs, education, and so on).

All these ideas are certain to modify our behavior and our perception of our environment, if the corresponding techniques are implemented and the philosophy accepted and appropriated by users.

■ 12.4 POSSIBLE TECHNICAL POSITIONING APPROACHES AND METHODS FOR THE FUTURE

The GPS has paved the way for the positioning revolution, as portable clocks did about two centuries ago for the time revolution. It is difficult to imagine what the future of positioning will be but let us remember some "fundamentals":

- There is a large number of applications that could take advantage of positioning, with as many performance specifications.

- There are currently numerous techniques of positioning, applying in almost all environments, but not having the same level of global performance.[12]

- GNSS (GPS indeed) have been designed to allow positioning in environments where no infrastructure was available. The extension of their use in cities, for example, has been driven by the goal to reduce the positioning infrastructure only to existing satellites. This is not the way GPS was thought of initially.

- The advent of huge telecommunication capabilities makes the future easier to imagine.

- Ubiquity is sought by large communities (such as positioning, computer science, telecommunication, and so on).

This book has given the overall limitations of the different positioning systems. It appears clearly that there is currently no single means of positioning that is acceptable for all applications; moreover, hybridization approaches are being thought of in order to fulfill all the specifications for many applications. This is even more the case for

---

[11]One could of course choose not to be reachable for such purposes or, on the contrary, to be included in many different groups in order to be aware of all events.
[12]By performance, one has to understand positioning performance, but also the ease of deploying the system, the complexity of the terminal, and so on.

applications concerning people, such as numerous location-based services. So the questions about future positioning systems are quite natural. Will the GNSS be extended to cover environments not yet available to these techniques? Will hybridization be developed to the point where it works transparently? Are networks of sensors the best solution?

For the next 10 to 15 years, it is probable that GNSS systems will have a growing part in our lives, but this is not so sure for the long-term future unless alternative solutions are found to overcome the real limitations of GNSS. The current methods, which consist of obtaining the absolute coordinate of a location with reference to a global frame with an accuracy indicator being given in meters, are perhaps not the only solution. For example, some positioning methods deal with so-called "symbolic" algorithms that rely on local positioning given in terms of rooms or zones. The accuracy is then an approximation of the number of individual rooms in which the user might be located. In addition, the specification that is being sought is the reliability of the positioning. It is of less importance to have an accurate position, but it is of uppermost importance to be able to rely on the position.

There is another way one could imagine the future of positioning. First, we can consider that for the near- and mid-term future, GNSS are almost ideal candidates for positioning where no infrastructure is available at all: at sea or in the desert, but also in the countryside where the coverage is very good. However, in cities or indoors, satellite-based positioning is not well suited, even considering all the efforts of the positioning communities to find a solution. In these environments, there are a lot of fixed objects distributed everywhere, that is, either mobile objects, such as personal computers, or fixed objects, such as doors, bus stops, or numbers on the front of houses. Usually, telecommunication features are associated with objects and people who need to exchange data, but one could easily imagine broadening the scope of telecommunication-equipped objects.[13]

Once an object is able to communicate it can also transmit its position, as long as it knows it. Depending on the range of the wireless transmission system used, all receiving electronic terminals located in the proximity of this object could, at least, know that they are not far away (at a maximum distance that is precisely the range of the telecommunication system of the object) from the object. This is indeed the basic concept of ubiquity in telecommunications; it allows the object to discover its environment. If the telecommunication capabilities grow sufficiently, this could be the real next step for positioning. Imagine one mobile terminal that cannot achieve GNSS positioning (because it is indoors for instance) but equipped with a communication capability. It is linked to another device that can transmit information gathered by its environment.[14] Gradually, the mobile terminal can obtain information about the geographical distribution of all the points in the network (this requires new approaches of data processing taking these aspects into account). As soon as one element in the network knows its own position, deductions can be carried out in order to determine the possible positions of other points. In such a way, it could be possible to obtain the positions of many points, knowing the exact locations of only a few.

---

[13]In a similar way in which RFID is intended to develop.
[14]In this context, "environment" means the "telecommunication environment."

Furthermore, it is imaginable that most fixed objects could be defined by their location, either indoors or outdoors. Moreover, one can easily imagine that some fixed objects could obtain their location through the use of GNSS-based systems. In such a case, the ultimate positioning system would be able to exchange data as well as detecting the proximity of two objects (transmitters). Only fixed objects would be equipped with GNSS (because of the reduced cost and the automatic configurability of the infrastructure). The location of a device is obtained through exchange with close objects, which, most of the time, know their own location. When this is not the case, more sophisticated algorithms are required (maybe inducing an increased uncertainty of the positioning).

An example of such a system could be the positioning of a pedestrian through his mobile phone, which is linked to the mobile phone, of the car driver going along the same street. As the car is equipped with a GNSS system, the mobile phone of the driver has an accurate location available. When passing close to the pedestrian, position data transfer is possible and gives information about the position to the pedestrian's mobile phone. The car is an obvious way to obtain the position, but one can think that bus stops, entrances of buildings, traffic lights, or also doors, windows, lights, and so on (for indoors) know their position. This idea brings us back to the initial aspects of GPS: positioning where no infrastructure is available and not in cities, buildings, and so on.

The situation is certainly a little different in developing countries where the type of infrastructure described above is probably not going to be available in the near to medium term. In such a case, satellite navigation systems are the enabling technology that will allow these countries not to be left behind. Applications would certainly be oriented more towards the transport domain (GNSS and SBAS are already envisaged to provide small airports with landing capabilities without requiring expensive radar, for example) or managing energy, resources, and population. Here, global positioning systems and their availability and coverage are probably the technology that will make numerous developments possible.

■ 12.5 CONCLUSION

As is often the case, the growth rate of some positioning-related applications (car navigation for instance), has led to the desire to have GNSS everywhere. This wish has rapidly been moderated by the real limitations of GNSS availability and coverage for numerous applications. Alternative or complementary positioning systems have to be found in order to develop the global domain of the location-based applications. Many everyday actions and behaviors are bound to be largely modified by the availability of the positions of both people and objects. The main goal of modern countries is almost always based on optimizing time (or rather reducing time wasted), and knowing the geography of one's environment is potentially an interesting plus. Brief ideas of such typical modifications are given in this chapter. It appears that current positioning systems nevertheless need further development in order to improve overall performance, mainly in availability and coverage. Current applications (car navigation, agriculture, leisure, geodesy, first location-based services, and so on) are perfect illustrations of this. They are limited to the performance of the positioning systems they are implementing. Current efforts are being carried out in order to broaden the spectrum of positioning systems: hybridization between GNSS and inertial, GSM/UMTS, WLAN, UWB, and so on, but also pseudolites and other radio beacons. It appears that when no infrastructure is deployed, GNSS have an incomparable advantage, and its development for the mid-term future is certain.

✅ 12.6 EXERCISES

Exercise 12.1 Describe the evolution of the time measurement systems and the corresponding modifications in society. In comparison, where would you put the current state of the development of positioning measurement systems?

Exercise 12.2 One of the first uses of time was collective, to allow people to join together in front of the church, for instance. Can you imagine a similar approach with modern positioning systems? Give details.

Exercise 12.3 What is your analysis of the current set of applications in the Location Based Services (LBS) domain? Give the reasons why they are not yet more widely spread.

Exercise 12.4 By analyzing a given application ("car navigation" or "personal friend finder," for example), try to extract features that are not yet provided by current positioning systems. Describe the technical means required in order to implement these features.

Exercise 12.5 Telecommunication is of prime importance in the development of positioning-based applications, mainly because the real need is to exchange positioning data. What are the main (high level) specifications required for these telecommunication transmissions? Can you describe how to use current GSM/UMTS, WPAN (Wireless Personal Area Networks, such as Bluetooth or UWB), WLAN (Wireless Local Area Networks such as WiFi), and WMAN (Wireless Metropolitan Area Networks such as WiMax) characteristics?

Exercise 12.6 List all the daily actions that could take advantage of positioning (consider your own case). You can, for example, split your time between a typical working day and weekend. From this list, make a classification of already possible applications and those that require further developments. Describe these developments.

■ BIBLIOGRAPHY

Abdel-Salam M. Natural disasters inference from GPS observations: case of earthquakes and tsunamis. In: ION GNSS 2005: Proceedings; September 2005. Long Beach (CA).

Fuller R, Grimm P. Tracking system for locating stolen currency. In: ION GNSS 2006: Proceedings; September 2006. Forth Worth (TX).

Heinrichs G. Personal localisation and positioning in the light of 3G wireless communications and beyond. In: IAIN2003: Proceedings; October 2003. Berlin; Germany.

Jensen ABO, Zabic M, Overø HM, Ravn B, Nielsen OA. Availability of GNSS for road pricing in Copenhagen. In: ION GNSS 2005: Proceedings; September 2005. Long Beach (CA).

Kaplan ED, Hegarty C. Understanding GPS: principles and applications. 2nd ed. Artech House; 2006. Norwood, MA, USA.

Pateli A, Fouskas K, Kourouthanassis P, Tsamakos A. On the potential use of mobile positioning technologies in indoor environments. In: 15th Bled Electronic Commerce Conference; Proceedings; June 2002. Bled; Slovenia.

Rohmer G, Dünkler R, Köhler S, von der Grün T, Franke N. A microwave based tracking system for soccer. In: ION GNSS 2005: Proceedings; September 2005. Long Beach (CA).

Schaefer RP, Lorkowski S, Brockfeld E. Using real-time FCD collection for trip optimisation in urban areas. The European Navigation Conference; GNSS 2003; Austria; Graz; 2003.

Woo D, Mariette N, Salter J, Rizos C, Helyer N. Audio nomad. In: ION GNSS 2006: Proceedings; September 2006. Forth Worth (TX).

INDEX

Acceleration
  definition, 124
Accelerometers, 305
  GNSS, 126
  non-GNSS positioning systems outdoor
    techniques, 124
Accuracy table, 329
Acquisition techniques
  GNSS, 163
Active Badge system, 289f
ADAS. *See* Advanced Driver Assistance
    Systems (ADAS)
ADC. *See* Analog-to-digital
    converter (ADC)
Advanced Driver Assistance Systems
    (ADAS), 366
Aeronautical Radio Navigation Service
    (ARNS), 58, 70, 151, 182
AGILE, 359
  classification, 360
A-GNSS. *See* Assisted-GNSS (A-GNSS)
A-GPS. *See* Assisted-GPS (A-GPS)
Air traffic
  GNSS systems, 388
AIS. *See* Automatic Identification
    System (AIS)
All-weather landing systems, 18
Altimeters
  non-GNSS positioning systems outdoor
    techniques, 126
Altitude problems
  ellipsoid, 46f
Amateur radio transmissions
  non-GNSS positioning systems outdoor
    techniques, 102
American GPS, 57
Analog-to-digital converter (ADC), 217

Ancient classical triangulation
  non-GNSS positioning systems outdoor
    techniques, 98
Angle measurements
  stars, 34
Angle of arrival (AOA), 109, 284
  based positioning system, 300
  disadvantage, 110
  measurements, 284
  positioning, 301
  power level measurements, 110
  principle, 109f
Antaris module
  associated chipset, 374
AOA. *See* Angle of arrival (AOA)
Arabian navigation
  navigation, 5
Argos system, 114, 116
  non-GNSS positioning systems outdoor
    techniques, 115–116
ARNS. *See* Aeronautical Radio Navigation
    Service (ARNS)
Arrival-related mathematics angle
  calculating position techniques,
    246–247
Artificial satellites
  navigation and positioning history,
    19–22
Ashtech GPS sensor, 139f
Assisted-GNSS (A-GNSS), 155t, 222, 303,
    312, 319
  configuration, 313f
  GSM-like techniques, 314
  indoor positioning, 312–313, 333
  light indoor positioning, 377
  positioning, 314t
  systems, 314

Global Positioning: Technologies and Performance. By Nel Samama
Copyright © 2008 John Wiley & Sons, Inc.

Assisted-GPS (A-GPS), 273
  FCC regulations, 361
Astrolabe, 7
Astronomical definitions, 43
Astronomical positioning
  accuracy, 12
  ancient times, 12
  navigation and positioning history, 11
Augmentation systems
  calculating position techniques, 258
Automatic Identification System
      (AIS), 348
  display screen, 349f
Automobile
  GNSS systems, 397
  positioning systems, 387, 394
  receivers, 372f
  vehicle displacement, 394

Barometers, 305
  non-GNSS positioning systems outdoor
      techniques, 126
Bat System, 288f
Battuta, Ibn, 5
Beidou
  non-GNSS positioning systems outdoor
      techniques, 121
Beidou Navigation and Positioning
      System (BNS)
  China, 89t
Binary offset carrier (BOC), 70
Binary phase shift keying, 67
Bipolar phase shift keying, 72
Bi-static radar, 102
  principle, 102f
Block IIR-M satellite, 136f
Bluetooth
  based GPS receivers, 354
  domain, 293
  local area telecommunication
      systems, 293
  positioning, 295t
  RSS, 291
BNS. *See* Beidou Navigation and
      Positioning System (BNS)
BOC. *See* Binary offset carrier (BOC)
BPSK. *See* Binary phase shift keying
Branly, Edouard, 13

C/A code generation, 206f

Calculating position techniques, 235–269
  arrival-related mathematics angle,
      246–247
  augmentation systems, 258
  coordinate system, 237–238
  hyperboloids analytical model, 243–245
  interoperability and integrity, 259–261
  least-square method, 248
  multipath, 262–268
  pseudo range errors impact on computed
      positioning, 255
  PVT solution calculations, 235–250,
      237–238
  quantified error estimation, 253–254
  satellite geometrical distribution,
      256–257
  satellite position computations, 251–252
  sphere intersection approach, 239–242
  time calculation, 251
  trilateration, 236
  velocity calculation, 249–250
Cambridge Positioning System
      (CPS), 112, 314
Cameras
  non-GNSS positioning systems outdoor
      techniques, 100
Carbon footprints, 385
Car navigation systems, 354
  access to services, 356
  GNSS capabilities, 357
  natural learning, 387
  next generation, 358
  positioning data, 387
  tourist information, 357
Carrier phase
  measurements, 226, 227f
  principle, 226
  technique, 223
Cartesian referential, 246
  ECEF, 45
Cartography
  forthcoming revolution, 388
  navigation and positioning history, 11
CDDS. *See* Commercial Data Distribution
      Service (CDDS)
CDMA. *See* Code division multiple access
      (CDMA)
Celestial Equatorial Referential, 42, 44f
Cell-ID
  angle of arrival, 300

concept, 108f
GSM networks, 112
positioning system, 300
timing advance technique, 113f
triangulation, 294
China
    BNS, 89t
    COMPASS global navigation satellite
        system, 40
    Galileo, 123
Chinese satellite based worldwide
        positioning system (COMPASS), 26
    constellation, 119
    global navigation satellite system, 40
    GPS systems, 123
    navigation satellite evolution, 122t
    non-GNSS positioning systems outdoor
        techniques, 121
Choke-ring antenna, 193f
Clocks
    GNSS-based indoor positioning,
        320–322
    H1 and H4, 32f
    religious practices, 382
Coastal navigation, 2
    fifteenth century, 30
Code division multiple access
        (CDMA), 68, 164
    FDMA, 178
    GLONASS, 173
    GPS, 23–24, 165
    techniques, 170
Code pulse related measurement approach,
        242
Cold War, 22, 38
Commercial activity synchronization, 383
Commercial Data Distribution Service
        (CDDS), 84
Commercial Service (CS), 76
Commercial tracking satellite
        services, 87t
COMPASS. *See* Chinese satellite based
        worldwide positioning system
        (COMPASS)
Compass bearing technique, 98t
Complexity table, 331t
Constellation geometry
    good, 257
    poor, 257
Constellations, 157t

Correlation function
    early minus late shape, 218
Correlation result, 263f
Cosmicheskaya Sistema Poiska Avarinykh
        Sudov (COSPAS-SARSAT)
    beacons, 118
    non-GNSS positioning systems outdoor
        techniques, 117–118
    system, 117f
COSPAS. *See* Cosmicheskaya Sistema
        Poiska Avarinykh Sudov
        (COSPAS-SARSAT)
Costa PLL, 219f
Coverage table, 330t
CPS. *See* Cambridge Positioning System
        (CPS)
CS. *See* Commercial Service (CS)
Cycling
    concept implementation, 340f
    global systems, 341

Davis cross staff, 8
Dead reckoning, 12, 34
Decca systems, 16, 35
    non-GNSS positioning systems outdoor
        techniques, 104–106
    United Kingdom, 106
Delay locked loop (DLL), 217, 219f
    early-late architecture, 217f
    PLL, 221
Difference measurement technique, 228f
Differential correction commercial satellite
        services, 87t
Differential phase measurement technique,
        318
Differential techniques, 82
Digital signal processor (DSP), 214
Dilution of precision (DOP), 235
    interoperability concerns, 260
    terms, 256
Direct distance measurements
    problems, 243
Distance measurement principle, 103f
Distance measuring equipment (DME), 106
DLL. *See* Delay locked loop (DLL)
DME. *See* Distance measuring equipment
        (DME)
DOP. *See* Dilution of precision (DOP)
Doppler, 209, 209f
    based positioning technique I and II, 115f

Doppler (*Continued*)
  GPS satellites, 363
  locating systems, 117
  PLL, 220, 251
  receiver stages, 150
Doppler Orbitography and Radiolocation
    Integrated by Satellite (DORIS), 118,
    119
Doppler shift, 210
  curve, 37f
  measurements, 37f, 39, 150
  receivers, 310
  satellites, 222, 250, 310
  Sputnik, 36
  TRANSIT system, 150
DORIS. *See* Doppler Orbitography and
    Radiolocation Integrated by Satellite
    (DORIS)
Double difference measurement technique,
    229f
DSP. *See* Digital signal processor (DSP)
Dual frequency method, 230
Dual-frequency repeater approach, 336

Early-late correlation result, 263f
  modified, 263–264f
Early positioning techniques
  world discovery, 30
Earth
  definition of graphical representation, 11
  representation changes, 383
Earth Centered Earth Fixed (ECEF), 237
  Cartesian referential, 45
  referential, 44f
Earth-Centered Inertial (ECI), 43, 44f
Earth climate
  meteorological observation, 365
Earthquakes
  analysis, 364
ECEF. *See* Earth Centered Earth Fixed
    (ECEF)
ECI. *See* Earth-Centered Inertial (ECI)
Edge correlators, 267
EGNOS. *See* European Geostationary
    Navigation Overlay Service (EGNOS)
Eighteenth century travels
  navigation, 9
Ellipsoid
  altitude, 238f
  altitude problems, 46f
  latitude, 238f

longitude, 238f
  representation, 45
eLoran, 16
  modulation scheme, 104f
Ephemeris
  satellites, 147
Equigravity field surface
  graphical representation, 42f
ESA. *See* European Space Agency (ESA)
Europe
  companies permit to pollute, 385
  E112, 360
  MCC, 158
  satellite based navigation systems, 58–59
  UTM reference grid, 49f
European Galileo constellation, 65
European Geostationary Navigation
    Overlay Service (EGNOS), 25, 83,
    157, 195
  CPS, 314
  integrity, 365
  integrity concept, 262
  message structure, 259t
  method implementation, 154
  SBAS, 195
European Space Agency (ESA), 76, 85
European Union
  emergency call, 274
  Galileo, 88t
  GLONASS constellation, 26
  GNSS, 82
  satellite navigation system, 97
EutelTRACS system, 84
  Eutelsat (Europe/Mediterranean /
    Middle East), 90t
Explorer 1 project, 21f

FCC. *See* Federal Communication
    Commission (FCC)
FDMA. *See* Frequency division multiple
    access (FDMA)
FEC. *See* Forward Error Correction (FEC)
Federal Communication Commission
  satellite based navigation systems, 58
Federal Communication Commission
    (FCC), 58
  regulations, 109
Field-Programmable Gate Array (FPGA),
    374
Fifteenth century
  coastal navigation, 30

Fifteenth century travels
  navigation, 6f
FLL. *See* Frequency lock loop (FLL)
Forthcoming revolution, 381–398
  cartography, 388
  daily lives, 389–394, 390–391
  evolution and time perception history,
    382
  first synthesis, 385
  future technical positioning approaches
    and methods, 395–396
  ideas, 394
  location-based services, 389
  nurse's day, 392
  objects, 393
  space perception change, 383–384
  time and space, 382–385, 386–388
  transportation, 386–387
Forward Error Correction (FEC), 171
Four-band architecture
  GLONASS, 337f
FPGA. *See* Field-Programmable Gate Array
    (FPGA)
France
  global receiving stations, 116
  Lambert projection, 51f
  Lambert zones, 52f
Frequency allocations
  band combinations, 367
  GNSS, 60
  regional division, 61
  United States, 59
  WRC'97 in L1 band, 60f
  WRC'97 in L2 band, 59f
Frequency division multiple access
    (FDMA), 71, 164
  CDMA, 178
Frequency lock loop (FLL), 222

GAGAN. *See* Geo Augmented Navigation
    (GAGAN)
Galileo, 58
  aspects, 64
  24-channel simulator, 375
  China, 123
  codes, 172, 203, 203t
  constellations, 77, 251, 283, 364
  correlation functions, 211
  E1 signal components, 181f
  E5 signal components, 180f
  European Union, 88t

frequency band occupation, 78f
functions, 172
fundamental techniques, 201
GNSS, 367
GPS, 142, 147, 151, 170, 183t–184t,
    190, 310
ground segment, 134f
ground tracks, 137
ICDs, 251
modeling, 191
multifrequency concept, 338
navigation message, 177, 179t, 260
navigation signals, 169f, 170t
navigation systems, 76–79
parameters, 81t, 143t, 170t,
    185t–186t
planning, 64
rescue repeaters, 169
satellites, 246, 259
service allocation *versus* frequencies,
    78f
specifications, 203t
systems, 123, 143t
three-dimensional indoor positioning,
    340
two-dimensional positioning, 337
Galileo Joint Undertaking (GJU), 63
Galileo Operating Company
    (GOC), 376
Galileo project. *See* AGILE
Galileo signals
  complexity, 168
  components, 180f
  correlation function, 266f, 267f
  cycling concept implementation, 340f
  modulation schemes, 78
  multifrequency approach, 339f
  spectrum, 80f
Galileo Supervisory Authority
    (GSA), 63
Galileo System Time (GST), 179t, 196
  PTF, 196
General conversion approach, 53f
Geo Augmented Navigation
    (GAGAN), 26, 120
  India, 89t
  non-GNSS positioning systems outdoor
    techniques, 121
Geo caching, 362
Geodesic networks
  development, 41

Geodesy
    modern geographical positioning
        systems, 350
    positioning techniques, 40–42
Geographical positioning, 1
Geographical referential systems, 44f, 45
GIOVE-A satellite, 138f
GIOVE-B satellite, 138
GJU. *See* Galileo Joint Undertaking (GJU)
Global Navigation Satellite Systems
        (GNSS), 57, 72, 97
    accelerometers, 126
    achievements, 100
    acquisition techniques, 163
    air traffic, 388
    architectures, 208–222
    atmosphere measurements, 362
    car navigation system capabilities, 275,
        357, 397
    carrier phase measurements, 225–226
    channel details, 217–222
    codes, 204–205, 223–224
    community goals, 194
    compatibility, 274
    components, 131
    constellations, 258, 259, 281, 283, 334,
        358
    coverage, 362
    effects, 151
    European Union, 82
    field measurements, 189
    frequency allocation, 60
    functionality, 156
    Galileo program, 367
    global coverage, 274
    global positioning, 275
    GPS, 58, 123, 164–168, 275
    high level approaches, 212
    hybridization approaches, 398
    indoor positioning, 283, 315, 358
    ITU, 58
    limitations, 258, 396
    manufacturers, 266
    measurements, 227–230, 231, 363
    mobile telecommunications, 100
    navigation message, 251
    networks, 278, 331
    odometers, 126
    ongoing developments, 346
    outdoor, 377

    outdoors, 315, 327
    performances, 131
    phase measurements, 223–224
    positioning, 95, 100, 156, 208, 302, 312t,
        396
    precise point positioning, 231
    promotion, 346
    proposal, 315f
    pseudolites, 283
    radio, 213–216
    relative techniques, 227–230
    repeaters, 279, 283
    SBAS, 131
    sensitivity, 312t
    sensors, 323
    server, 312
    stations, 363
    techniques, 154, 279, 281t, 282
    technology, 346, 352
    telecommunications, 145, 396
    terrestrial application, 352
    time, 251
    time references, 353
    time synchronization, 145
    TOA, 111
    transmission, 201–207
    triangulation method, 275
    troposphere analysis, 363
    UWB, 377
    vehicle, 275
Global Navigation Satellite Systems
        (GNSS) descriptions, 131–158
    comparison, 142
    differential approaches, 153–156
    error sources, 153
    ground segments, 132–133
    modernization, 151–152
    positioning parameters, 142–151,
        147–150
    SBAS system description, 157
    services, 140–141
    space segments, 134–138
    user (terminal) segments, 139
Global Navigation Satellite Systems
        (GNSS) indoor positioning, 309–341
    A-GNSS, 312–313, 333
    clock bias approach, 320–322
    evolutions, 333–340, 341
    HS-GNSS, 310–311, 333
    hybridization, 314

pseudolites, 315–318, 333
pseudo ranges approach, 323–327
recap tables and comparisons, 328–332
repeaters, 319–328, 334–340
Global Navigation Satellite Systems
  (GNSS) receivers, 124, 281, 302, 304
  architectures, 208–222
  civil engineering, 351f
  navigation purposes, 349
  positioning modules, 355
  power consumption, 311
  pseudo range, 327
  telecommunication systems, 352
  topographical measurements, 350
  TTFF, 313
Global Navigation Satellite Systems
  (GNSS) signals, 124, 163–197,
  206–207
  acquisition, 201–231
  codes, 171–178
  compatibility, 182
  developments, 186
  error budget estimation, 194
  error sources, 187–194
  features, 377
  frequencies, 174
  Galileo, 168–170
  GPS, GLONASS, 164–167
  indoor availability, 341
  ionized particle collisions, 363
  location-related errors, 193
  multiple access and modulations
    schemes, 178–179
  problems, 152–153
  propagation modeling, 363
  propagation-related errors, 190–192
  pseudo ranges impact, 188
  recap tables, 183–185
  refraction index, 362
  SBAS contribution, 195
  signal mathematical formulation,
    180–181
  time reference systems, 195–196
  time synchronization related errors, 189
  tracking, 201–231
  troposphere, 363
Global Positioning System (GPS), 17
  accuracy, 71t, 356
  architecture, 66f
  bi-directional links, 66

CDMA, 23–24, 165, 173
24-channel simulator, 375
civilian users, 347
COMPASS systems, 123
compatibility, 53
constellations, 67f, 163–164, 251, 283,
  316, 334, 364
correlation functions, 172, 211f, 218
coverage limitations, 350
definition, 150
development, 151, 350
find friend behavior, 370
four-band architecture, 337f
fundamental techniques, 201
Galileo, 123, 142, 147, 151, 183t–184t,
  190, 310
global performances, 68
global signals, 311
GLONASS, 65, 70, 72, 75t, 78, 80f, 142,
  149, 171, 173, 177, 183t–184t, 190
GNSS, 58, 123, 164–167, 275
ground segment, 132f
ICDs, 251
integrity, 365
manufacturers, 356
maps, 53
modeling, 191
modernization, 156
navigation and positioning history, 23
navigation message structure, 176f
new applications, 377
parameter comparison, 81t
parameters, 185t–186t
positioning revolution, 395
PPS, 71
principles, 54
prisoners, 366
program development phases, 148
pseudo range measurement, 254t
repeaters, 320
SA, 140, 152, 188
sector classification, 367
specifications, 348
SPS, 71, 140
success, 57
system parameters, 143t
technology, 397
TRANSIT applications, 283
two-band indoor architecture, 335f
two-dimensional positioning, 337

Global Positioning System
(GPS) (*Continued*)
unidirectional links, 66f
United States, 62, 82, 88t
UTC, 176
Global Positioning System (GPS) receivers,
212, 320, 322
civil engineering, 350
micro-barometers, 127
synchronization tool, 349
technical improvements, 377
world synchronization, 384
Global Positioning System (GPS) satellites,
25, 65–79, 142
Block IIR-M174, 174
Doppler measurements, 363
Galileo satellites, 259
GLONASS satellites, 195
navigation systems, 65–79
Global Positioning System (GPS)
signals, 167f
2003, 74f
2008, 74f
architecture, 68f
diagram, 73f
Galileo, 170
GLONASS signals, 166
L1, 206
L2, 206
LIC, 71f
reception, 309
SA application, 346
spectral representation, 69f
spectrum, 165f
structure, 70f, 71f, 166f
techniques, 68
Global Positioning System time (GPST),
189, 196
Global receiving stations
satellites, 116
Global system approaches, 35
Global theory, 246
GLONASS, 22, 209
CDMA, 173
codes, 223
constellations, 65, 163–164, 251, 283,
334
correlation functions, 211
development, 151
development phases, 24

developments, 75
European Union, 26
financial difficulties, 136
four-band architecture, 337f
frequency spectrum, 62
full operational capability (FOC), 135
fundamental techniques, 201
Galileo, 183t–184t
GG24 sensor, 139f
GNSS navigation signals, 164–167
goal achievements, 73
GPS, 24, 70, 72, 75t, 78, 80f, 142, 149,
171, 173, 177, 190
ground segment, 133f
ICDs, 251
modeling, 191
navigation and positioning history, 24
navigation message, 260
navigation message structure, 178t
parameter comparison, 81
parameters, 185t–186t
pilot tones, 176
quality, 172
renewal, 156
Russia, 62, 88t
satellite based navigation systems,
72–75
satellites, 195
SCC, 133
signals, 165f, 166, 167, 167f
spectrum, 165f
SPS, 141
structure, 167f
system parameters, 143t
two-band indoor architecture, 335f
UTC, 178
GLONASS-K, 137
GLONASS-M, 137
GLU. *See* Ground uplink stations (GLU)
GNSS. *See* Global Navigation Satellite
Systems (GNSS)
GNSS1, 80–84
GOC. *See* Galileo Operating Company
(GOC)
Goniometry, 107
GPS. *See* Global Positioning System (GPS)
GPST. *See* Global Positioning System time
(GPST)
Great Bear constellation, 121
Ground uplink stations (GLU), 158

GSA. *See* Galileo Supervisory Authority
    (GSA)
GSM, 282
  cell, 112, 304
  LORAN stations, 258
  mobile telecommunication networks, 277
  networks, 108, 112, 113, 331
  non-GNSS indoor positioning, 300
  prisoners, 366
  stations, 258
  techniques, 282
GST. *See* Galileo System Time (GST)
Gyroscope, 305
  accelerometers, 126
  non-GNSS positioning systems outdoor
    techniques, 125
  principle, 125f

Handheld receivers
  examples, 369
  meteorological forecast, 369
Hand Over Word (HOW), 242
Harrison, James, 1
Harrison, John, 11, 31
H1 clock, 32f
H4 clock, 32f
HDOP. *See* Horizontal DOP (HDOP)
Hieroglyphic inscriptions, 2
High Precision Service (HPS), 71
High sensitivity GNSS, 282, 312, 319
  GNSS-based indoor positioning,
    310–311, 333
  light indoor positioning, 377
  TTFF, 313
High sensitivity GPS, 273
  Assisted-GPS, 361
Homodyne conversion structure, 215f
Horizontal DOP (HDOP), 258
HOW. *See* Hand Over Word (HOW)
HPS. *See* High Precision Service (HPS)
Hybridization
  GNSS based indoor positioning, 314
  GNSS systems, 398
  non-GNSS indoor positioning, 302–303
Hyperbolic system approach, 16f
Hyperboloids analytical model
  calculating position techniques, 243–245

ICAO. *See* International Civil Aviation
  Organization (ICAO)

ICD. *See* Interface Control Document (ICD)
IGS. *See* International GNSS Service (IGS)
IGY. *See* International Geophysical
  Year (IGY)
ILS. *See* Instrument landing system (ILS)
India
  GAGAN, 89t
Indian Master Control Station
  (INMCS), 121
Indian Navigation Land Uplink Station
  (INLUS), 121
Indian Ocean
  commercial exchange routes, 4–5
Indian Reference Stations (INRES), 120
Indoor decision tables, 332t
Indoor local constellation
  pseudolite approach, 336
Indoor multipath configuration, 286f
Indoor navigation
  architectural purposes, 389
Indoor positioning
  non-GNSS indoor positioning,
    274–283
  problems, 333
Indoor pseudolite configuration, 316f
Indoor repeater based positioning system,
  323
Industrial revolutions, 383
Inertial positioning, 305t
Inertial systems, 304
  non-GNSS indoor positioning problem
    and main techniques, 304
Infrared positioning, 289
  global characteristics, 288
Infrared radiation (IR)
  definition, 287
  non-GNSS indoor positioning, 287–288
INLUS. *See* Indian Navigation Land Uplink
  Station (INLUS)
INMCS. *See* Indian Master Control Station
  (INMCS)
INRES. *See* Indian Reference Stations
  (INRES)
Instrument landing system (ILS), 106
Integrity concept
  MSAS, 262
  reliability indicator, 377
  WAAS, 262
Intelligent transport Systems (ITS), 387
  aspects, 387

Interface Control Document (ICD), 208
  Galileo constellations, 251
  GLONASS constellations, 251
  GPS constellations, 251
International Atomic Time (TAI), 195
International Civil Aviation Organization
    (ICAO), 261
International Geophysical Year (IGY), 19
International GNSS Service (IGS), 231
International spectrum conference
  satellite based navigation systems, 60
International Telecommunication Union
    (ITU), 58
  GNSS, 58
Ionized particle collisions
  GNSS signals, 363
Ionosphere propagation, 190
IR. *See* Infrared radiation (IR)
ITS. *See* Intelligent transport Systems (ITS)
ITU. *See* International Telecommunication
    Union (ITU)

Japan
  MSAS, 89t
  QZSS, 89t–90t
Japan Aerospace Exploration Agency
    (JAXA), 120
Javad
  multiconstellation products, 139
  TTGYG chip, 140f

Kalman filter parameters, 323
Kamal, 2, 3f
Kinematics mode, 231

LAAS. *See* Local area augmentation system
    (LAAS)
Lambert projection, 51f
  France, 51f
Lambert zones
  France, 52f
Landmark-based location evaluation, 33
Lasers
  non-GNSS positioning systems outdoor
      techniques, 99
  positioning system, 100
Latitude
  definition, 7
LBS. *See* Location based services (LBS)
L1C
  GPS signal structure, 71f

Least square method, 248
  calculating position techniques, 248
Levant, 4
LFSR. *See* Linear feedback shift registers
    (LFSRs)
Lighthouses
  navigation and positioning history, 11
  non-GNSS positioning systems outdoor
      techniques, 97
Light indoor positioning
  Assisted-GNSS, 377
  High-Sensitivity GNSS, 377
Linear feedback shift registers
    (LFSRs), 202
Linearization method
  satellites, 248
Line of Sight (LOS), 191
LNA. *See* Low noise amplifier (LNA)
Local area augmentation system (LAAS),
    83, 230, 316
Local area telecommunication systems
  non-GNSS indoor positioning, 291–299
Location-based approaches
  positioning systems, 392
Location based services (LBS)
  applications, 359f, 360f
  associated environments, 360f
  classification, 359f
  community, 384
  domain, 345
  evolution, 381
  forthcoming revolution, 389
  GNSS applications, 358
  modern geographical positioning
      systems, 358–359
  telecom services, 359
  transportation, 389
Location errors, 188
Longitude
  definition, 7
Long range navigation (LORAN), 23,
    36, 346
  antenna, 17f
  C coverage, 106f
  definition, 104
  GSM stations, 258
  non-GNSS positioning systems outdoor
      techniques, 104–106
  pulse shape, 105
  stations, 258
Long Term Ephemeris (LTE), 341

LORAN. *See* Long range navigation
    (LORAN)
LOS. *See* Line of Sight (LOS)
Low noise amplifier (LNA), 213
LTE. *See* Long Term Ephemeris (LTE)
Luminosity measurements
    non-GNSS positioning systems outdoor
        techniques, 101
Lux Trace system, 114

Magnetometers, 305
    non-GNSS positioning systems outdoor
        techniques, 126
Maps
    evolution, 8
    GPS compatibility, 53
Marconi, Guglielmo, 13
Maritime applications
    commercial activities, 348f
    modern geographical positioning
        systems, 346, 347
    rescue activities, 348f
Maritime currents, 3
Maritime receivers, 373f
Mass-market receivers
    professional ones, 372
Master control station (MCS), 121, 132
Matrix
    positioning approach, 112f
MCC
    Europe, 158
McClure, Frank, 22
MCS. *See* Master control station (MCS)
Medians, 41
Medieval maritime world, 4
Medium Earth Orbit, 77
Medium earth orbits (MEO)
    satellites, 137
MEO. *See* Medium earth orbits (MEO)
Mercator, 11
    projection, 11f
Meteorological forecast, 369
Microwave landing system (MLS), 106
    concept, 107f
Middle Ages
    navigation techniques, 6
Military maritime applications
    modern geographical positioning
        systems, 346
MLS. *See* Microwave landing system
    (MLS)

Mobile networks
    telecommunication, 278
Mobile phone, 371f
Mobile switching center (MSC), 111f
Mobile telecommunication networks
    GNSS, 100
    GSM, 277
    non-GNSS positioning systems outdoor
        techniques, 108–113
    UMTS, 277
Modern geographical positioning systems,
    345–377
    accuracy, 376
    applications chronological review,
        346–352
    application-specific mass-market
        receivers, 371
    atmospheric sciences, 362–363
    automobile navigation, 354–356
    availability and coverage, 377
    chipsets, 375
    civil engineering, 350
    classifications, 367
    commercial maritime
        applications, 347
    constellation simulators, 375
    under development, 366
    emergency calls, 360
    games, 362
    geodesy, 350
    individual applications, 353–362
    integrity, 377
    LBS, 358–359
    local guidance applications, 357
    mass-market handheld receivers,
        369–370
    military maritime applications, 346
    natural sciences, 364
    OEM receivers, 374
    prisoners, 365
    privacy issues, 368
    professional receivers, 372–373
    public regulatory forces, 365
    safety, 365
    scientific applications, 362–364
    security, 361
    tectonics and seismology, 364
    terrestrial applications, 352
    time-related applications, 348–349
    tourist information systems, 357
    TRANSIT, 346

Modulation schemes
  Galileo signals, 78
Mono-frequency receivers
  application fields, 350
MSAS. *See* Multitransport Satellite-Base
  Augmentation System (MSAS)
MSC. *See* Mobile switching center (MSC)
MTSAT. *See* Multifunctional transport
  satellite (MTSAT)
Multiconstellation, 338t
  multifrequency cycling method, 338
Multiconstellation products
  Javad, 139
Multifrequency approach
  Galileo signals, 339f
Multifrequency concept
  Galileo, 338
Multifrequency cycling, 338t
Multifunctional transport satellite
  (MTSAT), 120
Multipath evaluation curve, 192f
Multitransport Satellite-Base Augmentation
  System (MSAS), 83
  integrity, 262, 365
  Japan, 89t
  method implementation, 154
  United States, 195

National Geospatial Agency (NGA), 132
National Marine Electronics Association
  (NMEA), 374
Nautical navigation art
  logical evolution, 31
Navigation
  Arabian navigation, 5
  COMPASS, 122t
  data generation, 207
  eighteenth century travels, 9
  fifteenth century travels, 6f
  instruments, 2
  message, 166, 252
  origins, 1
  Phoenicians, 2
  Portuguese techniques, 5
  QZSS satellites, 119
  radio electric signals, 13–14
  signals, 119
  sixteenth century travels, 6
  three-dimensional display
    example, 388f

Navigation history, 1–26
  age of great navigators, 5–10
  artificial satellites, 19–22
  cartography, lighthouses, and
    astronomical positioning, 11
  first age of navigation, 1–4
  first terrestrial positioning systems,
    15–18
  radio age, 12–14
  real-time satellite navigation
    constellations, 23–26
Navigation land earth stations
  (NLES), 158
Navigation positioning
  data required, 175
  industrial communities, 279
  scientific communities, 279
Navigation satellite systems, 86t, 122t,
  186t–187t, 346
  positioning techniques, 36–38
  second generation, 39
  third generation, 40
Navigation techniques
  displacement measurement, 30
  Middle Ages, 6
  positioning, 29
  reflective mirrors, 8
NAVSTAR GPS, 24
NavTeq, 354
Navy Navigation Satellite System
  (NNSS), 22
Near-far effect, 334
Newton, Isaac, 8
NGA. *See* National Geospatial Agency
  (NGA)
NLES. *See* Navigation land earth stations
  (NLES)
NLOS. *See* Non line of sight (NLOS)
NMEA. *See* National Marine Electronics
  Association (NMEA)
NNSS. *See* Navy Navigation Satellite
  System (NNSS)
Non-GNSS indoor positioning problem and
  main techniques, 273–306
  global comparisons, 306
  hybridization, 302–303
  indoor environments, 285–286
  indoor positioning, 274–283
  inertial systems, 304
  infrared radiation (IR), 287–288

local area telecommunication systems, 291–299
measurements, 284
navigation application, 275
perceived needs, 276
pressure sensors, 289
radio frequency identification (RFID), 290
radio modules, 299
recap tables, 306
sensors network, 287–289
solutions, 279–282
techniques, 277–278, 284–286
ultrasound, 287
UMTS, 301
UWB, 296–297
wide-area telecommunication systems, 299–303, 300
WiFi (WLAN), 294–295
WiMax (WMAN), 298
WPAN, 296–297
Non-GNSS positioning system, 303
Non-GNSS positioning systems outdoor techniques, 95–127
accelerometers, 124
amateur radio transmissions, 102
Argos system, 115–116
barometers and altimeters, 126
Beidou and COMPASS, 121
cameras, 100
GAGAN, 121
gyroscopes, 125
ILS, MLS, VOR, DME, 107
large area without contact or wireless systems, 96
luminosity measurements, 101
magnetometers, 126
mobile telecommunication networks, 108–113
non-radio-based systems, 123–125
odometers, 125
optical systems, 97–101, 99
QZSS, 119–120
radio signals, 114
satellite radio systems, 117–118, 119
terrestrial radio systems, 101–115
WPAN, WLAN, WMAN, 114
Non-GPS devices
navigational capabilities, 370
Non line of sight (NLOS), 191
path, 285

Non-radio-based (NRB) systems
measurements, 284
non-GNSS positioning systems outdoor techniques, 123–125
Non-static radar, 102
NordNav R-30 software receiver, 374f
NRB. *See* Non-radio-based (NRB) systems

Observed time difference of arrival (OTDOA), 113
Octant, 9
Odometers, 305
GNSS, 126
non-GNSS positioning systems outdoor techniques, 125
OEM. *See* Original equipment manufacturer (OEM) receivers
OmniSTAR system, 84
United States (global coverage), 90t
OmniTRACS
United States (global coverage), 90t
Open cast mines, 351f
Open Service (OS), 76, 84
frequencies, 168
SoL, 169
Optical based calculation techniques
positioning techniques, 33–34
Optical positioning systems, 127
Optical systems
non-GNSS positioning systems outdoor techniques, 99
Orbital parameters, 252f
Original equipment manufacturer (OEM) receivers
modern geographical positioning systems, 374
OS. *See* Open Service (OS)
Oscillator drift considerations
PLL, 220
OTDOA. *See* Observed time difference of arrival (OTDOA)
Outdoor receiving antenna, 334

Parallels, 41
P code generation, 204f, 205f
PDA. *See* Personal Digital Assistant (PDA)
Pedestrian navigation module, 278
Pedestrian Navigation System (PNS), 281
Personal Digital Assistant (PDA), 354
receiver, 371f

Phase lock loop (PLL), 219
　aided DLL structure, 221
　DLL, 221
　Doppler considerations, 220
　Doppler measurements, 251
　oscillator drift considerations, 220
　structure, 220
Phoenicians
　navigation instruments, 2
Piezoelectric accelerometer, 124f
Pilot signals, 203
Pilot tones, 212
Plane projection form, 47f
PLL. *See* Phase lock loop (PLL)
PNS. *See* Pedestrian Navigation
　System (PNS)
Poldhu Station, 13
Pole Star, 3f, 34, 97
Polo, Marco, 5
Ponant, 4
Pop-up messages, 368
Portable positioning devices
　telecommunications, 384
Portolans, 4, 5f
Portuguese navigation techniques, 5
Position, velocity, and time (PVT)
　calculating position techniques, 235–250
Position error
　pseudo range errors, 255t
Position finding
　expanding mechanism, 237
Positioning
　availability, 382
　environmental applications, 386
　functionality, 156
　fundamentals, 395
　GNSS, 95, 100, 142–151, 156, 396
　history, 1–26
　hospital administrations, 393
　human evolution, 386
　hybridization approaches, 395
　Matrix, 112f
　navigation techniques, 29
　parameters, 142–151
　possible environments, 388
　predictive models for people, 395
　RFID-based detection, 393
　satellite based systems, 346
　technical solutions, 381
　techniques, 395

technology developments, 386
　telecom-based systems, 394
　telecommunication, 384
　transportation developments, 385
　transportation systems, 393
Positioning systems
　animal management, 364
　buses, 387
　location-based approaches, 392
　privacy issues, 368
　trains, 387–388
　weather forecast, 390
Positioning techniques, 29–54
　first age of navigation and longitude
　　problem, 30–32
　first navigation satellite systems, 36–38
　first optical-based calculation techniques,
　　33–34
　first terrestrial radio-based systems, 35
　forthcoming third generation navigation
　　satellite systems, 40
　geodesic systems, 53
　modern maps, 53
　navigation needs, 46–52
　reference systems, 42–44
　second generation navigation satellite
　　systems, 39
　world, 40–44
PPP. *See* Private Public Partnership (PPP)
Precise Positioning Service (PPS), 68, 254,
　365
　GPS, 71
　SA, 140
Precise Time Facilities (PTF)
　GST, 196
Pressure floor positioning, 290t
Pressure sensors
　non-GNSS indoor positioning problem
　　and main techniques, 289
Prisoners
　GPS systems, 366
　GSM systems, 366
　modern geographical positioning
　　systems, 365
Private Public Partnership (PPP), 63
PRN. *See* Pseudo random noise (PRN)
Professional receivers, 373f
PRS. *See* Public Regulated Service (PRS)
Pseudolite
　code techniques, 317f

GNSS-based indoor positioning, 333
indoor configuration, 318f
indoor positioning system, 319f, 333
phase-based techniques, 317
phase measurements, 333
positioning, 320t
repeaters, 328, 333
research programs, 328
synchronization problems, 317
Pseudo random noise (PRN), 202
code approach, 24
Pseudo range, 146
calculating position techniques, 255
errors, 255t
estimation of error sources, 194t
evolution, 324f
GNSS receivers, 327
GPS measurement, 254t
measurement configuration, 236f
theoretical aspects, 324
time differences, 340
PTF. *See* Precise Time Facilities (PTF)
Public-private partnership, 354
Public Regulated Service (PRS), 64, 76, 78,
79f, 122, 142, 169, 179, 365
PVT. *See* Precise Positioning Service (PPS)

Quantified error estimation
calculating position techniques, 253–254
Quasi-Zenith Satellite System (QZSS), 119
Japan, 89t–90t
navigation signals, 119
non-GNSS positioning systems outdoor
techniques, 119–120

Radar
non-GNSS positioning systems outdoor
techniques, 102–103
principle, 102
types, 102
Radio
electric signals, 13–15
GNSS signal acquisition and tracking,
213–216
goniometry, 15
modules, 299
navigation, 13–14
positioning systems, 127
systems, 97
time transfer, 14–15

Radio age
navigation and positioning history,
12–14
Radio Data System, 356
Radio frequency identification
(RFID), 290
detection positioning, 393
non-GNSS indoor positioning problem
and main techniques, 290
positioning, 290t
Radio Navigation Satellite Services
(RNSS), 59, 60
Radio signal strength, 108, 284
non-GNSS positioning systems outdoor
techniques, 114
time measurements, 35
RAIM. *See* Receiver Autonomous Integrity
Monitoring (RAIM)
Rapid static mode, 231
Real-time kinematics mode, 231
Real-time satellite navigation constellations
navigation and positioning history, 23, 24
Received signal strength, 278
Bluetooth system, 291
map, 292f
WLAN positioning, 291
Receiver Autonomous Integrity Monitoring
(RAIM), 261
Receivers, 214f
architecture, 215f
civilian users, 353
Doppler shift, 310
geometrical distribution, 257f
super-heterodyne architecture, 215
Reception, 201
Reflective mirrors
navigation techniques, 8
Regional augmentation systems, 230
Repeater positioning, 327t
global characteristics, 327
Repeaters
GNSS-based indoor positioning,
334–340
pseudolites, 328
research programs, 328
RFID. *See* Radio frequency identification
(RFID)
RnS
approach, 320
repeater configuration, 320f

RNSS. *See* Radio Navigation Satellite
    Services (RNSS)
Route guidance, 355
Route planning modules, 355
R1S repeater configuration, 320, 321f
RSS. *See* Radio signal strength; Received
    signal strength
Russia
    GLONASS, 62, 88t
    GLONASS constellation, 26

SA. *See* Selective availability (SA)
Safety
    modern geographical positioning
        systems, 365
Safety of life (SoL), 84, 141
    OS, 169
SAR. *See* Search and Rescue (SAR)
Saragossa treaty, 6
SARSAT. *See* Cosmicheskaya Sistema
    Poiska Avarinykh Sudov
    (COSPAS-SARSAT)
Satellite
    calculating position techniques, 256–257
    constellation representation, 137f
    displacement, 250
    Doppler shifts, 37f, 222, 250, 310
    ephemeris, 147
    geometrical distribution, 256–257, 257f
    global receiving stations, 116
    linearization method, 248
    MEO, 137
    position computations, 251–252
    radio systems, 117–118, 119
    TRANSIT system, 39
    transmitted frequency, 249
    urban canyons, 333
Satellite Based Augmentation System
    (SBAS), 120, 235
    coverage areas, 84f
    EGNOS, 195
    GNSS navigation signals, 195
    GNSS performances, 131
    GNSS system descriptions, 157
    WAAS, 157, 195, 258
Satellite based navigation systems, 57–91,
    86t, 88t–90t
    differential commercial services, 85–90
    European strategic, political and
        economic issues, 61–64

global positioning satellite systems,
    65–79
    GNSS1, 80–84
    strategic, economic and political aspects,
        58–61
Satellite based positioning systems, 347
    navigation, 346
    positioning, 346
    United States, 347
Satellite navigation, 21, 345
    classification of applications, 345
    European Union, 97
    instruments, 375
    signals, 310
Satellite-to-receiver radio link, 152f, 192f
SBAS. *See* Satellite Based Augmentation
    System (SBAS)
SCC. *See* System control center (SCC)
Search and Rescue (SAR), 76, 142
Search and Rescue Satellite aided tracking
    (SARSAT). *See* Cosmicheskaya
    Sistema Poiska Avarinykh Sudov
    (COSPAS-SARSAT)
Secant cone projection, 48f
Secant cylinder projection, 48f
Second generation navigation satellite
    systems
    positioning techniques, 39
Security
    GPS, 361
Seismology
    GPS, 364
Selective availability (SA)
    GPS, 140, 152
    GPS signals, 346
    GPS systems, 188
Sensors network
    non-GNSS indoor positioning, 287–289
Seven Wonders of the World, 97
Sextant, 9
Shovell, Clowdisley, 10
Signal acquisition
    GNSS signal acquisition and tracking,
        209–211
Signal architecture
    GPS, 68f
Signal generation diagram, 208f
Signal Hill station, 14f
Signal-related parameters
    GNSS system descriptions, 147–150

Sliding correlators, 267
Software, 201
    receiver concept, 216f
SoL. *See* Safety of life (SoL)
Spectrum allocation
    regional division, 61f
Sphere intersection approach
    calculating position techniques, 239–242
Sphere surfaces
    intersection, 144f
    intersection of three, 144f
Spirent GSS 7900 GNSS simulator, 375f
Spread spectrum
    natural interface, 174
SPS. *See* Standard Positioning
        Service (SPS)
Sputnik, 20
    Doppler shift, 36
Standard Positioning Service (SPS), 63,
      71, 254
    GLONASS, 141
    GPS, 71, 140
    PPS, 140
StarFire system, 84
    United States (global coverage), 90t
Stars
    angle measurements, 34
    non-GNSS positioning systems outdoor
        techniques, 97
Static geodesic mode, 230
Strobe correlators, 267
Super heterodyne
    conversion structure, 216f
    receiver architecture, 215
Synchronization
    errors, 188
    space, 391
    time, 391
System control center (SCC)
    GLONASS system, 133

TAI. *See* International Atomic Time (TAI)
Tangent cone projection, 48f
Tangent cylinder projection, 48f
Tangent plane projection, 48f
TDMA. *See* Time division multiple access
    (TDMA)
TDOA. *See* Time difference of arrival
    (TDOA)
TEC. *See* Total Electron Content (TEC)

Tectonics, 364
TeleAtlas, 354
Telecom-based systems
    positioning, 394
Telecommunication
    data transfer, 352
    GNSS, 145
    GNSS positioning, 396
    GNSS receivers, 352
    mobile networks, 278
    portable positioning devices, 384
    positioning availability, 389
    road services, 367
    synchronization, 383
    technique classification, 281t
Terrestrial positioning systems, 106
    navigation and positioning history,
        15–18
Terrestrial radio systems
    non-GNSS positioning systems outdoor
        techniques, 101–115
    positioning techniques, 35
Terrestrial satellite constellation, 315
Terrestrial systems, 346
Three-dimensional indoor positioning
    Galileo satellites, 340
Time
    calculation, 251
    environmental applications, 386
    GNSS synchronization, 145
    measurements, 35
    position, 386
    position technique calculation, 251
    radio electric signal transfer, 14–15
    radio signals, 35
    synchronization related errors, 189
Time difference of arrival (TDOA), 284
    approach, 111f
    method, 301
    positioning, 302t
Time division multiple access
    (TDMA), 164
    approach, 112
Time of arrival (TOA), 110, 284
    approach, 110f
    GNSS, 111
    positioning, 301t
Time To First Fix (TTFF), 176
    GNSS receivers, 313
    HS-GNSS, 313

TOA. *See* Time of arrival (TOA)
Tordesillas treaty, 6
Total Electron Content (TEC), 364
Tourism
    travelers, 277
TRANSIT, 37
    applications, 283
    architecture, 38f
    Doppler shift measurements, 150
    evolution, 309
    GPS, 346
    military availability, 347
    principles, 54
    program, 22
    satellites, 39
    specifications, 23
Transmission, 201
Transportation
    applications, 352f
    LBS, 389
Travelers
    tourism, 277
Triangulation
    carrier phase measurements, 99
    concept, 98
    method, 98, 99f
    non-GNSS positioning systems outdoor
        techniques, 98
Trilateration
    calculating position techniques, 236
    process, 225
Triple difference technique, 229f
Trireme, 2
Troposphere
    definition, 190
    GNSS signal propagation modeling, 363
TTFF. *See* Time To First Fix (TTFF)
Two-band indoor architecture
    GLONASS, 335f
    GPS, 335f
Two-dimensional positioning, 347
    Galileo, 337
    technique, 336
Typical receiver, 214f

Ultra-rapid products, 231
Ultrasound
    non-GNSS indoor positioning problem
        and main techniques, 287
    positioning, 288t

Ultra wide band (UWB)
    approach, 296
    GNSS, 377
    indoor positioning configuration, 297f
    indoor positioning system, 377
    non-GNSS indoor positioning problem
        and main techniques, 296–298
    positioning, 298t
    radar, 103
    technology, 273
UMTS. *See* Universal Mobile
    Telecommunication System (UMTS)
United Kingdom
    Decca system, 106
United States
    emergency number, 274, 360
    FCC, 361
    frequency allocation table, 59
    global receiving stations, 116
    GPS, 62, 82, 88t
    MSAS, 195
    replenishment policy, 66
    satellite-based positioning systems, 347
    WAAS, 88t–89t, 158, 195
United States (global coverage)
    OmniSTAR system, 90t
    OmniTRACS, 90t
    StarFire system, 90t
United States Naval Observatory, 133
Universal Mobile Telecommunication
    System (UMTS), 282
    mobile telecommunication networks, 277
    non-GNSS indoor positioning problem
        and main techniques, 301
Universal Time Coordinated (UTC), 195
    GLONASS, 178
    GPS, 176
Universal Transverse Mercator (UTM)
    coordinate, 50f
    Europe, 49f
    projection, 47, 50
    reference grid, 47
    world, 49f
Urban canyons
    satellites, 333
UTC. *See* Universal Time
    Coordinated (UTC)
UTM. *See* Universal Transverse
    Mercator (UTM)
UWB. *See* Ultra wide band (UWB)

Vanguard project, 20
VDOP. *See* Vertical DOP (VDOP)
Velocity
    calculation, 150
Velocity vector
    computation, 250
Vernal point
    definition, 43f
Vertical DOP (VDOP), 258

WAAS. *See* Wide Area Augmentation
    System (WAAS)
Wave propagation, 15
Weather forecast
    positioning systems, 390
Wide Area Augmentation System (WAAS),
    83, 120
    integrity, 365
    integrity concept, 262
    method implementation, 154
    SBAS, 195, 258
    United States, 88t–89t, 158, 195
Wide area reference station (WRS), 158
Wide-area telecommunication systems
    non-GNSS indoor positioning problem
        and main techniques, 299–303, 300
WiFi, 278, 291, 294–295
    communication, 371
    positioning, 296t
    positioning system, 295f
WiMax. *See also* Wireless metropolitan area
    networks (WMAN)
    non-GNSS indoor positioning problem
        and main techniques, 298
Wireless local area networks (WLAN), 299.
    See also WiFi

deployment, 273
    indoor positioning, 331
    indoors, 327
    non-GNSS positioning systems outdoor
        techniques, 114
    positioning, 280, 291
    RSS, 291
    technique classification, 281t
    telecommunication role, 299
Wireless metropolitan area networks
    (WMAN)
    non-GNSS indoor positioning problem
        and main techniques, 298
    non-GNSS positioning systems outdoor
        techniques, 114
    telecommunication role, 299
Wireless personal area networks (WPAN),
    61, 291. *See also* Bluetooth; Ultra wide
    band (UWB)
    non-GNSS positioning systems outdoor
        techniques, 114
    telecommunication role, 299
WLAN. *See* Wireless local area networks
    (WLAN)
WMAN. *See* Wireless metropolitan area
    networks (WMAN)
World
    geographical representation, 228
    UTM reference grid, 49f
WPAN. *See* Wireless personal area
    networks (WPAN)
Wristwatch receiver, 370f
WRS. *See* Wide area reference station
    (WRS)

X1 event generation, 205f

Printed and bound by CPI Group (UK) Ltd, Croydon, CR0 4YY